Statistical Methods for Health Sciences

SECOND EDITION

Statistical Methods for Health Sciences

SECOND EDITION

M.M. Shoukri, Ph.D.

University of Guelph
Ontario, Canada

C.A. Pause, Ph.D.

University of Western Ontario
London, Ontario, Canada

CRC Press
Boca Raton   London   New York   Washington, D.C.

Library of Congress Cataloging-in-Publication Data

Shoukri, M.M. (Mohamed M.)
    Statistical methods for health sciences / Mohamed M. Shoukri, Cheryl A. Pause. — 2nd ed.
        p.   cm.
    Includes bibliographical references and index.
    ISBN 0-8493-1095-4 (alk. paper)
    1. Epidemiology — statistical methods.   I. Pause, Cheryl A. II. Title.
RA652.2.M3S53 1998
610′.7′27—dc21
                                                                                98-40852
                                                                                CIP

THE AUTHORS

MOHAMED M. SHOUKRI

Received his BSc and MSc in applied statistics from Faculty of Economics and Political Sciences, Cairo University, Cairo, Egypt. He received his MSc and PhD from the Department of Mathematics and Statistics, University of Calgary, Alberta, Canada. He taught applied statistics at Simon Fraser University, the University of British Columbia, and the University of Windsor. He has published in the *Journal of the Royal Statistical Society* (series C), *Biometrics, Journal of Statistical Planning and Inference, The Canadian Journal of Statistics, Statistics in Medicine*, and other journals. He is a Fellow of the Royal Statistical Society of London, and an elected member of the International Statistical Institute. Presently, he is Professor of Biostatistics in the Department of Population Medicine at the University of Guelph, Ontario, Canada.

CHERYL A. PAUSE

Received her BSc and MSc from the Department of Mathematics and Statistics, McMaster University, Hamilton, Ontario, Canada. Following one year as a statistical research assistant at the Canada Centre for Inland Waters, Burlington, Ontario, she entered the doctoral program in Biostatistics at the University of Western Ontario, London, Ontario, and completed her thesis in 1998. During this time she was a statistical consultant in the Department of Population Medicine at the University of Guelph, Ontario. She taught statistics at Mohawk College, Hamilton, Ontario, and Sheridan College, Oakville, Ontario, and has published in the *Medical Teacher Journal*.

PREFACE TO THE FIRST EDITION

A substantial portion of epidemiologic studies, particularly in community medicine, veterinary herd health, field trials, and repeated measures from clinical investigations, produce data that are clustered and quite heterogeneous. Such clustering will inevitably produce highly correlated observations; thus, standard statistical techniques in non-specialized biostatistics textbooks are no longer appropriate in the analysis of such data. For this reason it was our mandate to introduce to our audience the recent advances in statistical modeling of clustered or correlated data that exhibit extra variation or heterogeneity.

This book reflects our teaching experiences of a biostatistics course in the University of Guelph's Department of Population Medicine. The course is attended predominantly by epidemiology graduate students; to a lesser degree we have students from Animal Science and researchers from disciplines which involve the collection of clustered and over-time data. The material in this text assumes that the reader is familiar with basic applied statistics, principles of linear regression and experimental design, but stops short of requiring a cognizance of the details of the likelihood theory and asymptotic inference. We emphasize the "how to" rather than the theoretical aspect; however, on several occasions the theory behind certain topics could not be omitted, but is presented in a simplified form.

The book is structured as follows: Chapter 1 serves as an introduction in which the reader is familiarized with the effect of violating the assumptions of homogeneity and independence on the ANOVA problem. Chapter 2 discusses the problem of assessing measurement reliability. The computation of the intraclass correlation as a measure of reliability allowed us to introduce this measure as an index of clustering in subsequent chapters. The analysis of binary data summarized in 2x2 tables is taken up in Chapter 3. This chapter deals with several topics including, for instance, measures of association between binary variables, measures of agreement and statistical analysis of medical screening test. Methods of cluster adjustment proposed by Donald and Donner (1987), Rao and Scott (1992) are explained. Chapter 4 concerns the use of logistic regression models in studying the effects of covariates on the risk of disease. In addition to the methods of Donald and Donner, and Rao and Scott to adjust for clustering, we explain the Generalized Estimating Equations, (GEE) approach proposed by Liang and Zeger (1986). A general background on time series models is introduced in Chapter 5. Finally, in Chapter 6 we show how repeated measures data are analyzed under the linear additive model for continuously distributed data and also for other types of data using the GEE.

We wish to thank Dr. A. Meek, the Dean of the Ontario Veterinary College, for his encouragement in writing this book; Dr. S.W. Martin, Chair of the Department of Population Medicine, for facilitating the use of the departmental resources; the graduate students who took the course "Statistics for the Health Sciences"; special thanks to Dr. J. Sargeant for being so generous with her data and to Mr. P. Page for his invaluable computing expertise. Finally, we would like to thank J. Tremblay for her patience and enthusiasm in the production of this manuscript.

M. M. Shoukri
V. L. Edge
Guelph, Ontario

July 1995

PREFACE TO THE SECOND EDITION

The main structure of the book has been kept similar to the first edition. To keep pace with the recent advances in the science of statistics, more topics have been covered. In Chapter 2 we introduced the coefficient of variation as a measure of reproducibility, and comparing two dependent reliability coefficients. Testing for trend using Cochran-Armitage chi-square, under cluster randomization has been introduced in Chapter 4. In this chapter we discussed the application of the PROC GENMOD in SAS, which implements the GEE approach, and "Multi-level analysis" of clustered binary data under the "Generalized Linear Mixed Effect Models," using Schall's algorithm, and GLIMMIX SAS macro. In Chapter 5 we added two new sections on modeling seasonal time series; one uses combination of polynomials to describe the trend component and trigonometric functions to describe seasonality, while the other is devoted to modeling seasonality using the more sophisticated ARIMA models. Chapter 6 has been expanded to include analysis of repeated measures experiment under the "Linear Mixed Effects Models," using PROC MIXED in SAS. We added Chapter 7 to cover the topic of survival analysis. We included a brief discussion on the analysis of correlated survival data in this chapter.

An important feature of the second edition is that all the examples are solved using the SAS package. We also provided all the SAS programs that are needed to understand the material in each chapter.

M.M. Shoukri, Guelph, Ontario
C.A. Pause, London, Ontario
July 1998

TABLE OF CONTENTS

Chapter 5

Chapter 6

Chapter 1

COMPARING GROUP MEANS WHEN THE STANDARD ASSUMPTIONS ARE VIOLATED

I. INTRODUCTION

A great deal of statistical experiments are conducted so as to compare two or more groups. It is understood that the word 'group' is a generic term used by statisticians to label and distinguish the individuals who share a set of experimental conditions. For example, diets, breeds, age intervals, methods of evaluations etc. are groups. In this chapter we will be concerned with comparing group means. When we have two groups, that is when we are concerned with comparing two means μ_1, and μ_2, the familiar Student t-statistic is the tool that is commonly used by most data analysts. If we are interested in comparing several group means; that is if the null hypothesis is $H_0:\mu_1=\mu_2=...=\mu_k$, a problem known as the "Analysis of Variance", we use the F-ratio to seek the evidence in the data and see whether it is sufficient to justify the above hypothesis.

In performing an "Analysis of Variance" (ANOVA) experiment, we are always reminded of three basic assumptions:

Assumption (1):   That the observations in each group are a random sample from a normally distributed population with mean μ_i and variance σ_i^2 (i=1,2,...k).

Assumption (2):   The variances $\sigma_1^2,...\sigma_k^2$ are all equal. This is known as the variance homogeneity assumption.

Assumption (3):   The observations within each group should be independent.

The following sections will address the effect of the violation of each of these three assumptions on the ANOVA procedure, how it manifests itself, and possible corrective steps which can be taken.

A. NON-NORMALITY

It has been reported (Miller, 1986, p.80) that lack of normality has very little effect on the significance level of the F-test. The robustness of the F-test improves with increasing the number of groups being compared, together with an increase in the group size n_i. The reason for this is a rather technical issue and will not be discussed here. However, the investigators should remain aware that although the significance level may be valid, the F-test may not be appreciably powerful. Transformations such as the square root, or the logarithm to improve normality can improve the power of the F-test when applied to non-normal samples. To provide the reader with the proper tool to detect the non-normality of the data, either before or after performing the proper transformation we need some notation to facilitate the presentation. Let y_{ij} denote the jth observation in the ith group, where j=1,2,...n_i and i=1,2,...k. Moreover suppose that $y_{ij} = \mu_i + e_{ij}$, where it is assumed that e_{ij} are identically, independently and normally distributed random variables with $E(e_{ij})=0$ and variance $(e_{ij})=\sigma^2$ (or $e_{ij} \sim$ iid $N(0,\sigma^2)$). Hence, $y_{ij} \sim$ iid $N(\mu_i,\sigma^2)$, and the assumption of normality of the data y_{ij} needs to be verified before using the ANOVA F-statistic.

Miller (1986, p.82) recommended that a test of normality should be performed for each group. One should avoid omnibus tests that utilize the combined groups such as a combined goodness-of-fit χ^2 test or a multiple sample Kolmogorov-Smirnov. A review of these tests is given in D'Agostino and Stephens (1986, chapter 9). They showed that the χ^2 test and Kolmogorov test have poor power properties and should not be used when testing for normality.

Unfortunately, the preceeding results are not widely known to nonstatisticians. Major statistical packages, such as SAS, perform the Shapiro-Wilk, W test for group size up to 50. For larger group size, SAS provides us with the poor power Kolmogorov test. This package also produces measures of skewness and kurtosis, though strictly speaking these are not the actual measures of skewness $\sqrt{b_1}$ and kurtosis b_2 defined as

$$\sqrt{b_1} = \frac{1}{n} \sum_{j=1}^{n} (y_j - \bar{y})^3 \Big/ \left[\frac{1}{n} \sum_{j=1}^{n} (y_j - \bar{y})^2 \right]^{3/2}$$

$$b_2 = \frac{1}{n} \sum_{j=1}^{n} (y_j - \bar{y})^4 \Big/ \left[\frac{1}{n} \sum_{j=1}^{n} (y_j - \bar{y})^2 \right]^{2} .$$

In a recent article by D'Agostino et al. (1990) the relationship between b_1 and b_2 and the measures of skewness and kurtosis produced by SAS is established. Also provided is a simple SAS macro which produces an excellent, informative analysis for investigating normality. They recommended that, as descriptive statistics, values of $\sqrt{b_1}$ and b_2 close to 0 and 3 respectively, indicate normality. More precisely, the expected values of these are 0 and 3(n-1)/(n+1) under normality.

To test for skewness, the null hypothesis, H_0: *underlying population is normal* is tested as follows (D'Agostino and Pearson 1973):

1. Compute $\sqrt{b_1}$ from the data.
2. Compute

$$u = \sqrt{b_1} \left\{ \frac{(n+1)(n+3)}{6(n-2)} \right\}^{1/2} ,$$

$$\beta_2(\sqrt{b_1}) = \frac{3(n^2+27n-70)(n+1)(n+3)}{(n-2)(n+5)(n+7)(n+9)}$$

$$w^2 = -1 + \left\{2\left[\beta_2(\sqrt{b_1})-1\right]\right\}^{1/2} \quad,$$

$$\delta = (ln\ w)^{-1/2} \quad, \quad and \quad \alpha = \left\{2/(w^2-1)\right\}^{1/2} \quad.$$

3. Compute

$$Z(\sqrt{b_1}) = \delta\ ln\ (u/\alpha + \{(u/\alpha)^2 +1\}^{1/2}) \quad.$$

$Z(\sqrt{b_1})$ has approximately standard normal distribution, under the null hypothesis.

To test for kurtosis, a two-sided test (for $\beta_2 \neq 3$) or one-sided test (for $\beta_2 > 3$ or $\beta_2 < 3$) can be constructed:

1. Compute b_2 from the data.

2. Compute

$$E(b_2) = \frac{3(n-1)}{n+1}$$

and

$$Var(b_2) = \frac{24n(n-2)(n-3)}{(n+1)^2(n+3)(n+5)}$$

3. Compute the standardized score of b_2. That is

$$x = \frac{(b_2 - E(b_2))}{\sqrt{Var(b_2)}}$$

4.   Compute

$$\sqrt{\beta_1(b_2)} = \frac{6(n^2-5n+2)}{(n+7)(n+9)} \sqrt{\frac{6(n+3)(n+5)}{n(n-2)(n-3)}}$$

and

$$A = 6 + \frac{8}{\sqrt{\beta_1(b_2)}} \left[\frac{2}{\sqrt{\beta_1(b_2)}} + \sqrt{1 + \frac{4}{\beta_1(b_2)}} \right]$$

5.   Compute

$$Z(b_2) = [2/(9A)]^{-1/2} \left\{ \left(1 - \frac{2}{9A}\right) - \left[\frac{1-2/A}{1+x\sqrt{2/(A-4)}}\right]^{1/3} \right\} .$$

$Z(b_2)$ has approximately standard normal distribution, under the null hypothesis.

D'Agostino and Pearson (1973) constructed a test based on both $\sqrt{b_1}$ and b_2 to test for skewness and kurtosis. This test is given by

$$K^2 = (Z(\sqrt{b_1}))^2 + (Z(b_2))^2 \qquad (1.1)$$

and has approximately a chi-square distribution with 2 degrees of freedom.

For routine evaluation of the statistic K^2, the SAS package produces measures of skewness (g_1) and kurtosis (g_2) that are different from $\sqrt{b_1}$ and b_2. After obtaining g_1 and g_2 from PROC UNIVARIATE in SAS, we evaluate $\sqrt{b_1}$ as

$$\sqrt{b_1} = \frac{(n-2)}{\sqrt{n(n-1)}} g_1 \qquad (1.2)$$

and

$$b_2 = \frac{(n-2)(n-3)}{(n+1)(n-1)} g_2 + \frac{3(n-1)}{n+1} . \qquad (1.3)$$

One then should proceed to compute $Z(\sqrt{b_1})$ and $Z(\sqrt{b_2})$ and hence K^2.

Note that when n is large, $\sqrt{b_1} \approx g_2$ and $b_2 \approx g_2$.

An equally important tool that should be used to assess normality is graphical representation of the data. A very effective technique is the empirical quantile-quantile plot (Wilk and

4

Gnanadesikan, 1968). It is constructed by plotting the quantiles of the empirical distribution against the corresponding quantiles of the normal distribution. If the quantiles plot is a straight line, it indicates that the data is drawn from a normal population.

A good complete normality test would consist of the use of the statistic K^2 and the normal quantile plot.

Note that if the empirical quantile-quantile plot follows a straight line one can be certain, to some extent, that the data is drawn from a normal population. Moreover, the quantile-quantile plot provides a succinct summary of the data. For example if the plot is curved this may suggest that a logarithmic transformation of the data points could bring them closer to normality. The plot also provides us with a useful tool for detecting outlying observations.

In addition to the quantile-quantile plot, a box plot (Tukey, 1977) should be provided for the data. In the box plot the 75th and 25th percentiles of the data are portrayed by the top and bottom of a rectangle, and the median is portrayed by a horizontal line segment within the rectangle. Dashed lines extend from the ends of the box to the adjacent values which are defined as follows. First the inter-quantile range IQR is computed as:

$$IQR = \text{75th percentile - 25th percentile.}$$

The upper adjacent value is defined to be the largest observation that is less than or equal to the 75th percentile plus (1.5) IQR. The lower adjacent value is defined to be the smallest observation that is greater or equal to the 25th percentile minus (1.5) IQR. If any data point falls outside the range of the two adjacent values, it is called an outside value (not necessarily an outlier) and is plotted as an individual point.

The box-plot displays certain characteristics of the data. The median shows the center of the distribution; and the lengths of the dashed lines relative to the box show how far the tails of the distribution extend. The individual outside values give us the opportunity to examine them more closely and subsequently identify them as possible outliers.

Example 1.1   120 observations representing the average milk production per day/herd in kilograms were collected from 10 Ontario farms (see; Appendix 1). All plots are produced by SAS.

A data summary via the box plot is shown in Figure 1.1 For illustrative purposes Figure 1.2 is the normal quantile plot of the data. Ignoring the identification by farm, there is a slight departure from the straight line that would be expected if the data were drawn from a normal population. The summary statistics produced by SAS UNIVARIATE PROCEDURE are given in Table 1.1.

Table 1.1
Output of the SAS UNIVARIATE PROCEDURE on the Original Data.

Variable = Milk

Moments

N	120	Sum Wgts	120
Mean	26.74183	Sum	3209.02
Std Dev	3.705186	Variance	13.7284
Skewness	**-0.54127**	**Kurtosis**	**0.345688**
USS	87448.76	CSS	1633.68
CV	13.85539	Std Mean	0.338236
T:Mean=0	79.06274	Prob>|T|	0.001
Sgn Rank	3630	Prob>|S|	0.001
Num ^= 0	120		
W:Normal	0.970445	Prob<W	0.1038

The result of D'Agostino et al (1990) procedure is summarized in Table 1.2.

Table 1.2
Results of a Normality Test on the Original Data Using a SAS Macro Written by D'Agostino et al (1990)

SKEWNESS	g_1=-0.54127	b_1=-0.53448	$Z(\sqrt{b_1})$=-2.39068	p=0.0168
KURTOSIS	g_2=0.34569	b_2=3.28186	$Z(\sqrt{b_2})$=0.90171	p=0.3672
OMNIBUS TEST	K^2=6.52846 (against chisq 2df)			p=0.0382

The value of the omnibus test (6.528), when compared to a chi squared value with 2 degrees of freedom (5.84) is found to be significant. Thus, based on this test, we can say that the data are not normally distributed.

In attempting to improve the normality of the data, it was found that the best results were obtained through squaring the milk yield values. This solved the skewness problem and resulted in a non significant K^2 value (Table 1.3). Combined with the normal probability plot (Figure 1.3) produced by the D'Agostino et al. (1990) macro, one can see that the data has been normalized.

Table 1.3
Results of a Normality Test on the Transformed Data (Milk^2)

SKEWNESS	g_1 = -0.11814	b_1 = 0.11666	Z = -0.54876	p = 0.5832
KURTOSIS	g_2 = -0.03602	b_2 = 2.91587	Z = 0.00295	p = 0.9976
OMNIBUS TEST	K^2 = 0.30115			p = 0.860

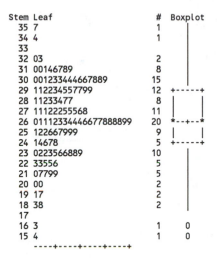

```
Stem Leaf                          #  Boxplot
  35 7                             1      |
  34 4                             1      |
  33                                      |
  32 03                            2      |
  31 00146789                      8      |
  30 001233444667889              15      |
  29 112234557799                 12  +------+
  28 11233477                      8  |      |
  27 11122255568                  11  |      |
  26 01112334446677888899         20  *--+--*
  25 122667999                     9  |      |
  24 14678                         5  +------+
  23 0223566889                   10      |
  22 33556                         5      |
  21 07799                         5      |
  20 00                           2      |
  19 17                           2      |
  18 38                           2      |
  17                                      |
  16 3                            1      0
  15 4                            1      0
     ----+----+----+----+
```

Figure 1.1. Stem leaf and box plot illustration of the milk production on 10 herds; the data are untransformed.

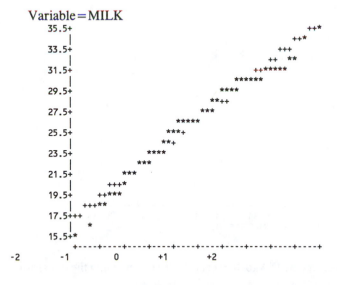

Figure 1.2 The normal quantile plot for the untransformed data of Example 1.1.

7

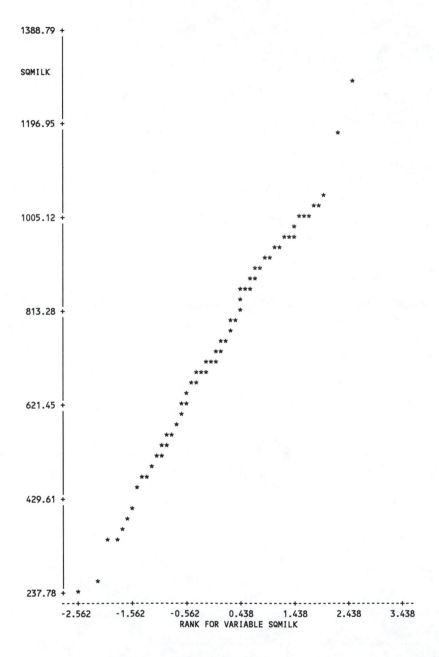

Figure 1.3 Normal plot produced by D'Agostino et al (1990) macro using the transformed milk yield data.

B.   HETEROGENEITY OF VARIANCES

The effect of unequal variances on the F-test of the ANOVA problem had been investigated by Box (1954a). He showed that in balanced experiments (i.e. when $n_1 = n_2 = \ldots = n_k$) the effect of heterogeneity is not too serious. When we have unbalanced experiments then the effect can be more serious, particularly if the large σ_i^2 are associated with the small n_i. The recommendation is to balance the experiment whenever possible, for then unequal variances have the least effect.

Detecting heterogeneity or testing the hypothesis $H_0 : \sigma_1^2 = \sigma_2^2 = \ldots = \sigma_k^2$, is a problem that has applications in the field of measurement errors, particularly reproducibility and reliability studies as will be seen in the next chapter. In this section we introduce some of the widely used tests of heterogeneity.

1.   Bartlett's Test

Bartlett (1937) modified the likelihood ratio statistic on $H_0 : \sigma_1^2 = \sigma_2^2 = \ldots \sigma_k^2$ by introducing a test statistic widely known as Bartlett's test on homogeneity of variances. This test statistic is given as

$$B = M \, / \, C \tag{1.4}$$

where

$$M = v \, \log s^2 - \sum_{i=1}^{k} v_i \, \log s_i^2 \quad ,$$

$$v_i = n_i - 1 \quad , \quad v = \sum_{i=1}^{k} v_i \quad , \quad s^2 = \frac{1}{v} \sum_{i=1}^{k} v_i \, s_i^2 \quad ,$$

$$s_i^2 = \frac{1}{v_i} \sum_{j=1}^{n_i} (y_{ij} - \overline{y}_i)^2 \quad , \quad \overline{y}_i = \frac{1}{n_i} \sum_{j=1}^{n_i} y_{ij}$$

and

$$C = 1 + \frac{1}{3(k-1)} \left[\left(\sum_{i=1}^{k} \frac{1}{v_i} \right) - \frac{1}{v} \right] \quad .$$

The hypothesis of equal variances is rejected if $B > \chi^2_{\alpha, k-1}$.

One of the disadvantages of Bartlett's test is that it is quite sensitive to departure from normality. In fact, if the distributions from which we sample are not normal, this test can have an actual size several times larger than its nominal level of significance.

2.   Levene's Test (1960)

The second test was proposed by Levene (1960). It entails doing a one-way ANOVA on the variables

$$L_{ij} = |\ y_{ij} - \bar{y}_i\ |\quad ,\quad (i=1,2,...k)\quad .$$

If the F-test is significant, homogeneity of the variances is in doubt. Brown and Forsythe (1974a) conducted extensive simulation studies to compare between tests on homogeneity of variances and reported the following.

(i) If the data are nearly normal, as evidenced by the previously outlined test of normality, then we should use Bartlett's test which is equally suitable for balanced and unbalanced data.

(ii) If the data are not normal, Levene's test should be used. Moreover, if during the step of detecting normality we find that the value of $\sqrt{b_1}$ is large, that is if the data tend to be very skewed, then Levene's test can be improved by replacing $\bar{y}_i$ by $\tilde{y}_i$, where $\tilde{y}_i$ is the median of the ith group. Therefore $L_{ij} = |y_{ij} - \tilde{y}_i|$, and an ANOVA is done on L_{ij}. The conclusion would be either to accept or reject $H_0 : \sigma_1^2 = ... = \sigma_k^2$. If the hypothesis is accepted, then the main hypothesis of equality of several means can be tested using the familiar ANOVA.

If the heterogeneity of variances is established, then it is possible to take remedial action by employing transformation on the data. As a general guideline in finding an appropriate transformation, first plot the k pairs $(\bar{y}_i, s_i^2)$ i=1,2...k as a means to detect the nature of the relationship between $\bar{y}_i$ and s_i^2. For example if the plot exhibits a relationship that is approximately linear we may try log y_{ij}. If the plot shows a curved relationship we suggest the square root transformation. Whatever transformation is selected, it is important to test the variance heterogeneity on the tranformed data to ascertain if the transformation has in fact stabilized the variance.

If attempts to stabilize the variances through transformation are unsuccessful, several approximately valid procedures to compare between group means in the presence of variance heterogeneity, have been proposed. The most commonly used are:

3. Welch's Statistic (1951) for Testing Equality of Group Means:

$$\mathrm{WL} = \frac{1}{k-1}\sum_{i=1}^{k}\frac{w_i\left(\bar{y}_i - \bar{\bar{y}}\right)^2}{c_1} \tag{1.5}$$

where

$$c_1 = 1 + \frac{2(k-1)}{(k^2-1)} \sum_{i=1}^{k} (n_i-1)^{-1} \left(1 - \frac{w_i}{w}\right)^2$$

$$w_i = n_i/s_i^2 \quad, \quad w = \sum_{i=1}^{k} w_i \quad, \text{ and}$$

$$\bar{\bar{y}} = \sum_{i=1}^{k} w_i \, \bar{y}_i \, / \, w$$

When all population means are equal (even if the variances are unequal), WL is approximately distributed as an F-statistic with k-1 and f_1 degrees of freedom, where

$$f_1 = \left[\frac{3}{(k^2-1)} \sum_{i=1}^{k} (n_i-1)^{-1} \left(1 - \frac{w_i}{w}\right)^2 \right]^{-1} \tag{1.6}$$

4.   **Brown and Forsythe Statistic (1974b) for Testing Equality of Group Means:**

$$BF = \sum_{i=1}^{k} n_i \left(\bar{y}_i - \bar{y}\right)^2 / c_2 \tag{1.7}$$

where

$$c_2 = \sum_{i=1}^{k} \left(1 - \frac{n_i}{N}\right) s_i^2$$

$$\bar{y} = \frac{1}{N} \sum_{i=1}^{k} \sum_{j=1}^{n_i} y_{ij} \quad, \quad \text{and } N = n_1 + n_2 + ... + n_k \quad.$$

Approximate critical values are obtained from the F-distribution with k-1 and f_2 degrees of freedom, where

$$f_2 = \left[\sum_{i=1}^{k} (n_i-1)^{-1} a_i^2 \right]^{-1}$$

and

$$a_i = \left(1 - \frac{n_i}{N}\right) s_i^2 / c_2$$

Remarks:

1. The approximations WL and BF are valid when each group has at least 10 observations and are not unreasonable for WL when the group size is as small as five observations per group.

2. The choice between WL and BF depends upon whether extreme means are thought to have extreme variance in which case BF should be used. If extreme means have smaller variance, then WL is preferred. Considering the structure of BF, Brown and Forsythe recommend the use of their statistic if some sample variances appear unusually low.

5. **Cochran's (1937) Method of Weighting for Testing Equality of Group Means:**

    To test the hypothesis $H_0: \mu_1 = \mu_2 = ... = \mu_k = \mu$, where μ is a specified value, Cochran suggested the weighted sum of squares

$$G = \sum_{i=1}^{k} w_i (\bar{y}_i - \mu)^2 \tag{1.8}$$

as a test statistic on the above hypothesis.

The quantity G follows (approximately) a χ_k^2 distribution. However, if we do not wish to specify the value of μ it can be replaced by $\bar{\bar{y}}$ and then G becomes

$$\hat{G} = \sum_{i=1}^{k} w_i (\bar{y}_i - \bar{\bar{y}})^2$$

$$= \sum_{i=1}^{k} w_i \bar{y}_i^2 - (\sum_{i=1}^{k} w_i \bar{y}_i)^2 / w \tag{1.9}$$

It can be shown that, under the null hypothesis of equality of means, $\hat{G}$ is approximately distributed as χ_{k-1}^2; the loss of 1 degree of freedom is due to the replacement by μ by $\bar{\bar{y}}$. Large values of $\hat{G}$ indicate evidence against the equality of the means.

Example 1.2. In this example we analyze the milk data (untransformed). The main objective of this example is to assess the heterogeneity of variances amongst the 10 farms. Further summary statistics are provided in Table 1.4.

Table 1.4
Summary Statistics for Each of the Ten Farms

Farm	Mean	n_i	Std. Dev.	s_i^2
1	30.446	12	1.7480	3.0555
2	29.041	12	2.9314	8.5931
3	24.108	12	3.3302	11.0902
4	27.961	12	3.2875	10.8077
5	29.235	12	1.7646	3.1138
6	27.533	12	2.1148	4.4724
7	28.427	12	1.9874	3.9498
8	24.473	12	1.8912	3.5766
9	24.892	12	1.7900	3.2041
10	21.303	12	3.8132	14.5405

The value of the Bartlett's statistic is $B = 18.67$, and from the table of chi-square at 9 degrees of freedom, the p-value ≈ 0.028. Therefore there is evidence of variance heterogeneity.

However, after employing the square transformation on the data (see summary statistics in Table 1.5) the Bartlett's statistic is $B = 14.62$ with p-value ≈ 0.102.

Hence, by squaring the data we achieve two objectives; first the data were transformed to normality; second, the variances were stabilized.

Table 1.5
Summary Statistics on the Transformed Data (Milk^2) for Each of the Ten Farms

Farm	Mean	n_i	Std. Dev.	s^2
1	929.75	12	109.020	11885.36
2	851.25	12	160.810	25.859.86
3	591.34	12	158.850	25233.32
4	791.72	12	195.020	38032.80
5	857.54	12	102.630	10532.92
6	762.18	12	116.020	13460.64
7	811.74	12	113.140	12800.66
8	602.22	12	92.774	8607.02
9	622.53	12	88.385	7811.91
10	467.13	12	165.080	27251.41

Example 1.3 The following data are the results of a clinical trial involving 4 groups of dogs assigned at random to each of 4 therapeutic drugs believed to treat compulsive behaviour. The scores given in Table 1.6 are measures of the severity of the disorder after 21 days of treatment. Before comparing the mean scores, we test the homogeneity of variances using Levene's test.

Table 1.6
Scores of 4 Drugs Given to Dogs in a Clinical Trial

Drug			Score			
1	5	10	2	10	10	4
2	1	2	5	2	5	
3	8	10	4	5	10	10
4	9	8	7	9	3	10

The estimated variances are $s_1^2=12.97$, $s_2^2=3.5$, $s_3^2=7.37$, $s_4^2=6.27$

Table 1.7 gives the location free scores $L_{ij}=\left|y_{ij}-\bar{y}_i\right|$

Table 1.7
The Location Free Score $L_{ij}=|y_{ij}-\bar{y}_i|$ for the Data of Table 1.6

Drug			Score (L_{ij})			
1	1.83	3.17	4.83	3.17	3.17	2.83
2	2	1	2	1	2	
3	0.17	2.170	3.83	2.13	2.17	2.17
4	1.33	0.33	0.67	1.33	4.67	2.33

In Table 1.8 we provide the result of the ANOVA on L_{ij}; the F-statistic = 2.11 with P-value = 0.133. This indicates that there is no significant difference between the variances of the 4 groups even though visually the estimated variances seem quite different from each other.

Table 1.8
ANOVA of L_{ij}

Source	df	Sum Square	Mean Square	F Value	Pr > F
Model	3	8.47824	2.82608	2.11	0.1326
Error	19	25.4350	1.33868		
Corrected Total	22	33.9132			

14

Levene's test could not detect this heterogeneity because the sample sizes and the number of groups are small. This lack of power is a common feature of many nonparametric tests such as Levene's.

C.   NON-INDEPENDENCE

A fundamental assumption in the analysis of group means is that the samples must be independent, as are the observations within each sample. For now, we shall analyze the effect of within sample dependency with emphasis on the assumption that the samples or the groups are independent of each other.

Two types of dependency and their effect on the problem of comparing means will be discussed in this section. The first type is caused by a sequence effect. The observations in each group may be taken serially over time in which case observations that are separated by shorter time intervals may be correlated because the experimental conditions are not expected to change, or because the present observation may directly affect the succeeding one. The second type of correlation is caused by blocking effect. The n_i data points $y_{i1} \ldots y_{in_i}$ may have been collected as clusters. For example the y's may be collected from litters of animals, or from "herds". The possibility that animals or subjects belonging to the same cluster may respond in a similar manner (herd or cluster effect) creates a correlation that should be accounted for. The cluster effect can be significantly large if there are positive correlations between animals within clusters or if there is a trend between clusters. This within herd correlation is known as the intracluster correlation and will be denoted by ρ.

Before we examine the effect of serial correlation on comparing group means, we shall review its effect on the problem of testing one mean, that is its effect on the t-statistic.

The simplest time sequence effect is of a serial correlation of lag 1. Higher order lags are discussed in Albers (1978). Denoting $y_1 \ldots y_n$ as the sample values drawn from a population with mean μ and variance σ^2, we assume that

$$\text{Cov} (y_i , y_{i+j}) \quad = \quad \begin{cases} \rho^j \sigma^2 & j = 0, 1 \\ 0 & \text{otherwise} \end{cases}$$

When n is large, the variance of $\bar{y}$ is given approximately as

$$var\ (\bar{y}) = \frac{\sigma^2}{n} [1 + 2\rho] \quad,$$

15

and since the parameters σ^2 and ρ are unknown we may replace them by their moment estimators

$$\hat{\sigma}^2 = s^2 = \frac{1}{n-1} \sum_{i=1}^{n} (y_i - \bar{y})^2$$

and

$$\hat{\rho} = \frac{1}{n-1} \sum_{i=1}^{n-1} (y_i - \bar{y})(y_{i+1} - \bar{y}) / s^2 \ . \tag{1.10}$$

Note that the var $(\bar{y})$ in the presence of serial correlation is inflated by the factor $d=1+2\rho$. This means that if we ignore the effect of ρ the variance of the sample mean would be under estimated. The t-statistic thus becomes

$$t = \frac{\sqrt{n} \, (\bar{y}-\mu)}{s\sqrt{1+2\hat{\rho}}} \ . \tag{1.11}$$

To illustrate the unpleasant effect of $\hat{\rho}$ on the p-value, let us consider the following hypothetical example. Suppose that we would like to test $H_0: \mu=30$ versus $H_1: \bar{y}>30$. The sample information is, $\bar{y}=32$, $n=100$, $s=10$, and $\hat{\rho}=0.30$. Therefore, from equation (1.11), $t=1.58$, and the p-value ≈ 0.06. If the effect of $\hat{\rho}$ were ignored, then $t=2.0$ and the p-value ≈ 0.02.

This means that significance will be spuriously declared if the effect of $\hat{\rho}$ is ignored.

Remark: It was recommended by Miller (1986) that a preliminary test of $\rho=0$ is unnecessary, and that it is only the size of $\hat{\rho}$ that is important.

We now return to the problem of comparing group means. If the observations within each group are collected in time sequence, we compute the serial correlations $\hat{\rho}_i$ ($i=1,2,\ldots k$). Few papers researching the effect of serial correlation on the ANOVA F statistic are available. We shall report on their quantitative merits in Chapter 6, which is devoted to the analysis of repeated measurements. However, the effect of correlation can be salvaged by using the serial correlation estimates together with one of the available approximate techniques. As an illustration; let $y_{i1}, y_{i2}, \ldots y_{in_i}$ be the observations from the ith group and let $\bar{y}_i$, and s_i^2 be the sample means and variances respectively. If the data in each sample are taken serially, the serial correlation estimates are $\hat{\rho}_1 \ \hat{\rho}_2, \ldots, \hat{\rho}_k$. Since the estimated variance of $\bar{y}_i$ is

$$v_i = v\hat{a}r(\bar{y}_i) = \frac{s_i^2}{n_i} \left[1 + 2\hat{\rho}_i \right]$$

then $w_i = v_i^{-1}$ can be used to construct a pooled estimate for the group means under the hypothesis $H_0: \mu_1 = ... = \mu_k$. This estimate is

$$\bar{\bar{y}} = \sum_{i=1}^{k} w_i \, \bar{y}_i \, / \, w \quad , \quad w = \sum_{i=1}^{k} w_i \, .$$

The hypothesis is thus rejected for values of

$$\hat{G} = \sum_{i=1}^{k} w_i \, (\bar{y}_i - \bar{\bar{y}})^2$$

exceeding χ^2_{k-1} at the α-level of significance.

To account for the effect of intraclass correlation a similar approach may be followed. Procedures for estimating the intraclass correlation will be discussed in Chapter 2.

Example 1.4   The data are the average milk production per cow per day, in kilograms, for 3 herds over a period of 10 months (Table 19). For illustrative purposes we would like to use Cochran's statistic $\hat{G}$ to compare between the herd means. Because the data were collected in serial order one should account for the effect of the serial correlation.

Table 1.9
Milk Production Per Cow Per Day in Kilograms for 3 Herds Over a Period of 10 Months.

Farm				Months						
	1	2	3	4	5	6	7	8	9	10
1	25.1	23.7	24.5	21	19	29.7	28.3	27.3	25.4	24.2
2	23.2	24.2	22.8	22.8	20.2	21.7	24.8	25.9	25.9	25.9
3	21.8	22.0	22.1	19	18	17	24	24	25	19

The estimated variances and serial correlations of the 3 herds are: $s_1^2 = 9.27$, $s_2^2 = 3.41$, $s_3^2 = 6.95$, $\hat{\rho}_1 = 0.275$, $\hat{\rho}_2 = .593$, and $\hat{\rho}_3 = 0.26$. Therefore $v_1 = 1.36$, $v_2 = 0.75$, and $v_3 = 1.06$. Since $\hat{G} = 6.13$, the p-value < 0.05, one concludes that the hypothesis of no difference in mean milk production cannot be accepted.

Remarks on SAS programming:

We can test the homogeneity of variances (HOVTEST) of the groups defined by the MEANS effect within PROC GLM. This can be done using either Bartlett's test (1.4) or Levene's test L_{ij} by adding the following options to the MEANS statement :

MEANS group | HOVTEST = BARTLETT; or
                        HOVTEST = LEVENE (TYPE = ABS);

Chapter 2

STATISTICAL ANALYSIS OF MEASUREMENTS RELIABILITY

"The case of interval scale measurements"

I. INTRODUCTION

The verification of scientific hypotheses is only possible by the process of experimentation, which in turn often requires the availability of suitable instruments to measure materials of interest with high precision.

The concepts of "precision" and "accuracy" are often misunderstood. To correctly understand such concepts, modern statistics provides a quantitative technology for experimental science by first defining what is meant by these two terms, and second, developing the appropriate statistical methods for estimating indices of accuracy of such measurements. The quality of measuring devices is often described by *reliability, repeatability (reproducibility) and generalizability*, so the accuracy of the device is obviously an integral part of the experimental protocol.

The reader should be aware that the term "device" refers to the means by which the measurements are made, such as the instrument, the clinician or the interviewer. In medical research, terms such as patient *reliability, interclinician agreement, and repeatability* are synonymous with a group of statistical estimators which are used as indices of the accuracy and precision of biomedical measurements. To conduct a study which would measure the reliability of a device, the common approach is to assume that each measurement consists of two parts, the true unknown level of the characteristic measured (blood pressure, arm girth, weight etc.) plus an error of measurement. In practice it is important to know how large or small the error variance, or the imprecision, is in relation to the variance of the characteristic being measured. This relationship between variances lays the ground rules for assessing reliability. For example, the ratio of the error variance to the sum of the error variance plus the variance of the characteristic gives, under certain conditions, a special estimate of reliability which is widely known as the intraclass correlation (denoted by ρ). There are numerous versions of the intraclass correlation, the appropriate form for the specific situation being defined by the conceptual intent of the reliability study.

The main objective of this chapter is to provide the readers with guidelines on how to use the intraclass correlation as an index of reliability.

II. MODELS FOR RELIABILITY STUDIES

An often encountered study for estimating ρ assumes that each of a random sample of k patients is evaluated by n clinicians.

Three different cases of this kind of study will be investigated. For each of these, a description of the specific mathematical models that are constructed to evaluate the reliability of measurements is given.

A. CASE 1: THE ONE-WAY RANDOM EFFECTS MODEL

This model has each patient or subject being measured (evaluated) by the same device.

Let y_{ij} denote the jth rating on the ith patient (i=1,2,...k ; j=1,2,...n). It is quite possible that each patient need not be measured a fixed number of times n, in which case we would have unbalanced data where $j=1,2,...n_i$. In general we assume the following model for y_{ij}

$$y_{ij} = \mu + g_i + e_{ij} \qquad (2.1)$$

where μ is the overall population mean of the measurements; g_i is the true score of the ith patient; and e_{ij} is the residual error. It is assumed that g_i is normally distributed with mean zero and variance σ_g^2 and independent of e_{ij}. The error e_{ij} is also assumed to be normally distributed with mean zero and variance σ_e^2. The variance of y_{ij} is then given by $\sigma_y^2 = \sigma_g^2 + \sigma_e^2$.

Since

$$\text{Cov } (y_{ij}, y_{il}) = \sigma_g^2 \qquad i=1,2,...k ; j \neq l=1,2,...n_i$$

the correlation between any pair of measurements on the same patient is

$$\rho = \frac{cov(y_{ij}, y_{il})}{\sqrt{var(y_{ij}), \; var(y_{il})}}$$

$$= \frac{\sigma_g^2}{\sigma_g^2 + \sigma_e^2} \; .$$

This model is known as the components of variance model, since interest is often focused on estimation of the variance components σ_g^2 and σ_e^2. The ANOVA corresponding to (2.1) is shown in Table 2.1. In this table,

$$N = \sum_{i=1}^{k} n_i \, , \quad \bar{y}_i = \frac{1}{n_i} \sum_{j=1}^{n_i} y_{ij} \, , \quad \text{and} \quad \bar{y} = \frac{1}{N} \sum_{i}^{k} \sum_{j}^{n_i} y_{ij} \, ,$$

and

$$n_o = \frac{1}{k-1} \left[N - \sum_{i=1}^{k} n_i^2 / N \right] .$$

Since unbiased estimators of σ_e^2 and σ_g^2 are given respectively by $\hat{\sigma}_e^2 = MSW$ and $\hat{\sigma}_g^2 = (MSB-MSW)/n_o$, it is natural to define the ANOVA estimator of ρ as

$$r_1 = \hat{\sigma}_g^2 / (\hat{\sigma}_g^2 + \hat{\sigma}_e^2) = \frac{MSB-MSW}{MSB+(n_o-1)MSW} . \tag{2.2}$$

This estimator of reliability is biased.

Table 2.1
The ANOVA Table Under the One-way Random Effects Model.

Source of Variation	D.F.	S.S.	M.S.	E(M.S.)
Between Patients	K-1	$SSB = \sum_{i=1}^{K} n_i(\bar{y}_i - \bar{y})^2$	MSB=SSB/(K-1)	$(1+(n_0-1)\rho)\sigma_y^2$
Within Patients	N-K	$SSW = \sum_{i=1}^{K} \sum_{j=1}^{n_i} (y_{ij} - \bar{y}_i)^2$	MSW=SSW/(N-K)	$(1-\rho)\sigma_y^2$
Total	N-1	$SST = \sum_{i=1}^{K} \sum_{j=1}^{n_i} (y_{ij} - \bar{y})^2$		

When we have moderately large number of subjects (patients, slides, ... etc) a reasonably good approximation to the variance of r_1 derived by Smith (1956) is given by

$$v(r_1) = \frac{2(1-\rho)^2}{n_o^2} \left\{ \frac{[1+\rho(n_o-1)]^2}{N-k} + \frac{(k-1)(1-\rho)[1+\rho(2n_o-1)]+\rho^2\lambda(n_i)}{(k-1)^2} \right\}$$ (2.3)

where

$$\lambda(n_i) = \sum n_i^2 - 2N^{-1} \sum n_i^3 + N^{-2} \left(\sum n_i^2 \right)^2 \ .$$

Donner and Wells (1985) showed that an accurate approximation of the confidence limits on ρ for moderately large k, is given by:

$$\left\{ r_1 - Z_{\alpha/2}\sqrt{\hat{v}(r_1)}, \ r_1 + Z_{\alpha/2}\sqrt{\hat{v}(r_1)} \right\}$$

where $\hat{v}(r_1)$ is defined in (2.3), with r_1 replacing ρ, and $Z_{\alpha/2}$ is the two-sided critical value of the standard normal distribution corresponding to the α-level.

Note that the results of this section apply only to the estimate of reliability obtained from the one-way random effects model.

B. CASE 2: THE TWO-WAY RANDOM EFFECTS MODEL

In this model, a random sample of n clinicians is selected from a larger population and each clinician evaluates each patient.

When the n raters, clinicians or devices that took part in the reliability study are a sample from a larger population, the one-way random effects model should be extended to include a rater's effect, so that

$$y_{ij} = \mu + g_i + c_j + e_{ij} \ .$$ (2.4)

This is the so-called two-way random effects model, where the quantity c_j, which characterizes the additive effect of a randomly selected rater, is assumed to vary normally about a mean of zero with variance σ_c^2. The three random variables g,c, and e are assumed mutually independent. The variance of y_{ij} is

$$var(y_{ij}) = \sigma_g^2 + \sigma_c^2 + \sigma_e^2 \ ,$$

and the covariance between two measurements on the same patient, taken by the lth and j^{th} raters is $Cov(y_{ij}, y_{il}) = \sigma_g^2$.

22

Hence the intraclass correlation to be used as an appropriate measure of reliability under this model is,

$$R = \frac{\sigma_g^2}{\sigma_g^2 + \sigma_c^2 + \sigma_e^2} .$$ (2.5)

Table 2.2

The ANOVA Table Under the Two-way Random Effects and Mixed Effects Models

Source of Variation	D. F.	S.S.	M.S.	Raters Random (Case 2)	Raters Fixed (Case 3)
Patients	k-1	$n\sum (\bar{y}_{i.} - \bar{y})^2$	PMS	$\sigma_e^2 + n\sigma_g^2$	$\sigma_e^2 + n\,\sigma_g^2$
Raters	n-1	$k\sum (\bar{y}_{j} - \bar{y})^2$	CMS	$\sigma_e^2 + k\sigma_c^2$	$\sigma_e^2 + \dfrac{k}{n-1}\sum_{i=1}^{n} d_i^2$
Error	(k-1)(n-1)		MSW	σ_e^2	σ_e^2
Total		$\sum\sum (y_{ij} - \bar{y})^2$			

The unbiased variance components estimates of σ_g^2, σ_c^2, and σ_e^2 are given respectively as

$$\hat{\sigma}_g^2 = \frac{PMS - MSW}{n}$$

$$\hat{\sigma}_c^2 = \frac{CMS - MSW}{k}$$

$$\hat{\sigma}_e^2 = MSW .$$

23

An estimator of reliability is then

$$r_2 = \frac{k(PMS-MSW)}{k(PMS) + n(CMS) + (nk-n-k)MSW} \tag{2.6}$$

a formula that was derived by Bartko (1966).

C. CASE 3: THE TWO-WAY MIXED EFFECTS MODEL

What typifies this model is that each patient is measured by each of the same n raters, who are the only available raters. Furthermore, under this model, the raters taking part in the study are the only raters about whom inferences will be made. In the terminology of the analysis of variance, the raters effect is assumed fixed. The main distinction between Case 2 (where clinicians are considered random) and Case 3 (where clinicians are considered fixed) is that under Case 2 we wish to generalize the findings to other raters within a larger group of raters, whereas in the Case 3 situation we are only interested in the group that took part in the study.

Following Fleiss (1986), y_{ij} is written as

$$y_{ij} = \mu + g_i + d_j + e_{ij} \tag{2.7}$$

Here, $d_1, d_2, \ldots d_n$ are no longer random, rather they are assumed to be a set of fixed constants, and

$$\sum_{t=1}^{n} d_t = 0 \ .$$

The assumptions on g_i and e_{ij} are the same as those of Cases 1 and 2. The ANOVA for this case is provided by Table 2.2.

Under this model, the appropriate measure of reliability from Fleiss (1986), is given as

$$r_3 = \frac{k(PMS-MSW)}{k(PMS) + (n-1)CMS + (n-1)(k-1)MSW} \tag{2.8}$$

Fleiss (1986) describes the following sequential approach to estimating r_3.

1- Test for clinician to clinician variation to see if clinicians differ significantly from one another. To test this hypothesis (H_0: $d_1 = d_2 = .. = d_n = 0$) one would compare the ratio $F = CMS/MSW$ to tables of the F distribution with $(n-1)$ and $(n-1)(k-1)$ degrees of freedom. Acceptance of the null hypothesis would imply the absence of inter-clinician bias, thus one could estimate the reliability using (equation 2.8). If $F > F_{(n-1),(n-1)(k-1),\alpha}$ then the null hypothesis is rejected indicating that differential measurement bias exists.

2-   When the above hypothesis is rejected, the clinician or the rater responsible for most of the significance of the differences among the raters should be determined. There is no doubt that the estimated reliability will be higher if this rater is removed.

For example, if the jth rater is the one in doubt we form the following contrast

$$L = \bar{y}_j - \frac{1}{n-1} (\bar{y}_{.1} + ... \bar{y}_{j-1} + \bar{y}_{j+1} + ... + \bar{y}_{.n})$$

with a standard error

$$SE(L) = \left[\frac{n\, MSW}{k(n-1)} \right]^{1/2} .$$

The jth rater is considered non-conforming if $\dfrac{L}{SE(L)}$ exceeds $| t_{(n-1)(k-1),\alpha/2} |$, in which case

 a recommended strategy is to drop this rater, construct a new ANOVA, and estimate reliability using (2.8).

Example 2.1

In this example, the following data will be analyzed as Case 1, Case 2 and Case 3 situations. An analytical chemist wishes to estimate with high accuracy the critical weight of chemical compound. She has 10 batches and 4 different scales. Each batch is weighed once using each of the scales. A SAS program which outlines how to create the output for the data described in Table 2.3 is given below.

Table 2.3
Critical Weights of a Chemical Component Using 4 Scales and 10 Batches.

Scale	1	2	3	4	5	6	7	8	9	10
1	55	44.5	35	45	50	37.5	50	45	65	58.5
2	56	44.5	33	45	52	40	53	50	66	56.5
3	65	62	40	37	59	50	65	50	75	65
4	58.5	45	35.5	37	55	40	48	47	79	57

EXAMPLE SAS PROGRAM

```
data chemist;
input scale batch y;
cards;
1 1  55
1 2  44.5
1 3  35
....

4 7  48
4 8  47
4 9  79
4 10 57
;

/** one way **/
proc glm data=chemist;
class batch;
model y=batch;
run;

/** two way - random mixed **/
proc glm data=chemist;
class batch scale;
model y=batch scale;
run;

/** two way dropping scale 3 **/
data drop3;
set chemist;
if scale=3 then delete;

proc glm data=drop3;
class batch scale;
model y=batch scale;
run;
```

As a Case 1 Situation:

Treating this as a Case 1 situation by ignoring the scale effect (which means that the order of the measurements of weights 'within' a batch does not matter), the measurement reliability is computed as r_1 (see equation 2.2).

Extracts from the ANOVA tables in the SAS OUTPUT are given here:

Source	DF	Sum Squares	Mean Square	F value	Pr>F
Model	9	3913.881250	434.875694	14.60	0.0001
Error	30	893.562500	29.785417		
C Total	39	4807.443750			

26

Therefore, r_1 = [434.875694 - 29.785417]/[434.875694 + (4-1)(29.785417)]
                = 0.773

This is an extremely good reliability score based on the 'standard values' of comparison which are: excellent (> .75), good (.40, .75) and poor (< .40).

As a Case 2 Situation:

The ANOVA OUTPUT for the two-way random effects model that results is,

Source	DF	Sum Squares	Mean Square	F Value	Pr>F
Model	12	4333.0000	361.083333	20.55	0.0001
BATCH	9	3913.88125	434.875694	24.75	0.0001
SCALE	3	419.11875	139.706250	7.95	0.0006
Error	27	474.118750	17.571991		
C Total	39	4807.44375			

Therefore, r_2 (equation 2.6) is computed as,
r_2 = 10[434.8757 - 17.572] /[(10)(434.8757) + (4-1)(139.70625) + (4-1)(10-1)(17.571991)]
    = 0.796

Again, this value indicates an excellent measure of reliability, but note that the scale effect is highly significant (p=0.0006).

As a Case 3 Situation:

First of all we can test whether Scale 3 is significantly 'deviant', by using the contrast,

$$L = \bar{y}_{.3} - \frac{1}{3} (\bar{y}_{.1} + \bar{y}_{.2} + \bar{y}_{.4})$$

with standard error,

$$SE(L) = \left| \frac{(n)(MSW)}{k(n-1)} \right|^{\frac{1}{2}}$$

Thus, L = 7.35, SE(L) = 1.5307 and L/SE(L) = 4.802.

Since 4.802 exeeds the value of t = 2.052 (27 df at α = 0.05), there is reason to remove this 'deviant' scale.

Now, by removing the 'deviant' scale (scale 3) we can recalculate the reliability measure under the fixed effects model.

The ANOVA for the two-way mixed effects model that results is,

Source	DF	Sum Squares	Mean Square	F Value	Pr>F
Model	11	2865.45833	260.496212	22.85	0.0001
BATCH	9	3913.88125	434.875694	27.79	0.0001
SCALE	2	13.9500	6.975000	0.61	0.5533
Error	18	205.216667	11.400926		
C Total	29	3070.6750			

Note that after removing scale 3 the F-value of the scale effect is no longer significant. In calculating r_3 (equation 2.8),

$$r_3 = 10[316.834 - 11.4009]/[(10)(316.8343 + (3-1)(6.975) + (10-1)(3-1)(11.4009)] = 0.911$$

We can see that the removal of the 3rd scale has resulted in a considerable improvement in the reliability score, thus, if the scales are considered a fixed effect, it would be wise to drop the third scale.

D. COVARIATE ADJUSTMENT IN THE ONE-WAY RANDOM EFFECTS MODEL

Suppose that each of k patients is measured n_i times. In a practical sense, not all measurements on each patient are taken simultaneously, but at different time points. Since the measurements may be affected by the time point at which they are taken, one should account for the possibility of a time effect. In general, let x_{ij} denote the covariate measurement of the ith patient at the jth rating. The one-way analysis of covariance random effects model is given by

$$y_{ij} = \mu + g_i + \beta(x_{ij} - \bar{x}) + e_{ij}$$

where

$$\bar{x} = \frac{1}{N} \sum_i^k \sum_j^{n_i} x_{ij} \quad i=1,2,...k$$
$$j=1,2,...n_i$$

and

$$N = \sum_{i=1}^k n_i \quad .$$

28

Note that the assumptions for g_i and e_{ij} as described in section A are still in effect. This model is useful when an estimate of the reliability ρ, while controlling for a potentially confounding variable, such as time, is desired.

Define:

$$\bar{x}_i = \frac{1}{n_i} \sum_j x_{ij} \quad , \quad \bar{y}_i = \frac{1}{n_i} \sum_j y_{ij}$$

$$\bar{y} = \frac{1}{N} \sum_i \sum_j y_{ij} \quad ,$$

$$E_{yy} = \sum_i \sum_j (y_{ij} - \bar{y}_i)^2 \quad ,$$

$$E_{xx} = \sum_i \sum_j (x_{ij} - \bar{x}_i)^2$$

$$E_{xy} = \sum_i \sum_j (y_{ij} - \bar{y}_i)(x_{ij} - \bar{x}_i)$$

$$T_{yy} = \sum_i \sum_j (y_{ij} - \bar{y})^2$$

$$T_{xx} = \sum \sum (x_{ij} - \bar{x})^2$$

$$T_{xy} = \sum \sum (y_{ij} - \bar{y})(x_{ij} - \bar{x})$$

From Snedecor and Cochran (1980), the mean squares between patients and within patients, MSB^x and MSW^x respectively, are given as:

$$MSB^x = \frac{1}{k-1}\left[T_{yy} - \frac{T_{xy}^2}{T_{xx}} - \left(E_{yy} - \frac{E_{xy}^2}{E_{xx}}\right)\right]$$

$$MSW^x = \frac{1}{N-k-1}\left[E_{yy} - \frac{E_{xy}^2}{E_{xx}}\right]$$

Stanish and Taylor (1983) showed that the estimated intraclass correlation which controls for the confounder x is given as

$$r_1^x = \frac{MSB^x - MSW^x}{MSB^x + (m_o-1)MSW^x}$$

where

$$m_0 = \frac{1}{k-1}\left[N - \frac{\sum n_i^2}{N} - \frac{\sum n_i^2(\bar{x}_i-\bar{x})^2}{T_{xx}}\right].$$

III. COMPARING THE PRECISIONS OF TWO METHODS
(GRUBBS, 1948 and SHUKLA, 1973)

In this section we will be concerned with the comparison of the performance of two measuring methods, where every sample is measured twice by both methods. This means that each patient will provide four measurements. The data layout is given in Table 2.4; the replications from the two methods are obtained from the same patient. This produces estimates of variance components and indices of reliability that are correlated, and hence, statistical inference procedures need to be developed to deal with this situation.

Table 2.4
Measurements from Two Methods Using k Patients

Patient

	1	2	.	.	.	.	.	k
Method(1)	x_{11}	x_{21}	.	.	.	.	.	x_{k1}
	x_{12}	x_{22}	.	.	.	.	.	x_{k2}
Method(2)	y_{11}	y_{21}	.	.	.			y_{k1}
	y_{12}	y_{22}	.	.	.			y_{k2}

Let
$$x_{ij} = \alpha + \mu_i + \xi_{ij} \qquad i=1,2,...k$$
$$y_{ij} = \alpha + \mu_i + \eta_{ij} \qquad j=1,2$$

where x_{ij} is the jth replication on the ith patient obtained by the first method and y_{ij} is the jth replication on the ith patient obtained by the second method.

Let

$$x_i = \frac{x_{i1}+x_{i2}}{2}, \quad \text{and} \quad y_i = \frac{y_{i1}+y_{i2}}{2}.$$

Here, μ_i is the effect of the ith patient, ξ_{ij} and η_{ij} are the measurement errors.

Define the estimates of the variances and covariance as,

$$S_{xx} = \frac{1}{k-1} \sum_{i=1}^{k} (x_i - \bar{x})^2,$$

$$S_{yy} = \frac{1}{k-1} \sum_{i=1}^{k} (y_i - \bar{y})^2,$$

and

$$S_{xy} = \frac{1}{k-1} \sum_{i=1}^{k} (x_i - \bar{x})(y_i - \bar{y}).$$

Assume that $\mu_i \sim N(0,\sigma^2)$, $\xi_{ij} \sim N(0,\sigma_\xi^2)$, and $\eta_{ij} \sim N(0,\sigma_\eta^2)$ and that μ_i, ξ_{ij} and η_{ij} are mutually independent. Thus,

$$var(x_i) = \sigma^2 + \frac{1}{2} \sigma_\xi^2$$

$$var(y_i) = \sigma^2 + \frac{1}{2} \sigma_\eta^2$$

and

$$cov(x_i, y_i) = \sigma^2.$$

Grubbs (1948) showed that the MLE of σ^2, σ_ξ^2, σ_η^2, are given respectively as

$$\hat{\sigma}^2 = s_{xy}$$

$$\hat{\sigma}_\xi^2 = 2(s_{xx} - s_{xy})$$

$$\hat{\sigma}_\eta^2 = 2(s_{yy} - s_{xy}) \ .$$

If we let $u_i = x_i + y_i$
$v_i = x_i - y_i$

then the Pearson's correlation between u_i and v_i is

$$r = \frac{s_{xx} - s_{yy}}{\left[(s_{xx} + s_{yy} - 2s_{xy})(s_{xx} + s_{yy} + 2s_{xy})\right]^{1/2}} \ . \tag{2.9}$$

The null hypothesis by which one can test if the two methods are equally precise is $H_0 : \sigma_\xi^2 = \sigma_\eta^2$, versus the alternative $H_1 : \sigma_\xi^2 \neq \sigma_\eta^2$.

Shukla (1973) showed that H_0 is rejected whenever

$$t_o = r\sqrt{\frac{k-2}{1-r^2}} \tag{2.10}$$

exceeds $|t_{\alpha/2}|$ where $t_{\alpha/2}$ is the cut-off point found in the t-table at $(1-\alpha/2)\,100\%$ confidence, and $(k-2)$ degrees of freedom.

Note that we can obtain estimates of reliability for method (1) and method (2) using the one-way ANOVA. Clearly such estimates are correlated since the same group of subjects is evaluated by both methods. A test of the equality of the intraclass correlations ρ_1 and ρ_2 must account for the correlation between $\hat{\rho}_1$ and $\hat{\rho}_2$.

Alsawalmeh and Feldt (1994) developed a test statistic on the null hypothesis $H_o : \rho_1 = \rho_2$.

They proposed rejecting H_0 for extreme values of

$$T = \frac{1 - \hat{\rho}_1}{1 - \hat{\rho}_2},$$

where $\hat{\rho}_j$ is the one way ANOVA of ρ_j.

The statistic T, is approximately distributed as an F random variable with d_1 and d_2 degrees of freedom, where,

$$d_2 = \frac{2M}{M-1}, \quad d_1 = \frac{2d_2^3 - 4d_2^2}{(d_2 - 2)^2(d_2 - 4)V - 2d_2^2}$$

$$M = \frac{E_1}{E_2} + \frac{E_1}{E_2^3}V_2 - \frac{C_{12}}{E_2^2}$$

$$E_j = \frac{v_j}{v_j - 2} - \frac{(1 - \rho_j)}{(k - 1)} \qquad j = 1,2$$

$$V_j = \frac{2v_j^2(C_j + v_j - 2)}{C_j(v_j - 2)^2(v_j - 4)} - \frac{2(1 - \rho_j)}{k - 1}$$

$$v_j = \frac{2(k - 1)}{1 + \rho_j^2}, \quad C_j = (k - 1), \quad C_{12} = \frac{2}{(k - 1)}\rho_{12}^2,$$

and,

$$V = \left(\frac{E_1}{E_2}\right)^2 \left[\frac{V_1}{E_1^2} + \frac{V_2}{E_2^2} - \frac{2C_{12}}{E_1 E_2}\right]$$

The parameter ρ_{12} is the correlation between individual measurements x_j and y_j from the two methods. An estimate of ρ_{12} is given by,

$$\hat{\rho}_{12} = \frac{\hat{\sigma}^2}{\left[(\hat{\sigma}^2 + \hat{\sigma}_\xi^2)(\hat{\sigma}^2 + \hat{\sigma}_\eta^2)\right]^{1/2}}$$

Unfortunately, the test is only valid when k>50.

Example 2.2 Comparing the precision of two methods:

The following data (Table 2.5) were provided by Drs. Viel and J. Allen of the Ontario Veterinary College in investigating the reliability of readings of bacterial contents from 15 samples of nasal swabs taken from 15 animals with respiratory infection.

Table 2.5
Bacterial Counts Taken by Two Raters on 15 slides; Two Readings For Each Slide

Rater	Reading	\multicolumn{15}{c}{SLIDE}
		1	2	3	4	5	6	7	8	9	10	11	12	13	14	15
1	1	52	68	106	98	40	87	98	98	122	66	80	94	98	105	93
	2	62	62	104	98	46	77	107	101	113	75	89	99	98	77	105
2	1	72	68	102	71	77	78	89	75	105	71	92	88	82	74	116
	2	93	63	61	79	44	90	98	88	109	71	82	73	92	72	98

Let X represent the mean values of readings 1 and 2 for rater 1 and Y represent the mean values of readings 1 and 2 for rater 2.

Then, $s_{xx} = 411.817$ $s_{yy} = 172.995$ $s_{xy} = 186.037$

and, by *Grubb's theorem*,

34

$$\hat{\sigma}^2_\xi = 2(s_{xx} - s_{xy}) = 2(411.81667 - 186.0369) = 451.156$$
$$\hat{\sigma}^2_\eta = 2(s_{yy} - s_{xy}) = 2(172.99524 - 186.0369) = -26.083 \approx 0$$

so, we can see that the variances of the random error are not equal.

Now, to test that the two methods are equally precise, that is to test the null hypothesis;

$$H_0 : \sigma^2_\xi = \sigma^2_\eta,$$

we define $U_i = X_i + Y_i$ and $V_i = X_i - Y_i$.

Then the correlation between U_i and V_i is, $r = .529$, and the t-statistic is $t_0 = 2.249$ where,

$$t_0 = r\sqrt{\frac{k-2}{1-r^2}}.$$

Since t_0 exceeds $t_{13,.025} = 2.160$, we conclude that there is significant difference between the precision of the two raters.

IV. CONCORDANCE CORRELATION AS A MEASURE OF REPRODUCIBILITY

In this section our focus will remain on the statistical analysis of evaluating agreements between two measuring devices, or alternately, "split halves", or repeated measures using the same instrument. It is quite common among clinicians and chemists, to use linear regression techniques for the relative calibration of measuring instruments. They are particularly fond of using Pearson product-moment correlation as a measure of agreement between pairs of measurements. Linear regression is not appropriate when both measures are subject to error because product-moment correlation coefficients are measures of association and are not suitable as indices of agreement. Other commonly used methods of validation include, the paired t-test, the coefficient of variation, and the simultaneous test of slope (=1) and intercept (=0) of the least squares regression line. None of these methods alone can fully assess the desired reproducibility characteristics. For example, to evaluate the blood cell counter for hematology analysis in a laboratory, it is desirable to have duplicates of the same blood sample measured by the counter at different times (usually at most 1 day apart). When we sketch the scatter plot of the first measurement against the second measurement of the red blood cell counts for all blood samples available we hope to see, within tolerance limits on the error, that the measurements fall on a 45° line through the origin (0,0). The Pearson product-moment correlation measures a linear relationship but fails to detect any departure from the 45° line through the origin. Figures 2.1a, 1b and 1c illustrate this situation.

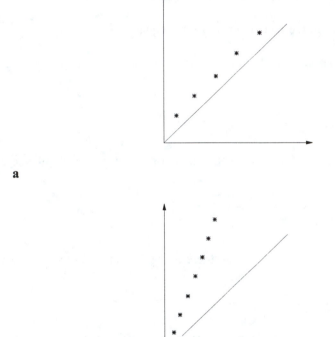

a

b

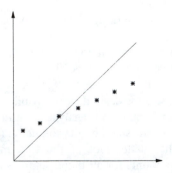

c

Figure 2.1   Cases when the Pearson's product moment correlation attains its maximum value of 1, even though the agreement is poor.

To show how the paired t-test falls short in the detection of poor agreement let us consider the following numerical examples:

Example 2.3

x : 3	4	5	6	7
y : 5	5	5	5	5

$$t = \frac{\sqrt{n}\ \bar{d}}{S_d} = 0 \qquad \text{(p-value} = 0.50).$$

Here, even though there is virtually no agreement, we fail to reject the hypothesis $H_0:\mu_x=\mu_y$.

Example 2.4

x : 4	5	6	7	8
y : 8.5	8.6	8.7	8.8	8.9

The Pearson's product correlation (r) is 1, even though the paired t-test on the hypothesis $H_0:\mu_x=\mu_y$ is

$$t = -4.24 \ , \ with \ p \approx 0.0001 \ .$$

Thus, despite the fact that the two sets of values do not agree, the value of r implies that there is total agreement. The worst is yet to come!

Suppose now we test the simultaneous hypothesis H_0:intercept (α) = 0 and slope (β) = 1. From Leugrans (1980), the test statistic on the above hypothesis is

$$F = (MSE)^{-1} \left[k\ \hat{\alpha}^2 + (\hat{\beta}-1)^2 \sum_{i=1}^{k} x_i^2 + 2\ k\ \bar{x}\ \hat{\alpha}(\hat{\beta}-1) \right] \ ,$$

where $\hat{\alpha}$ and $\hat{\beta}$ are respectively the least squares estimates of the intercept and slope. The MSE is the residual mean square which can be obtained from the ANOVA of the regression line $y=\alpha+\beta x+\epsilon$. The hypothesis $H_0:\alpha=0$, $\beta=1$ is rejected when F exceeds $F_{\alpha,2,k-2}$. From the definition of the test statistic, we may fail to reject the hypothesis if data are very scattered (see Figure 2.2) . This means that the more the data are scattered, or MSE is large (less agreement), the less chance one could reject the above hypothesis.

On the other hand, based on the above statistic F, we can reject a highly reproducible technique, if MSE is very small, which is also true when a paired t-test is used (see Figure 2.3)

To avoid the deficiencies of the paired t-test and Pearson's product correlation in the context of reproducibility, Lin (1989) introduced a measure of reproducibility which he termed, "Concordance Correlation Coefficient" denoted by ρ_c. We shall show in a later chapter that ρ_c is analogous to Cohen's weighted kappa which is used as a measure of agreement between two clinicians when the response categories are ordinal.

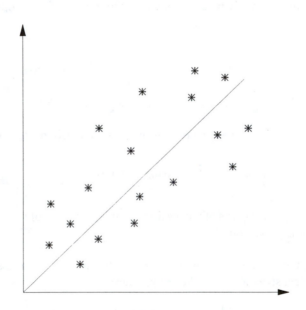

Figure 2.2 H_0:α=0; β=1 is not rejected even though data are very scattered.

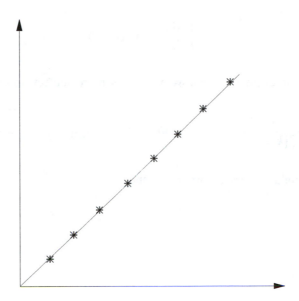

Figure 2.3 H_0: $\alpha=0$; $\beta=1$ is rejected where the technique is in fact highly reproducible.

The concordance correlation coefficient of Lin is estimated by:

$$\rho_c = \frac{2s_{xy}^{*}}{s_x^{*2} + s_y^{*2} + (\bar{x}-\bar{y})^2}$$

(2.11)

where

$$\bar{x} = \frac{1}{k}\sum_{i=1}^{k}x_i \quad , \quad \bar{y} = \frac{1}{k}\sum_{i=1}^{k}y_i$$

$$s_x^{*2} = \frac{1}{k}\sum_{i=1}^{k}(x_i-\bar{x})^2$$

$$s_y^{*2} = \frac{1}{k}\sum_{i=1}^{k}(y_i-\bar{y})^2$$

$$s_{xy}^* = \frac{1}{k} \sum_{i=1}^{k} (x_i - \bar{x})(y_i - \bar{y}) \ .$$

In the case of large samples (that is when we have 30 or more subjects to measure), the standard error of ρ_c is given by

$$SE(\rho_c) = \left\{ \frac{1}{k-2} \left[\left((1-r^2)\rho_c^2(1-\rho_c^2)/r^2 \right) + \left(4\rho_c^3(1-\rho_c)u^2/r \right) - 2\rho_c^4 \ u^4/r^2 \right] \right\}^{1/2}$$

where r is Pearson's product moment correlation, and

$$u = (\bar{x} - \bar{y}) / \sqrt{s_x^* s_y^*} \ .$$

It should be noted that:

i) $\rho_c = 0$ if and only if $r=0$
ii) $\rho_c = r$ if and only if $\mu_x = \mu_y$ and $\sigma_x = \sigma_y$
iii) $\rho_c = \pm 1$ if and only if $r = \pm 1$, $\sigma_x = \sigma_y$, and $\mu_x = \mu_y$

   This concordance correlation ρ_c evaluates the degree to which pairs fall on the 45° line through the origin (0,0). Any departure from this line would produce $\rho_c < 1$ even if $r=1$, an illustration of this being found in Example 2.5.

Example 2.5

x :	4	5	6	7	8
y :	8.5	8.6	8.7	8.8	8.9

$\bar{x} = 6, \bar{y} = 8.7, s_x^2 = 2, s_y^2 = 0.02 \ s_{xy} = 0.2$, and hence $r = 1$. This gives $\rho_c = 0.043$, which is a very poor agreement, even though Pearson's correlation attains its maximum value of 1. The estimated standard error of ρ_c is 0.0259.

Remarks:

1. Before the concordance correlation is computed, we recommend plotting half the difference between the two measurements $\frac{(x_i - y_i)}{2}$ against their mean $\frac{(x_i + y_i)}{2}$.

This plot can be useful in detecting systematic bias, outliers and whether the variance of the measurements is related to the mean. This approach was first suggested by Bland and Altman (1986).

2. Assessing measurement reliability is a vast area of research that attracts the attention of statisticians, quality assurance engineers, chemists, clinicians, and researchers in numerous other disciplines. The literature on this topic is quite voluminous and interesting; we refer the reader to the recent review paper by Dunn (1992). Judging by the work that has already been published, our account of some of the methods of evaluating reliability is by no means exhaustive. Indeed, we hope to have shed some light on the complex nature of the particular designs that need to be used and their importance in correctly analyzing the errors produced by measuring devices.

V. ASSESSING REPRODUCIBILITY USING THE COEFFICIENT OF VARIATION

Quan and Shih (1996) noted that the intraclass correlation coefficient is study population based since it involves between-subject variation. This may cause problems in comparing results from different studies. For example, a clinical test may be shown to be more reproducible just by applying it to a more heterogeneous population (σ_g^2 is large) than a homogeneous population even when the within-subject variation is the same. To avoid this problem, they considered the within-subject coefficient of variation (WCV) as an alternative measure of reliability; the smaller the WCV, the better the reproducibility. Under the one-way model of 2.1, the WCV is defined as

$\delta = \frac{\sigma_e}{\mu}$ and is estimated by:

$$WCV = \frac{(MSW)^{1/2}}{\bar{y}} \qquad (2.12)$$

where $\bar{y}$ and *MSW* are defined in part A, Case 1, of this chapter. To construct confidence limits on δ, we need the standard error of WCV. From Quan and Shih (1996), the estimated standard error of WCV (assuming that the data are normally distributed) is :

$$S\hat{E}(WCV) = \left[\left(\frac{k\hat{\sigma}_e^2}{(\bar{y})^4}\right)\left(\frac{N\hat{\sigma}_e^2 + \left(\sum_{i=1}^{k} n_i^2\right)\hat{\sigma}_g^2}{N^2}\right) + \left(\frac{k\hat{\sigma}_e^2}{2(\bar{Y})^2 (N-k)}\right)\right]^{1/2}$$ (2.13)

Example 2.6

The following data (Table 2.6) are the systolic blood pressure scores of 10 patients. Three measurements were taken on each patient.

Table 2.6
Blood Pressure Scores Taken on 10 Patients.

				Patient					
1	2	3	4	5	6	7	8	9	10
121	130	140	117	111	133	130	140	132	111
120	129	121	112	117	130	129	141	128	120
119	127	119	111	115	131	128	140	129	129

$N=30$, $k=10$, $\sum_{i=1}^{10} n_i^2 = 90$, $\bar{y} = 125.33$, $MSW = \hat{\sigma}_e^2 = 24.63$, $\hat{\sigma}_g^2 = 61.57$.

Therefore, WCV=0.04; its standard error is 0.02 with a 95% confidence interval on δ of (0.001, 0.08).

Note that $\hat{\rho} = \dfrac{61.57}{61.57 + 24.63} = 0.72$.

Remark:

The intraclass correlation coefficient is definitely a good measure of reliability. In situations where the reliability of two measuring devices are being compared, based on two independent samples, we recommend comparing their WCV only if the sample means of all measurements taken by the first method differ substantially from that of the second method.

Chapter 3

STATISTICAL ANALYSIS OF CROSS CLASSIFIED DATA

I. INTRODUCTION

There are two broad categories of investigative studies that produce statistical data; the first is *designed controlled experiments* and the second is *observational studies*. Controlled experiments are conducted to achieve the following two major objectives: i) To define with absolute clarity and appreciable power the effects of the treatment (or combination of treatments) structure, and ii) to ensure maximum control on the experimental error. There are several advantages of studies run under controlled experiments, the most important being that it permits one to disentangle a complex causal problem by proceeding in a stepwise fashion. As was suggested by Schlesselman (1982), within an experimental procedure the researcher can decompose the main problem and explore each component by a series of separate experiments with comparatively simple causal assumptions.

Observational studies are those which are concerned with investigating relationships among characteristics of certain populations, where the groups to be compared have not been formed by random allocation. A frequently used example to illustrate this point is that of establishing a link between smoking cigarettes and contracting lung cancer. This cannot be explored using random allocation of patients to smoking and non-smoking conditions for obvious ethical and practical reasons. Therefore, the researcher must rely on observational studies to provide evidence of epidemiological and statistical associations between a possible risk factor and a disease.

One of the most common and important questions in observational investigation involves assessing the occurrence of the disease under study in the presence of a potential risk factor. The most frequently employed means of presenting statistical evidence is the 2x2 table. The following section is devoted to assessing the significance of association between disease and a risk factor in a 2x2 table.

II. MEASURES OF ASSOCIATION IN 2X2 TABLES

In this section, an individual classified as *diseased* will be denoted by D, and by $\bar{D}$ if *not diseased*. Exposure to the risk factor is denoted by E and $\bar{E}$ for *exposed* and *unexposed* respectively. Table 3.1 illustrates how a sample of size n is cross classified according to the above notation.

Table 3.1
Model 2x2 Table

	Disease		
Exposure	D	$\bar{D}$	Total
E	n_{11}	n_{12}	$n_{1.}$
$\bar{E}$	n_{21}	n_{22}	$n_{2.}$
Total	$n_{.1}$	$n_{.2}$	n

There are, in practice, several methods of sampling by which the above table of frequencies can be obtained. The three most common methods are the cross sectional (historical cohort study), the cohort study (prospective design) and the case control study (retrospective design). These are described here with regards to the 2x2 table analysis.

A. CROSS SECTIONAL SAMPLING

This method calls for the selection of a total of n subjects from a large population after which the presence or absence of disease and the presence or absence of exposure is determined for each subject. Only the sample size can be specified prior to the collection of data. With this method of sampling, the issue of association between disease and exposure is the main concern. In the population from which the sample is drawn the unknown proportions are denoted as in Table 3.2.

Table 3.2
Model 2x2 Table
For Cross-Sectional Sampling

	Disease		
Exposure	D	$\bar{D}$	Total
E	p_{11}	p_{12}	$p_{1.}$
$\bar{E}$	p_{21}	p_{22}	$p_{2.}$
Total	$p_{.1}$	$p_{.2}$	1

Disease and exposure are independent if and only if $p_{ij} = p_{i.}p_{.j}$ $(i,j = 1,2)$. Assessing independence based on the sample outcome is determined by how close the value of n_{ij} is to $e_{ij} = n \hat{p}_{i.}\hat{p}_{.j}$ (the expected frequency under independence), where

$$\hat{p}_{i.} = \frac{n_{i.}}{n} \quad and \quad \hat{p}_{.j} = \frac{n_{.j}}{n}$$

44

are the maximum likelihood estimators of $p_{i.}$ and $p_{.j}$ respectively. There are two commonly used measures of distance between n_{ij} and e_{ij}. The first is the Pearson χ^2 statistic, where

$$\chi^2 = \sum_{ij} \sum \frac{(n_{ij} - e_{ij})^2}{e_{ij}} \ ,$$

which with Yate's continuity correction becomes,

$$\chi^2 = \sum_{ij} \sum \frac{(|n_{ij} - e_{ij}| - 1/2)^2}{e_{ij}} \ .$$

The hypothesis of independence is rejected for values of χ^2 that exceed $\chi_{\alpha}^2{}_{,1}$ (the cut off value of chi-square at α-level of significance and 1 degree of freedom). The second is Wilk's statistic,

$$G^2 = 2 \sum_{ij} \sum n_{ij} \left[\ln n_{ij} - \ln e_{ij} \right] .$$

This statistic is called the likelihood-ratio chi-squared statistic; the larger the value of G^2, the more evidence there is against the null hypothesis of independence.

When independence holds, the Pearson χ^2 statistic and the likelihood ratio statistic G^2 have asymptotic (i.e. in large samples) chi-squared distribution with 1 degree of freedom. It is not a simple matter to describe the sample size needed for the chi-square distribution to approximate well the exact distributions of χ^2 and G^2. Two items of note in this regard are that for a fixed number of cells (the case being discussed here), χ^2 converges more quickly than G^2, and the chi-square approximation is usually poor for G^2 when $n < 20$. Further guidelines regarding sample size considerations and the validity of χ^2 and G^2 are given in Agresti (1990; page 246).

The most commonly used measures of association between disease and exposure are the relative risk and the odds ratio. To explain how such measures are evaluated, changes in notation in the 2x2 table (Table 3.1) would be appropriate and are found in Table 3.3.

Table 3.3
Model 2x2 Table

Exposure	Disease		Total
	D	$\overline{D}$	
E	$n_{11}{=}y_1$	$n_{12}{\equiv}n_1{-}y_1$	n_1
$\overline{E}$	$n_{21}{=}y_2$	$n_{22}{\equiv}n_2{-}y_2$	n_2
Total	y.	n-y.	n

The following estimates obtained using the entries in Table 3.3, are of prime importance to epidemiologists and health officials.

- Estimated risk of disease among those exposed to the risk factor:

$$Pr[D|E] = \frac{y_1}{n_1} \equiv \hat{p}_1$$

- Estimated risk of disease among those not exposed to the risk factor:

$$Pr[D|\overline{E}] = \frac{y_2}{n_2} \equiv \hat{p}_2$$

- The risk of disease for those exposed to the risk factor relative to those not exposed is called the *relative risk* (RR).

$$RR = \frac{\left(\dfrac{y_1}{n_1}\right)}{\left(\dfrac{y_2}{n_2}\right)}$$

The relative risk represents how many times more (or less) likely disease occurs in the exposed group as compared with the unexposed. For RR > 1, a "positive" association is said to exist, and for RR < 1, there is a negative association.

- The fourth extensively used estimate is the *odds ratio*. Note that, if "T" is defined as an "event", then the odds of this event would be written as

$$\frac{Pr[T]}{1-Pr[T]}$$

Now, by defining T as "disease among the exposed", and in using this format, the odds of T would be

$$odds_E(T) = \frac{Pr\ [disease\ |\ exposed]}{1-Pr\ [disease\ |\ exposed]}$$

$$= \frac{\left(\dfrac{y_1}{n_1}\right)}{\dfrac{(n_1-y_1)}{n_1}} = \frac{y_1}{n_1-y_1}$$

Similarly we can take the event T to signify "disease among the unexposed", then

$$odds_{\overline{E}}(T) = \frac{Pr\ [disease\ |\ unexposed]}{1-Pr\ [disease\ |\ unexposed]}$$

$$= \frac{\left(\dfrac{y_2}{n_2}\right)}{\dfrac{(n_2-y_2)}{n_2}} = \frac{y_2}{n_2-y_2}$$

The odds ratio, denoted by ψ, is the ratio of these two odds, thus

$$\psi = \frac{odds_E(T)}{odds_{\overline{E}}(T)} = \frac{y_1(n_2-y_2)}{y_2(n_1-y_1)} \tag{3.1}$$

The odds ratio is a particularly important estimator in at least two contexts. One is that in the situation of rare diseases, the odds ratio approximates relative risk, and secondly, it can be determined from either cross sectional, cohort, or case-control studies as will be illustrated later in this chapter.

B. COHORT AND CASE-CONTROL STUDIES

In a cohort study, individuals are selected for observation and followed over time. Selection of subjects may depend on the presence or absence of exposure to risk factors that are believed to influence the development of the disease.

This method entails choosing and studying a predetermined number of individuals, n_1 and n_2, who are exposed and not exposed, respectively. This method of sampling forms the basis of

47

prospective or cohort study, and retrospective studies. In prospective studies, n_1 individuals with, and n_2 individuals without, a suspected risk factor are followed over time to determine the number of individuals developing the disease. In retrospective case-control studies, n_1 individuals are selected due to having the disease (cases) and n_2 non-diseased (controls) individuals would be investigated in terms of past exposure to the suspected antecedent risk factor.

The major difference between the cohort and case-control sampling methods is in the selection of study subjects. A cohort study selects individuals who are initially disease free and follows them over time to determine how many become ill. This would determine the rates of disease in the absence or presence of exposure. By contrast, the case-control method selects individuals on the basis of presence or absence of disease.

Recall that the odds ratio has been defined in terms of the odds of disease in exposed individuals relative to the odds of disease in the unexposed. An equivalent definition can be obtained in terms of the odds of exposure conditional on the disease, so that the odds of exposure among diseased and not diseased are:

$$odds_D(E) \;=\; \frac{Pr[E \mid D]}{1 - Pr[E \mid D]} \;=\; \frac{y_1}{y_2}$$

and

$$odds_{\bar{D}}(E) \;=\; \frac{Pr[E \mid \bar{D}]}{1 - Pr[E \mid \bar{D}]} \;=\; \frac{n_1 - y_1}{n_2 - y_2} \;.$$

The odds ratio of exposure in diseased individuals relative to the nondiseased is

$$\hat{\Psi} \;=\; \frac{odds_D(E)}{odds_{\bar{D}}(E)} \;=\; \frac{y_1(n_2 - y_2)}{y_2(n_1 - y_1)} \tag{3.2}$$

Thus the exposure odds ratio defined by equation (3.2) is equivalent to the disease odds ratio defined by equation (3.1). This relationship is quite important in the design of case-control studies.

III. STATISTICAL INFERENCE ON ODDS RATIO

Cox (1970) indicated that the statistical advantage of the odds ratio (OR) is that it can be estimated from any of the study designs which were outlined in the previous section (prospective cohort study, cross-sectional survey, and retrospective case-control study).

A problem that is frequently encountered when an estimate of OR is constructed is the situation where $n_{12}\, n_{21} = 0$ in which case Ψ is undefined. To allow for estimation under these conditions, Haldane (1956) suggested adding a correction term $\delta = \frac{1}{2}$ to all four cells in the 2x2 tables, to modify the estimator proposed earlier by Woolf (1955). The OR estimate is then given by

$$\psi_H = \frac{(n_{11}+.5)(n_{22}+.5)}{(n_{12}+.5)(n_{21}+.5)} .$$

Adding $\delta=\frac{1}{2}$ to all cells gives a less biased estimate than if it is added only as necessary, such as when a zero cell occurs (Walter 1985). Another estimator of ψ was given by Jewell (1984,1986) which is,

$$\psi_J = \frac{n_{11}n_{22}}{(n_{12}+1)(n_{21}+1)} .$$

The correction of $\delta=1$ to the n_{12} and n_{21} cells is intended to reduce the positive bias of the uncorrected estimator ψ and also to make it defined for all possible tables. Walter and Cook (1991) conducted a large scale Monte-Carlo simulation study to compare among several point estimators of the OR. Their conclusion was that for sample size $n \geq 25$, ψ_J has lower bias, mean-square error, and average absolute error, than the other estimators included in the study. Approximate variances of ψ, ψ_H and ψ_J are given by:

$$var(\psi) = (\psi)^2 \left| \frac{1}{n_{11}} + \frac{1}{n_{12}} + \frac{1}{n_{21}} + \frac{1}{n_{22}} \right| ;$$

that of ψ_H is

$$var(\psi_H) = (\psi_H)^2 \left| \frac{1}{n_{11}+.5} + \frac{1}{n_{12}+.5} + \frac{1}{n_{21}+.5} + \frac{1}{n_{22}+.5} \right| ,$$

and that of ψ_J is

$$var(\psi_J) = (\psi_J)^2 \left| \frac{1}{n_{11}} + \frac{1}{n_{22}} + \frac{1}{n_{12}+1} + \frac{1}{n_{21}+1} \right| .$$

Before we deal with the problem of significance testing of the odds ratio, there are several philosophical points of view concerning the issue of "statistical" significance as opposed to "scientific" significance. It is known that the general approach to testing the association between disease and exposure is verified by contradicting the null hypothesis $H_0:\psi=1$. The p-value of this test is a summary measure of the consistency of the data with the null hypothesis. A small p-value is evidence that the data are not consistent with the null hypothesis (in this case implying a significant association). As was indicated by Oakes (1986) the p-value should be considered only a guide to interpretation. The argument regarding the role played by the p-value in significance tests dates back to Fisher's work (1932). He indicated that the null hypothesis cannot be affirmed as such but is possibly disproved. On the other hand scientific inference is concerned with measuring the magnitude of an effect, regardless of whether the data are consistent with the null hypothesis. Therefore, the construction of a confidence interval on ψ is very desirable as an indication of whether or not the data contain adequate information to be consistent with the H_0, or to signal departures from the H_0 that are of scientific importance.

A. SIGNIFICANCE TESTS

The standard chi-square test for association in a 2x2 table provides an approximate test on the hypothesis $H_0 : \psi = 1$. Referring to the notation in Table 3.3, the χ^2 statistic is given by

$$\chi^2 = \frac{n(|y_1(n_2 - y_2) - y_2(n_1 - y_1)| - \frac{n}{2})^2}{n_1 n_2 y_.(n - y_.)} \tag{3.3}$$

Under H_0, the above statistic has an approximate chi-square distribution with one degree of freedom, and is used in testing the two-sided alternative. However, if there is interest in testing the null hypothesis of no association against a one-sided alternative, the standard normal approximation

$$Z_\alpha = \pm\sqrt{\chi^2}$$

can be used. In this case, the positive value is used for testing the alternative, $H_1 : \psi > 1$; the negative values test $H_1 : \psi < 1$. Quite frequently the sample size may not be sufficient for the asymptotic theory of the chi-square statistic to hold. In this case an exact test is recommended. The evolution of Fisher's Exact Test will now be illustrated. First let $p_{11} \equiv p_1$ and $p_{21} \equiv p_2$ be the proportion of diseased individuals in the population of exposed and unexposed respectively. Then, two independent samples, n_1 and n_2 are taken from the exposed and unexposed population. Referring to Table 3.3, it is clear that y_1 and y_2 are independent binomially distributed random variables with parameters (n_1, p_1) and (n_2, p_2). Hence their joint probability distribution is:

$$p(y_1, y_2) = \binom{n_1}{y_1} \binom{n_2}{y_2} p_1^{y_1} q_1^{n_1 - y_1} p_2^{y_2} q_2^{n_2 - y_2} \tag{3.4}$$

Under the transformation $y_. = y_1 + y_2$, equation (3.4) can be written as:

$$p(y_., y_1 | \psi) = \binom{n_1}{y_1} \binom{n_2}{y_. - y_1} \psi^{y_1} p_2^{y_.} q_2^{n_2 - y_.} q_1^{n_1}$$

where $\psi = p_1 q_2 / q_1 p_2$ is the population odds ratio parameter. Clearly $\psi = 1$ if and only if $p_1 = p_2$. Conditional on the sum $y_.$, the probability distribution of y_1 is

$$p(y_1 | y_., \psi) = c(y_., n_1, n_2; \psi) \binom{n_1}{y_1} \binom{n_2}{y_. - y_1} \psi^{y_1} \tag{3.5}$$

where

$$c^{-1}(y_{.},n_1,n_2;\psi) = \sum_x \binom{n_1}{x}\binom{n_2}{y_{.}-x}\psi^x$$

Under the hypothesis; $\Psi=1$, (3.5) becomes the hypergeometric distribution

$$p(y_1|y_{.},1) = \frac{\binom{n_1}{y_1}\binom{n_2}{y_{.}-y_1}}{\binom{n_1+n_2}{y_{.}}}. \tag{3.6}$$

The exact p-value of the test on the hypothesis $\Psi=1$ is calculated from (3.6) by summing the probabilities of obtaining all tables with the same marginal totals, with observed y_1 as extreme as that obtained from the sample.

Example 3.1

One of the researchers in VMI conducted a clinical trial on two drugs used for the treatment of diarrhea in calves. Colostrum deprived calves are given a standard dose of an infectious organism (strain B44 E.coli) at two days of age and then therapy is instituted as soon as the calves begin to show diarrhea. The following data were obtained from the trial

	Died	Lived
Drug (1)	7	2
Drug (2)	3	5

Since,

$$P(y_1|y_{.},1) = \frac{\binom{9}{y_1}\binom{8}{10-y_1}}{\binom{17}{10}}$$

we have

y_1	0	1	2	3	4	5	6	7	8	9
$P(y_1|y_{.},1)$	0	0	.0018	.0345	.1814	.363	.302	.104	.013	.0004

Hence the P-value of Fisher's exact test is
$$P = 0 + .0004 + .013 + .0345 + .0018 + .104 \approx .15$$

and therefore, no significant association between treatment and mortality can be justified by the data.

From the properties of the hypergeometric distribution, if $E(y_1 \mid y_\cdot, 1)$ and $Var(y_1 \mid y_\cdot, 1)$ denote the mean and variance of (3.6), then

$$E(y_1 \mid y_\cdot, 1) = \frac{n_1 y_\cdot}{n_1 + n_2} \tag{3.7}$$

and

$$Var(y_1 \mid y_\cdot, 1) = \frac{n_1 n_2 y_\cdot (n_1 + n_2 - y_\cdot)}{(n_1 + n_2)^2 (n_1 + n_2 - 1)} \; . \tag{3.8}$$

As an approximation to the tail area required to test $H_0: \Psi = 1$ we refer

$$Z = \frac{|y_1 - E(y_1 \mid y_\cdot, 1)| - \frac{1}{2}}{\sqrt{Var(y_1 \mid y_\cdot, 1)}} \tag{3.9}$$

to the table of standard normal distribution.

From the above example $y_1 = 7$, $E(y_1 \mid y_\cdot, 1) = 5.29$, and $Var(y_1 \mid y_\cdot, 1) = 1.03$. Hence $z = 1.19$, $p \approx .12$ indicating no significant association.

B. INTERVAL ESTIMATION

To construct an approximate confidence interval on ψ, it is assumed that when n is large, then $(\hat{\psi} - \psi)$ follows a normal distribution with mean 0 and variance $Var(\hat{\psi})$. An approximate $(1-\alpha)$ 100% confidence interval is

$$\hat{\psi} \pm Z_{\alpha/2} \sqrt{Var(\hat{\psi})} \; .$$

To avoid asymmetry, Woolf (1955) proposed constructing confidence limits on $\beta = \ln \Psi$. He showed that

$$Var(\hat{\beta}) = Var(\hat{\psi})/(\hat{\psi})^2 = \frac{1}{y_1} + \frac{1}{n_1 - y_1} + \frac{1}{y_2} + \frac{1}{n_2 - y_2} \tag{3.10}$$

and the lower and upper confidence limits on ψ are given respectively by

$$\psi_L = \psi \ exp \left[-Z_{\alpha/2} \ \sqrt{Var(\hat{\beta})} \right]$$

$$\psi_U = \psi \ exp \left[+Z_{\alpha/2} \ \sqrt{Var(\hat{\beta})} \right]$$

C. ANALYSIS OF SEVERAL 2x2 CONTINGENCY TABLES

Consider k pairs of mutually independent binomial variates y_{i1} and y_{i2} with corresponding parameters p_{i1} and p_{i2} and sample sizes n_{i1} and n_{i2}, where $i=1,2,...k$. This information has a k 2x2 table representation as follows.

		Disease		
		D	$\bar{D}$	Total
Exposure	E	y_{i1}	$n_{i1}-y_{i1}$	n_{i1}
	$\bar{E}$	y_{i2}	$n_{i2}-y_{i2}$	n_{i2}
	Total	$y_{i.}$	$n_i-y_{i.}$	n_i

There is a considerable literature on the estimation and significance testing of odds ratios in several 2x2 tables (Thomas and Gart, 1977; Fleiss, 1979; Gart and Thomas, 1982). The main focus of such studies was to address the following questions:

(i)   Does the odds ratio vary considerably from one table to another?

(ii)   If no significant variation among the k odds ratios is established, is the common odds ratio statistically significant?

iii)   If no significant variation among the *k* odds ratios is established, how can we construct confidence intervals on the common odds ratio after pooling information from all tables?

Before addressing these questions, the circumstances under which several 2x2 tables are produced will now be explored in more detail. One very important consideration is the effect of confounding variables. In a situation where a variable is correlated with both the disease and the exposure factor, 'confounding' is said to occur. Failure to adjust for this effect would bias the estimated odds ratio as a measure of association between the disease and exposure variables.

If we assume for simplicity, that the confounding variable has several distinct levels, then one way to control for its confounding effect is to construct a 2x2 table for the disease and exposure variable, at each level of the confounder. This procedure is known to epidemiologists as "stratification". Example 3.2 illustrates this idea.

Example 3.2

The following data (Table 3.4) are from a case-control study on enzootic pneumonia in pigs by Willeberg, (1980). Under investigation is the effect of ventilation systems where exposure (E) denotes those farms with fans.

Table 3.4
Classification of Diseased and Exposed Pigs in Enzootic Pneumonia Study
(Willeberg, 1980).

	Disease		
	D	$\bar{D}$	
(Fan) E	91	73	164
(No Fan) $\bar{E}$	25	60	Total
Total	116	133	249

$$\hat{\psi} = 2.99$$
$$\ln\hat{\psi} = 1.095$$

$$\sqrt{var(\hat{\beta})} = \left[\frac{1}{91} + \frac{1}{73} + \frac{1}{25} + \frac{1}{60}\right]^{\frac{1}{2}} = 0.285$$

and 95% confidence limits are:

$$\hat{\psi}_L = 2.99\,e^{-(1.96)(0.285)} = 1.71$$
$$\hat{\psi}_U = 2.99\,e^{(1.96)(0.285)} = 5.23$$

A factor which is not taken into account in analyzing the data as a whole is the size of the farms involved in the study. In attempting to filter out the effect of farm size (if it is present) on the disease, two groups are formed, large and small farms. By stratifying this possible confounder into large and small farms, the table that results (Table 3.5) produces the odds ratios found in Table 3.6. These subgroup analyses are aimed at evaluating the association between the disease and the risk factor (ventilation) at this level. Now it is evident that the relationship between the disease and the exposure factor is not clear; and this could be due to the possible confounding effect of farm size.

Table 3.5
Stratification of Farm Sizes

		Fan	Disease D	$\bar{D}$
Farm Size	Large	E	61	17
		$\bar{E}$	6	5
	Small	E	30	56
		$\bar{E}$	19	55

Table 3.6
Odds Ratios and Confidence Intervals After Stratification by Farm Size

Size	ψ_i	$\hat{\beta}_i$	$SE(\hat{\beta}_i)$	C.I. on ψ
large	2.99	1.095	0.665	(.67, 13.2)
small	1.55	0.438	0.349	(.74, 3.26)

For farms of large size, the estimated odds ratio of disease-risk association is $\psi = 2.99$ which is identical to the estimate obtained from Table 3.4. However, it is not statistically significant ($p \approx .10$, two-sided). It follows then that pooling the data from large and small farms to form a single 2x2 table can produce misleading results. Therefore the subgroup-specific odds ratio may be regarded as descriptive of the effects. Now, in the context of multiple tables, the three questions posed previously will be addressed.

1. Test of Homogeneity

This is a reformulation of the question "Does the odds ratio vary considerably across tables?" which tests the hypothesis H_0: $\Psi_1 = \Psi_2 \ldots \Psi_k = \Psi$. Woolf (1955) proposed a test that is based on the estimated log odds ratios ($\hat{\beta}_i$) and their estimated variances.

Since the estimated variance of ($\hat{\beta}_i$) is

$$Var\ (\hat{\beta}_i) = w_i^{-1} = \frac{1}{y_{i1}} + \frac{1}{n_{i1}-y_{i1}} + \frac{1}{y_{i2}} + \frac{1}{n_{i1}-y_{i2}}, \qquad (3.11)$$

an estimate of $\ln \Psi$ is constructed as

55

$$\hat{\beta} = \ln\hat{\psi} = \frac{\sum_{i=1}^{k} w_i \hat{\beta}_i}{\sum_{i=1}^{k} w_i}$$

Furthermore, it can be shown that

$$var\ (\hat{\beta}) = \frac{1}{\sum_{i=1}^{k} w_i} \quad .$$

To assess the homogeneity (i.e. constancy of odds ratios) the statistic

$$\chi_w^2 = \sum_{i=1}^{k} w_i (\hat{\beta}_i - \hat{\beta})^2$$

has approximately a chi-square distribution with k-1 degrees of freedom. Large values of χ_w^2 is an evidence against the homogeneity hypothesis.

Example 3.3

This example applies Woolf's Method to the summary estimates from Table 3.6 to test for a difference between the odds ratios of the two groups. The chi-square value is calculated as follows:

$$\hat{\beta}_1 = 1.095 \quad \hat{\beta}_2 = 0.438$$

$$w_1 = 2.26 \quad w_2 = 8.20, \quad \sum w_i = 10.46$$

$$\hat{\beta} = \frac{\sum w_i \beta_i}{\sum w_i} = 0.58 \tag{3.12}$$

$$\chi_w^2 = \sum_{i=1}^{k} w_i\ (\hat{\beta}_i - \hat{\beta})^2 = 0.764 \tag{3.13}$$

From the value of the χ^2 we can see that there is no significant difference between the odds ratios of the two groups.

Remark (test for interaction):

If we intend to find a summary odds ratio from several 2x2 tables, it is useful to test for interaction. Consider the summary calculations in Table 3.6. The odds ratios for the two strata are 2.99 and 1.55. If the underlying odds ratios for individual strata are really different, it is questionable if a summary estimate is appropriate. In support of this, the implication of a significant value of χ_w^2 should be stressed. A large value of χ_w^2 indicates that the homogeneity of ψ_i's is not supported by the data and that there is an interaction between the exposure and the stratification variable. Thus, in the case of a significant value of χ_w^2 one should not construct a single estimate of the common odds ratio.

2. Significance Test of Common Odds Ratio

Recall that inference on the odds ratio can be based on the total $y_{.i}$ conditional on the marginal total of the 2x2 table. From Cox and Snell (1989), the inference on the common odds ratio Ψ is based on the conditional distribution of

$$T = \sum_{i=1}^{k} y_{.i}$$

given the marginal total of all tables.

Now, we know that under $H_0:\psi=1$ the distribution of $y_{.i}$ (the total of diseased in the ith table) for a particular i is given by the hypergeometric distribution (3.6) and the required distributions of T is the convolution of k of these distributions. It is clear that this is impracticable for exact calculation of confidence limits. However, we can test the null hypothesis that $\Psi=1$ by noting from (3.7) and (3.8) that the mean and variance of T are given respectively as:

$$E(T|\psi=1) = \sum_{i=1}^{k} \frac{n_{i1} y_{.i}}{n_{i1}+n_{i2}}$$

$$Var(T|\psi=1) = \sum_{i=1}^{k} \frac{n_{i1} n_{i2} y_{.i}(n_{i1} + n_{i2} - y_{.i})}{(n_{i1} + n_{i2})^2 (n_{i1} + n_{i2} - 1)}$$

where $y_{.i}$ is the observed total number of diseased in the ith table. A normal approximation, with continuity correction, for the distribution of T will nearly always be adequate. Cox and Snell (1989) indicated that the approximation is good even for a single table and will be improved by convolution. The combined test of significance from several 2x2 contingency tables is done by referring

$$\chi_1^2 = \frac{\left(|T - E(T|\psi=1)| - \frac{1}{2}\right)^2}{Var(T|\psi=1)} \tag{3.14}$$

to the chi-square table with one degree of freedom.

Another form of the one-degree of freedom chi-square test on $H_0: \psi = 1$ was given by Mantel and Haenszel (1959) as

$$\chi_{mh}^2 = \frac{\left(\sum_{i=1}^{k} \frac{y_{i1}(n_{i2}-y_{i2}) - y_{i2}(n_{i1}-y_{i1})}{n_i}\right)^2}{\sum_{i=1}^{k} \frac{n_{i1}n_{i2}y_{.i}(n_i - y_{.i})}{n_i^2(n_i-1)}} \tag{3.15}$$

where, $n_i = n_{i1} + n_{i2}$, $y_{.i} = y_{i1} + y_{i2}$.

In Example 3.4 we consider the data in Table 3.5 and calculate both Cox and Snell, and Mantel and Haenszel formulations.

Example 3.4

As a test of significance of the common odds ratio using the data in Table 3.5, we have T=91, $E(T|\Psi=1) = 85.057$, and $Var(T|\Psi=1) = 10.3181$. Hence $\chi_1^2 = 2.872$, and the combined odds ratio is non-significant.

Using Mantel and Haenszel's χ_{mh}^2 formula we got $\chi_{mh}^2 = 2.882$ which is quite similar to the value obtained using the Cox and Snell method with similar conclusions.

3.   Confidence Interval on the Common Odds Ratio

Mantel and Haenszel (1959) suggested a highly efficient estimate of the common odds ratio from several 2x2 tables as:

$$\Psi_{mh} = \frac{\sum_{i=1}^{k} \frac{y_{i1}(n_{i2}-y_{i2})}{(n_{i1}+n_{i2})}}{\sum_{i=1}^{k} \frac{(n_{i1}-y_{i1})y_{i2}}{(n_{i1}+n_{i2})}}. \tag{3.16}$$

Note that $\hat{\beta}_{mh} = \ln \hat{\psi}_{mh}$ can be regarded as a special form of weighted means, based on the linearizing approximation near $\beta_{mh} = 0$.

This equivalence has been used to motivate various estimates of the variance of $\hat{\beta}_{mh}$. Robins et al.(1986) found that an estimated variance is

$$Var(\hat{\beta}_{mh}) = \frac{1}{2}\left\{ \sum \frac{A_i C_i}{C_.^2} + \sum \frac{(A_i D_i + B_i C_i)}{(C. D.)} + \sum \frac{B_i D_i}{D_.^2} \right\} \tag{3.17}$$

where

$$A_i = (y_{i1} + n_{i2} - y_{i2}) / (n_{i1} + n_{i2})$$
$$B_i = (y_{i2} + n_{i1} - y_{i1}) / (n_{i1} + n_{i2})$$
$$C_i = y_{i1} (n_{i2} - y_{i2}) / (n_{i1} + n_{i2})$$
$$D_i = y_{i2} (n_{i1} - y_{i1}) / (n_{i1} + n_{i2})$$

and

$$C. = \Sigma C_i , \quad D. = \Sigma D_i$$

Using the results from the previous example (Example 3.4), we find that $\hat{\psi}_{mh} = 1.76$ and that the 95% confidence limits for $\hat{\psi}_{mh}$ are (0.96, 3.24).

Hauck (1979) derived another estimator of the large sample variance of $\hat{\psi}_{mh}$ as

$$\hat{V}_{mh} = \frac{\hat{\psi}_{mh}^2 \left(\sum_{i=1}^{k} \hat{w}_i^2 \, \hat{b}_i \right)}{\left(\sum \hat{w}_i \right)^2} \tag{3.18}$$

where

$$\hat{w}_i = (n_{i1}^{-1} + n_{i2}^{-1})^{-1} \, \hat{p}_{i2}(1 - \hat{p}_{i1})$$

$$\hat{b}_i = \frac{1}{n_{i1} \, \hat{p}_{i1} \, \hat{q}_{i1}} + \frac{1}{n_{i2} \, \hat{p}_{i2} \, \hat{q}_{i2}}$$

$$\text{and} \quad \hat{p}_{ij} = \frac{y_{ij}}{n_{ij}} \quad j=1,2$$

$$\hat{q}_{ij} = 1 - \hat{p}_{ij}$$

IV. ANALYSIS OF MATCHED PAIRS (ONE CASE AND ONE CONTROL)

The *matching* of one or more control to each case means that the two are paired based on their similarities with respect to some characteristic(s). Pairing individuals can involve attributes such as age, weight, farm covariates, parity, hospital of admission and breed. These are just a few examples.

It is important to note that since cases and controls are believed to be similar on the matching variables, their differences with respect to disease may be attributed to different extraneous variables. It was pointed out by Schlesselman (1982; page 105) that "if the cases and controls differ with respect to some exposure variable, suggesting an association with the study disease, then this association cannot be explained in terms of case-control differences on the matching variables". The main objective of matching is in removing the bias that may affect the comparison between the cases and controls. In other words, matching attempts to achieve comparability on the important potential confounding variables (a confounder is an extraneous variable that is associated with the risk factor and the disease of interest). This strategy of matching is different from what is called "adjusting", in that adjusting attempts to correct for differences in the cases and controls during the data analysis step, as opposed to matching, which occurs at the design stage.

In this section we investigate the situation where one case is matched with a single control, where the risk factor is dichotomous. As before we denote the presence or absence of exposure to the risk factor by E and $\bar{E}$ respectively. In this situation responses are summarized by a 2x2 table which exhibits two important features. First, all probabilities or associations may show a symmetric pattern about the main diagonal of the table. Second, the marginal distributions may differ in some systematic way.

Subsections A and B will address the estimation of the odds ratio and testing the equality of the marginal distributions under the matched case-control situation.

A. ESTIMATING THE ODDS RATIO

Suppose that one has matched a single control to each case, and that the exposure under study is dichotomous. Denoting the presence or absence of exposure by E and $\bar{E}$ respectively there are four possible outcomes for each case-control pair.

To calculate ψ_{mh} for matched pair data, in which each pair is treated as a stratum, we must first change the matched pair table to its unpaired equivalent resulting in four possible outcomes (Tables 3.7, 3.8, 3.9, and 3.9a).

Table 3.7

outcome (1)   unpaired equivalent

control

	E	$\overline{E}$
E	1	0
$\overline{E}$	0	0

case

	case	control
E	1 (y_1)	1 (n_1-y_1)
E	0 (y_2)	0 (n_2-y_2)

$n=2$

$$y_1(n_2\text{-}y_2) = 0, \text{ and } y_2(n_1\text{-}y_1) = 0$$

Table 3.8

outcome (2)   unpaired equivalent

control

	E	$\overline{E}$
E	0	0
$\overline{E}$	1	0

case

	case	control
E	0	1
$\overline{E}$	1	0

$$y_1(n_2\text{-}y_2) = 0, \text{ and } y_2(n_1\text{-}y_1) = 1$$

Table 3.9

outcome (3)   unpaired equivalent

control

	E	$\overline{E}$
E	0	1
$\overline{E}$	0	0

case

	case	control
E	1	0
$\overline{E}$	0	1

$$y_1(n_2\text{-}y_2) = 1, \text{ and } y_2(n_1\text{-}y_1) = 0$$

Table 3.9a

outcome (4)   unpaired equivalent

control

	E	$\bar{E}$
E	0	0
$\bar{E}$	0	1

case (for first table, row labels E, $\bar{E}$)

	case	control
E	0	0
$\bar{E}$	1	1

$$y_1(n_2-y_2) = y_2(n_1-y_1) = 0$$

Since,

$$\Psi_{mhp} = \frac{\sum y_{i1}(n_{i2}-y_{i2})/n_i}{\sum y_{i2}(n_{i1}-y_{i1})/n_i}$$

and $n_i=2$, then,

$$\Psi_{mhp} = \frac{\text{number of pairs with } y_{i1}(n_{i2}-y_{i2}) = 1}{\text{number of pairs with } y_{i2}(n_{i1}-y_{i1}) = 1}$$

If the matched pairs table was,

control

	f	g
case	h	i

then the odds ratio estimate would be $\Psi_{mhp} = g/h$.

It was shown (see Fleiss, 1981) that the variance of the odds ratio estimate is

$$Var(\Psi_{mhp}) = \left(\frac{g}{h}\right)^2 \left(\frac{1}{g} + \frac{1}{h}\right) .$$

To test the hypothesis $H_0 : \Psi_{mhp} = 1$, the statistic

$$\chi^2_{(1)} = \frac{(g-h)^2}{g+h}$$

has an asymptotically chi-square distribution with one degree of freedom. A better approximation includes a correction factor, such that the statistic becomes:

$$\chi^2_{(1)} = (|g-h| - 1)^2/(g+h) \quad .$$

The implication of a large value of this χ^2 is that the cases and the controls differ with regards to the risk factor.

B. TESTING THE EQUALITY OF MARGINAL DISTRIBUTIONS

Suppose that the 2x2 table is summarized as

Control

		E	E	
	E	p_{11}	p_{12}	$p_{1.}$
Case	E	p_{21}	p_{22}	$p_{2.}$
		$p_{.1}$	$p_{.2}$	

When $p_{1.} = p_{.1}$, then $p_{2.} = p_{.2}$, subsequently there is marginal homogeneity. From the above table, if $(p_{1.} - p_{.1}) = (p_{2.} - p_{.2})$, then marginal homogeneity is equivalent to symmetry of probabilities off the main diagonal, that is, $p_{12} = p_{21}$. Naturally, a test of marginal homogeneity should be based on

$$D = \hat{p}_{1.} - \hat{p}_{.1} = \frac{(g-h)}{n} \quad .$$

From Agresti (1990) (page 348), it can be shown that

$$Var(D) = \frac{1}{n} \left[p_{1.}(1-p_{1.}) + p_{.1}(1-p_{.1}) - 2(p_{11}p_{22} - p_{12}p_{21}) \right] \qquad (3.19)$$

Note that the dependence in the sample marginal probabilities resulting from the matching of cases and controls, is reflected in the term $2(p_{11}p_{22} - p_{12}p_{21})$. Moreover, dependent samples usually show

63

positive association between responses, that is

$$\Psi = p_{11}p_{22} / p_{12}p_{21} > 1 \quad \text{or} \quad p_{11}p_{22} - p_{12}p_{21} > 0.$$

From (3.19) this positive association implies that the variance of D is smaller than when the samples are independent. This indicates that matched studies which produce dependent proportions can help improve the precision of the test statistic D.

For large samples, $D = \hat{p}_{1.} - \hat{p}_{.1} = (g-h)/n$ has approximately a normal sampling distribution. An $(1-\alpha)$ 100% confidence interval on $p_{1.} - p_{.1}$ is

$$D \pm Z_{\alpha_{/2}} \sqrt{\hat{Var}(D)}$$

where

$$\hat{Var}(D) = \frac{1}{n} \left[\hat{p}_{1.}(1-\hat{p}_{1.}) + \hat{p}_{.1}(1-\hat{p}_{.1}) - 2(\hat{p}_{11}\hat{p}_{22} - \hat{p}_{12}\hat{p}_{21}) \right]$$

The statistic

$$Z = \frac{D}{\sqrt{\hat{var}(D)}}$$

is a test on the hypothesis $H_0: p_{1.} = p_{.1}$.

Under H_0 it can be shown that the estimated variance is,

$$\hat{Var}(D|H_0) = \frac{g+h}{n^2}$$

The statistic simplifies to

$$Z_0 = \frac{g-h}{(g+h)^{1/2}} \tag{3.20}$$

the square of Z_0 gives the one-degree of freedom chi-square test on

$$H_0: \psi = 1 , \tag{3.21}$$

this being McNemar's test, an example of which is given below.

Example 3.5

To illustrate McNemar's test the following hypothetical data from a controlled trial with matched pairs is used. The null hypothesis being tested is $H_0: \Psi_{mhp} = 1$.

Control

		E	$\bar{E}$
Case	E	15	20
	$\bar{E}$	5	85

$$\hat{\psi} = \frac{20}{5} = 4$$

$$SE(\hat{\psi}) = \frac{20}{5} \left[\frac{(20+5)}{(20)(5)} \right]^{1/2} = 2$$

$$\chi^2_{(1)} = \frac{(|20-5|-1)^2}{25} = 7.84$$

Based on the calculated value of the χ^2, there is no reason to accept the null hypothesis which means that there is a significant difference ($p < 0.05$) in the rate of exposure among the cases and the controls.

V. STATISTICAL ANALYSIS OF CLUSTERED BINARY DATA

In the analysis of binary data we are always reminded that, for the statistical inference on the proportion or the odds ratios to be valid, the individual responses must be independent from each other. For example in a randomized controlled trial, individuals are assigned to the treatment and control groups at random. Moreover, within each group (treatment and control) the response collected from an individual is assumed to be independent of the responses of other individuals in the same group as well as the other groups. There are situations, however, where randomization of individuals into groups may not be feasible. Studies carried out by community health researchers, teratologists, and epidemiologists often use clusters of individuals as units of randomization. A main feature of cluster randomization studies is that inferences are often intended to apply at the individual level, while randomization is at the cluster level.

We would like to point out that what is meant by the term *cluster* depends on the nature of the study. For example, repeated measurements of individuals over time, as in longitudinal studies that include n independent individuals, is a form of randomization by cluster. Here each individual is a cluster and the number of over-time measurements is the cluster size.

Another example arising from veterinary epidemiology is where investigators want to test if current bacterial infection in the cow's udder affects the invasion by other pathogenic bacteria. Each cow is treated as a cluster with the four quarters (teats) as the cluster size. In genetic epidemiology where the concern is with the possibility that a certain disease is genetically controlled, nuclear families or sibships are the clusters that constitute the sampling frame.

Although different in their specific objectives and their significance to the scientific community, the above examples share an important characteristic. That is, responses of individuals within a cluster cannot be regarded as statistically independent and tend to exhibit intracluster correlation. This means that measurements on individuals within a cluster tend to be more alike than measurements collected from individuals in different clusters. Standard methods of analysis for binary data tend to be inadequate when used with clustered data; in particular, variances of the estimated parameters are under-estimated and statistical tests do not maintain their nominal error rates. In this section we present the methodologies available to account for the effect of cluster sampling on the chi-square statistic when it is used to test the homogeneity of several proportions in nested randomized experiments. Other applications such as estimating the variance of the estimated common odds ratio and the Mantel-Haenszel chi-square test for independence in a series of 2x2 tables are also presented. Before we discuss the methods, it is worthwhile to clarify the nature of interdependence within a cluster.

Let n_i be the size of the ith cluster $i=1,2,\ldots k$ and y_{ij} be the response of the jth individual in the ith cluster. Since y_{ij} is binary, we assume that $P[y_{ij}=1] = p_i$ for all $j=1,2,\ldots n_i$, and that the correlation ρ between any pair y_{ij}, y_{il} has the same value for any i and $j \neq \ell$. In other words all individuals within the cluster are equally correlated. The parameter ρ is a measure of intraclass correlation (Crowder, 1978), first formally suggested by Landis and Koch (1977) as the common correlation. Under the common correlation model we have:

$\Pr (y_{ij} = y_{i\ell} = 1) = p_i^2 + \rho p_i(1-p_i)$
$\Pr (y_{ij} = y_{i\ell} = 0) = (1-p_i)^2 + \rho p_i(1-p_i)$
$\Pr (y_{ij} = 1, y_{i\ell} = 0) = \Pr (y_{ij} = 0, y_{i\ell} = 1) = p_i(1-p_i)(1-\rho)$

If $y_i = \sum_j y_{ij}$ is the total number of diseased individuals in the ith cluster, then

$$Var(y_i) = \sum_j Var(y_{ij}) + \sum_{l \neq j} \sum Cov(y_{ij}, y_{il})$$

$$= n_i p_i (1-p_i) + n_i (n_i - 1) p_i (1-p_i) \rho$$

$$= n_i p_i (1-p_i)[1+(n_i-1)\rho]$$

The variance of y_i under the assumption of independence is $n_i p_i(1-p_i)$.

In the absence of independence the variance is inflated by the factor $[1+(n_i-1)\rho]$ which is denoted by ϕ_i. Therefore, for $\rho > 0$, ignoring the cluster effect would result in underestimating the variance of y_i.

A. TESTING HOMOGENEITY

In this section, our chief objective is to make inferences about the unknown overall proportions of affected individuals in conceptual groups denoted by p_i ($i=1...I$), where I is the number of groups (treatments). Within each of these groups (treatment), we randomize k_i clusters. Let y_{ij} be the number of positive responses (affected, diseases, etc.) among the n_{ij} units in the jth cluster ($j=1,2,...k_i$) within the ith treatment ($i=1,2,...I$).

Note that under this setup

$$y_{ij} = \sum_{l=1}^{n_{ij}} y_{ijl}$$

where $y_{ijl} = \begin{cases} 1 & \text{if } l^{th} \text{ individual in } j^{th} \text{ cluster receiving } i^{th} \text{ treatment is affected} \\ 0 & \text{else} \end{cases}$

and the association between y_{ijl} and $y_{ijl'}$ is measured by the common intraclass correlation ρ.

From Fleiss (1981), the homogeneity hypothesis is given by H_0: $p_1=p_2=...=p_I$. Under H_0 the statistic

$$\chi^2 = \sum_{i=1}^{I} \frac{n_i(\hat{p}_i-\hat{p})^2}{[\hat{p}(1-\hat{p})]} \tag{3.22}$$

has asymptotic distribution of χ^2 with I-1 degrees of freedom (when used without accounting for the clustering effect) where

$$\hat{p}_i = \frac{y_i}{n_i}$$

$$y_i = \sum_{j=1}^{k_i} y_{ij} \ ,$$

$$n_i = \sum_{j=1}^{k_i} n_{ij} \ , \quad \text{and}$$

$$\hat{p} = \frac{\displaystyle\sum_{i=1}^{I} y_i}{\displaystyle\sum_{i=1}^{I} n_i} \ .$$

To illustrate the use of the statistic (3.22) in a special case where the responses are assumed to be independent we shall assess the data of Whitmore (reported in Elston (1977)). In this study, 12 strains of mice were treated with a carcinogen and the numbers with and without tumours were noted. The results are as found in Table 3.10.

Table 3.10
Number of Tumours Observed in Rats Treated with Carcinogens

Strain	1	2	3	4	5	6	7	8	9	10	11	12
No. with tumours	26	27	35	18	33	11	11	13	13	5	5	2
No. without tumours	1	3	14	9	20	11	11	15	22	19	30	24

We would like to test the null hypothesis of no difference among the strains with respect to the incidence of tumours.

Direct computations give $\hat{p} = \dfrac{199}{378} = 0.526$, and $\chi^2 = 103.85$. The critical value of chi-square with 11 degrees of freedom at α-level of 0.05 is 19.67. Thus, we would conclude, based on the data, that the strains differ significantly.

In the presence of within-cluster correlation, χ^2 given by (3.22) no longer provides a valid test on H_0 but may be adjusted to account for such correlation. Of the several methods which can be used to correct the χ^2 statistic, two will be presented here.

1. Donner's Adjustment

Donner (1989) suggested weighting the χ^2 statistic by a factor $D_i = 1+(c_i-1)\rho$, where ρ is the intraclass correlation, assumed common across groups, with

$$c_i = \sum_{j=1}^{k_i} \frac{n_{ij}^2}{n_i} \quad .$$

His suggested adjusted chi-square statistic is

$$\chi_D^2 = \sum_{i=1}^{I} \frac{n_i(\hat{p}_i - \hat{p})^2}{[\hat{D}_i \, \hat{p}(1-\hat{p})]}$$

where $\hat{D}_i = 1+(c_i-1)\hat{\rho}$ and $\hat{\rho}$ is the analysis of variance estimator of ρ given by:

$$\hat{\rho} = \frac{MSB - MSW}{MSB + (n_0 - 1)MSW}$$

where

$$MSW = (N - M)^{-1} \sum_{i=1}^{I} \sum_{j=1}^{k_i} \frac{y_{ij}(n_{ij} - y_{ij})}{n_{ij}}$$

$$MSB = (M - I)^{-1} \sum_{i=1}^{I} \sum_{j=1}^{k_i} \frac{(y_{ij} - n_{ij}\hat{p}_i)^2}{n_{ij}}$$

$$M = \sum_{i=1}^{I} k_i \quad , \quad N = \sum_{i=1}^{I} n_i \quad ,$$

and

$$n_0 = \frac{\left(N - \sum_i c_i\right)}{(M-I)} \quad .$$

Donner proposed that the asymptotic distribution of χ_D^2 is chi-square with I-1 degrees of freedom.

Returning again to the rat tumour data (Table 3.10), recall that there was clearly a highly significant strain effect. This may be due to the fact that the between strain variation is much larger than the within strain variation. In other words, this could indicate the presence of high correlation between binary responses within the strain. Thus, one could question the validity of the chi-square statistic used (3.22). To detect departure from the assumption of independence, Tarone (1979) constructed a one-degree of freedom chi-square test on the null hypothesis that the data are independently binomially distributed versus a correlated binomial alternative. His test statistic is given by

$$\chi_T^2 = \frac{\left[(\hat{p}\hat{q})^{-1} \sum_{i=1}^{I} (y_i - n_i\hat{p})^2 - N \right]^2}{(2 \sum_{i=1}^{I} n_i^2 - N)}$$

Since $\hat{p} = 0.526$, $N = 378$, $\sum n_i^2 = 13022$ *and* $\chi_T^2 = 315.11$ the binomial hypothesis is rejected. Hence the statistic (3.22) must be adjusted to account for the effect of within strain correlation.

To illustrate Donner's adjustment on the rat tumour data, we used the SAS program given below to produce the MSW, MSB and n_o, which in turn are used to calculate the $\hat{\rho}$. Looking at the SAS output derived from this program, the (highlighted) values for the MSW, MSB and n_o are 2.35, 0.187 and 31.23 respectively. Therefore, $\hat{\rho}$ is equal to 0.27 and χ_D^2 is equal to 11.33, which is no longer significant.

```
data  donner;
input strain tumour count;
cards;
1 1   26
1 0    1
2 1   27
2 0    3
3 1   35
3 0   14
....
12 1   2
12 0 24
;
proc freq;
tables strain*tumour/chisq;
weight count;
run;

proc glm;
class strain;
freq count;
model tumour = strain;
random strain;
run;
```

OUTPUT:

General Linear Models Procedure
Class Level Information

Class   Levels   Values
STRAIN   12   1 2 3 4 5 6 7 8 9 10 11 12
Number of observations in data set = 378

Dependent Variable: TUMOUR
Frequency: COUNT

Source	DF	Sum of Squares	Mean Square	F Value	Pr > F
Model	11	25.89374083	**2.35397644**	12.61	0.0001
Error	366	68.34170890	**0.18672598**		
Corrected Total	377	94.23544974			

R-Square	C.V.	Root MSE	TUMOUR Mean
0.274777	82.08071	0.43211802	0.52645503

General Linear Models Procedure

Source   Type III Expected Mean Square

STRAIN   Var(Error) + **31.232** Var(STRAIN)

2. Rao and Scott's Adjustment

Recently Rao and Scott (1992) proposed a more robust procedure to account for the intracluster dependence, which assumes no specific model for the intracluster correlation as required by Donner's adjustment. Their suggested procedure is based on the concepts of "design effect" and effective sample size widely used in sample surveys (Kish, 1965, p.259). Their approach is summarized as follows: Since $\hat{p}_i = y_i/n_i$ may be written as the ratio of two sample means

$$\hat{p}_i = \frac{\bar{y}_i}{\bar{n}_i} \quad , \quad \text{with}$$

$$\bar{y}_i = \sum_{j=1}^{k_i} \frac{y_{ij}}{k_i} \quad , \quad \text{and}$$

71

$$\bar{n}_i = \sum_{j=1}^{k_i} \frac{n_{ij}}{k_i} \quad,$$

we have a variance of the $\hat{p}_i$, as

$$v_i = \frac{k_i}{n_i^2(k_i-1)} \sum_{j=1}^{k_i} (y_{ij} - n_{ij}\,\hat{p}_i)^2$$

for large k_i. Moreover, since

$$\frac{\hat{p}_i(1-\hat{p}_i)}{n_i}$$

is the variance of $\hat{p}_i$ under the binomial assumptions, then

$$d_i = \frac{n_i v_i}{[\hat{p}_i(1-\hat{p}_i)]}$$

represents the inflation factor due to clustering and is known in sample surveys as the 'design effect'. The ratio $\tilde{n}_i = n_i/d_i$ is called the effective sample size.

Transforming the aggregate data (y_i, n_i) to ($\tilde{y}_i, \tilde{n}_i$), i=1,2,...I where $\tilde{y}_i = y_i/d_i$ and, treating $\tilde{y}_i$ as binomial random variable with parameters ($\tilde{n}_i, \tilde{p}_i$), the adjusted χ^2 statistic is

$$\tilde{\chi}^2 = \sum_{i=1}^{I} \frac{\tilde{n}_i(\hat{p}_i - \tilde{p})^2}{\tilde{p}(1-\tilde{p})}$$

where

$$\tilde{p} = \frac{\sum_{i=1}^{I} \tilde{y}_i}{\sum_{i=1}^{I} \tilde{n}_i}$$

Under H_0, $\tilde{\chi}^2$ is asymptotically distributed as a χ^2 variable with I-1 degrees of freedom, unlike χ^2 given in (3.22).

Example 3.6 **"Hypothetical Drug Trial"**

The data are the result of a drug trial aimed at comparing the effect of 4 antibiotics against "Shipping Fever" in calves. Calves arrive in trucks at the feedlot and are checked upon arrival for signs of the disease. Animals that are confirmed as cases (from the same truck) are randomized as a group to one of the 4 treatment regimes.

The following table (Table 3.11) gives the number of treated animals within a 2 week period and the number of deaths at the end of the 2 weeks.

Table 3.11
A Hypothetical Drug Trial to Compare the Effect of
Four Antibiotics Against Shipping Fever in Calves

Treatment								
1	(a)	30	25	25	32	12	10	10
	(b)	3	0	4	10	0	0	1
2	(a)	30	30	15	15	20	19	
	(b)	1	1	1	1	4	0	
3	(a)	30	35	30	10	25		
	(b)	5	7	9	1	2		
4	(a)	40	10	25	20	19	25	
	(b)	10	1	1	2	1	2	

(a) = number treated   (b) = mortality

We would like to test the hypothesis that the mortality rates are the same for all treatments. Let p_i be the mortality rate of the ith treatment. Hence the null hypothesis would be $H_0: p_1 = p_2 = \ldots = p_I$.
Since,

$$\hat{p}_i = \frac{\sum\limits_{i=1}^{I} y_i}{\sum\limits_{i=1}^{I} n_i}$$

where, $n_1 = 144$, $n2 = 129$, $n_3 = 130$, $n_4 = 139$ and $\hat{p}(1-\hat{p}) = 0.1236(0.8764) = 0.1083$,

73

then the uncorrected χ^2 is given by:

$$\chi^2 = 0.002606 + 4.520 + 4.47 + 0.00217 = 8.995$$

Comparing this to a χ_3^2 value of 7.81, the null hypothesis is rejected at a p level of 0.029. This implies that there is a significant difference in the proportions of dead animals for the four different antibiotics.

To estimate the intracluster correlation, the ANOVA results obtained from SAS are summarized in the following table:

Treatment	SSB	(df)	SSW	(df)	M_i
1	1.915	6	13.835	137	144
2	.504	5	7.00	123	129
3	.763	4	18.807	125	130
4	.974	5	13.947	133	139

MSB = 0.208 and MSW = .103, n_0 = 21.934 and thus $\hat{\rho}$ = 0.044.

Since $c_1 = 24.43$, $c_2 = 23.34$, $c_3 = 28.85$, $c_4 = 26.708$, then,
$D_1 = 2.023$, $D_2 = 1.98$, $D_3 = 2.22$, $D_4 = 2.13$.

To facilitate the computations, we summarize the results as follows:

Treatment	D_i	$\hat{P}_i$	n_i	$\dfrac{n_i(\hat{p}_i - \bar{p})^2}{D_i\,\bar{p}(1-\bar{p})}$
1	2.023	0.13	144	0.0013
2	1.98	0.06	129	2.28
3	2.22	0.18	130	2.008
4	2.13	0.12	139	0.001

Hence $\chi_D^2 = 4.29$, which indicates that there is no significant difference between the mortality rates.

Rao and Scott's Method:

The computations are summarized as follows:

Drug 1: $\hat{P}_1 = \dfrac{18}{144} = 0.125$

cluster	y_{1j}	n_{1j}	$(y_{1j} - n_{1j}\hat{P}_1)^2$
1	3	30	0.563
2	0	25	9.766
3	4	25	0.766
4	10	32	36.00
5	0	12	2.25
6	0	10	1.563
7	1	10	0.063
Total	18	144	50.96

$v_1 = 0.003$, $d_1 = 3.775$,
$\tilde{y}_1 = 4.768$, $\tilde{n}_1 = 38.141$

Drug 2: $\hat{P}_2 = \dfrac{8}{129} = 0.062$

cluster	y_{2j}	n_{3j}	$(y_{2j} - n_{2j}\hat{P}_2)^2$
1	1	30	0.74
2	1	30	0.74
3	1	15	0.005
4	1	15	0.005
5	4	20	7.616
6	0	19	1.388
Total	8	129	10.495

$v_2 = 0.0008$, $d_2 = 1.678$,
$\tilde{y}_2 = 4.767$, $\tilde{n}_2 = 76.863$

Drug 3: $\hat{P}_3 = \dfrac{24}{130} = 0.184$

cluster	y_{3j}	n_{3j}	$(y_{3j} - n_{3j}\hat{P}_3)^2$
1	5	30	0.27
2	7	35	0.31
3	9	30	12.11
4	1	10	0.716
5	2	25	6.76
Total	24	130	20.17

$v_3 = 0.0015$, $d_3 = 1.285$,
$\tilde{y}_3 = 18.676$, $\tilde{n}_3 = 101.161$

Drug 4: $\hat{P}_4 = \dfrac{17}{139} = 0.122$

cluster	y_{4j}	n_{4j}	$(y_{4j} - n_{4j}\hat{P}_4)^2$
1	10	40	26.09
2	1	10	0.05
3	1	25	4.234
4	2	20	0.198
5	1	19	1.752
6	2	25	1.118
Total	17	139	33.442

$v_4 = 0.0021$, $d_4 = 2.69$,
$\tilde{y}_4 = 6.32$, $\tilde{n}_4 = 51.68$

Hence,

$$\chi^2_{RS} = \frac{.6601}{0.129(1-.129)} = 5.883,$$

which again does not exceed the $\chi^2_{3,0.05}$ value of 7.81, so we have no reason to reject the null hypothesis. This implies that there is no significant difference between the mortality rates.

From this example we can see the importance of accounting for intracluster correlations as the resulting conclusion about the equality of the proportions is different when the correlation is taken into account.

B. INFERENCE ON THE COMMON ODDS RATIO

Consider a series of 2x2 tables wherein the notation has been changed to accommodate multiple tables, groups and clusters. Explicitly, n_{ijt} is the size of the tth cluster in the jth group ($j=1$ for *exposed* and $j=2$ for *unexposed*) from the ith table, and k_{ij} is the number of cluster in the jth group from the ith table. To further clarify this altered notation, Table 3.12 describes the data layout for the ith stratum.

This can be reduced to the 2x2 format as shown in Table 3.13.

Table 3.12

Stratum (i)		cluster (t)			
		1	2	3 ...... k_{ij}	
Exposed $j=1$	cluster size	n_{i11}	n_{i12}	n_{i13}	$n_{i1k_{ij}}$
	number of deaths	y_{i11}	y_{i12}	y_{i13}	$y_{i1k_{ij}}$
Unexposed $j=2$	cluster size	n_{i21}	n_{i22}	n_{i23}	$n_{i2k_{ij}}$
	number of deaths	y_{i21}	y_{i22}	y_{i23}	$y_{i2k_{ij}}$

Table 3.13

		D^+	D^-	
$j=1$	E^+	y_{i1}	$y_{i1}-n_{i1}$	n_{i1}
$j=2$	E^-	y_{i2}	$y_{i2}-n_{i2}$	n_{i2}

where , $\quad y_{ij} = \sum_{t=1}^{k_{ij}} y_{ijt}$ and $n_{ij} = \sum_{t=1}^{k_{ij}} n_{ijt}$ $\quad$ ($j=1,2$; $i=1,2,...k$).

Now, because the sampling units are clusters of individuals the χ^2_{mh} statistic used to test $H_0: \psi = 1$ would not be appropriate. To adjust this statistic for the clustering effect we introduce two procedures, one proposed by Donald and Donner (1987), and the other by Rao and Scott (1992).

1. Donald and Donner's Adjustment

Because we have 2k rows in Table 3.12, an intraclass correlation, ρ, is first estimated from each row. Let $\hat{\rho}_{ij}$ be the estimate of ρ_{ij}, from the jth row in the ith table. Under a common correlation model, it is reasonable to construct an overall estimate of ρ as

$$\hat{\rho}_A = \frac{1}{2k} \sum_{i=1}^{k} \sum_{j=1}^{2} \hat{\rho}_{ij}$$

Let $\hat{D}_{ijt} = 1 + (n_{ijt} - 1)\hat{\rho}_A$ be the correction factor for each cluster, and let B_{ij} be the weighted average of such correction factors, where the weights are the cluster sizes themselves.

77

Hence

$$B_{ij} = \frac{\sum\limits_{t=1}^{k_{ij}} n_{ijt} \hat{D}_{ijt}}{\sum\limits_{t=1}^{k_{ij}} n_{ijt}} \quad .$$

Donald and Donner (1987) suggested that the Mantel Haenszel test statistic on $H_0 : \psi = 1$, adjusted for the clustering effect is given by

$$\chi^2_{mhc} = \frac{\left(\sum\limits_{i=1}^{k} \dfrac{y_{i1}(n_{i2}-y_{i2}) - y_{i2}(n_{i1}-y_{i1})}{n_{i1}B_{i1} + n_{i2}B_{i2}} \right)^2}{\sum\limits_{i=1}^{k} \dfrac{n_{i1}n_{i2}y_{.i}(n_{i.}-y_{.i})}{(n_{i1}B_{i1} + n_{i2}B_{i2})n_i^2}} \quad .$$

Since cluster sampling affects the variance of the estimated parameters, the variance $\hat{V}_{mh}$ of ψ_{mh} is no longer valid. Donald and Donner defined the cluster variant $\hat{b}_i^*$ of the $\hat{b}_i$ contained in Hauck's formula (3.18)

$$\hat{b}_i^* = \frac{B_{i1}}{n_{i1} \hat{P}_{i1} \hat{q}_{i1}} + \frac{B_{i2}}{n_{i2} \hat{P}_{i2} \hat{q}_{i2}} \quad .$$

The corrected variance, $\hat{V}_{mhc}$, is:

$$\frac{\psi^2_{mh} \left(\sum\limits_{i=1}^{k} w_i^2 \, \hat{b}_i^* \right)}{\left(\sum w_i \right)^2} \quad .$$

Hence an $(1-\alpha)$ 100% confidence limits on ψ_{mh} after adjusting for clustering is

$$\psi_{mh} \pm Z_{\alpha/2} \sqrt{\hat{V}_{mhc}} \quad .$$

2. Rao and Scott's Adjustment

This adjustment requires the computation of the variance inflation factors d_{ij} using the cluster-level data (y_{ijt}, n_{ijt}), where $t=1,2,...k_{ij}$. The inflation factor is computed in two steps. First,

$$v_{ij} = \frac{k_{ij}}{(k_{ij}-1)n_{ij}^2} \sum_{t=1}^{k_{ij}} \left(y_{ijt}-n_{ijt}\,\hat{p}_{ij}\right)^2$$

where

$$\hat{p}_{ij} = \frac{y_{ij}}{n_{ij}} \quad ,$$

and then,

$$d_{ij} = \frac{n_{ij}\,v_{ij}}{\hat{p}_{ij}\,\hat{q}_{ij}} .$$

An asymptotically (as $k_{ij} \to \infty$ for each i and j) valid test of $H_0: \psi=1$ is obtained by replacing (y_{ij}, n_{ij}) by ($\tilde{y}_{ij},\tilde{n}_{ij}$), where $\tilde{y}_{ij} = y_{ij}/d_{ij}$ and ,$\tilde{n}_{ij} = n_{ij}/d_{ij}$.

To construct an asymptotic confidence interval on ψ_{mh}, RS suggested replacing (y_{ij},n_{ij}) by ($\tilde{y}_{ij},\tilde{n}_{ij}$) in ψ_{mh} to get

$$\tilde{\psi}_{mhc} = \frac{\displaystyle\sum_{i=1}^{k} \frac{\tilde{y}_{i1}\left(\tilde{n}_{i2} - \tilde{y}_{i2}\right)}{\left(\tilde{n}_{i1} + \tilde{n}_{i2}\right)}}{\displaystyle\sum_{i=1}^{k} \frac{\left(\tilde{n}_{i1} - \tilde{y}_{i1}\right)\tilde{y}_{i2}}{\left(\tilde{n}_{i1} + \tilde{n}_{i2}\right)}}$$

as the clustered adjusted point-estimator of ψ_{mh}. Similarly Hauck's variance estimator for ψ_{mh} becomes

$$\tilde{V}_{mhc} = \frac{\tilde{\psi}_{mhc}^2 \left(\displaystyle\sum_{i=1}^{k} \tilde{w}_i^2\,\tilde{b}_i\right)}{\left(\displaystyle\sum_{i=1}^{k} \tilde{w}_i\right)^2}$$

where,

$$\tilde{w}_i = \left(\tilde{n}_{i1}^{-1} + \tilde{n}_{i2}^{-1} \right)^{-1} \tilde{p}_{i2} \left(1 - \tilde{p}_{i1} \right)$$

$$\tilde{b}_i = \frac{d_{i1}}{n_{i1}\tilde{p}_{i1}\tilde{q}_{i1}} + \frac{d_{i2}}{n_{i2}\tilde{p}_{i2}\tilde{q}_{i2}} \quad .$$

$$\tilde{P}_{ij} = \frac{\tilde{y}_{ij}}{\tilde{n}_{ij}} \qquad j=1,2$$

$$\tilde{q}_{ij} = 1 - \tilde{p}_{ij}$$

Example 3.7

The following data are the result of a case-control study to investigate the association between Bovine-Leukemia-virus (BLV) infection and Bovine-Immunodeficiency virus (BIV) infection (Table 3.14). Each BLV$^+$ cow as matched on sex and age (within two years) with a BLV$^-$ cow from a different herd. The Pedigree relatives of a BLV$^+$ cow constituted clusters of BIV$^+$ or BIV$^-$ while the pedigree relatives of a BLV$^-$ cow constituted clusters of BIV$^+$ or BIV$^-$.

A region-stratified (unmatched) analysis is conducted to test the above hypothesis using,

(a)   the Mantel-Haenszel one-degree-of-freedom chi-square test on the significance of the common odds ratio and a 95% confidence interval using Hauck's variance formula and,

(b)   adjusting the above chi-square test and the variance expression for clustering using Donald and Donner's procedure and Rao and Scott's procedure.

Table 3.14

Case-Control Study Results for Investigation of Association Between Bovine-Leukemia-Virus Infection and Bovine-Immunodeficiency Infection

Region 1				Region 2			
BLV+ COWS		CONTROLS		BLV+ COWS		CONTROLS	
BIV+	BIV-	BIV+	BIV-	BIV+	BIV-	BIV+	BIV-
1	4	0	4	7	1	0	2
1	5	0	4	6	1	0	0
1	2	0	3	0	0	1	6
2	4	0	2	1	1	0	6
0	1	0	4	0	3	0	2
2	0	0	7	0	1	1	0
1	1	0	3	1	1	1	0
1	2	1	1				
2	1	2	5				
3	0	0	3				
1	1	1	2				
2	4	0	6				
1	1	0	4				

From Table 3.14 we construct the following 2 (2x2) tables

	Region 1			Region 2	
	BIV+	BIV-		BIV+	BIV-
BLV+	18	26		15	8
BLV-	4	48		3	16

The MH common odds ratio is given by:

$$\Psi_{MH} = \frac{\dfrac{(18)(48)}{96} + \dfrac{(15)(16)}{42}}{\dfrac{(4)(26)}{96} + \dfrac{(3)(8)}{42}} = 8.89$$

and Hauck's variance is 17.85.

Now, the 1 degree of freedom chi squared test of the common odds ratio will be shown using three methods:

- uncorrected
- Donald and Donner's adjustment
- Rao and Scott's adjustment.

Uncorrected Chi Squared Test of Common Odds Ratio:

H_0: no association between BIV and BLV status in pedigree.

Using (3.14) we have

$$\chi^2 = \frac{\left(|33 - 19.94| - \dfrac{1}{2} \right)^2}{6.86} = 24.863$$

Thus, since 24.863 exceeds the $\chi^2_{(1, 0.05)}$ value of 3.84, the null hypothesis is rejected implying that there is a strong association between BIV and BLV status.

Donald and Donner's Adjustment

H_0: no association between BIV and BLV status in pedigree.
(Here the chi square is adjusted by an inflation factor, D_i.)

The χ^2_{MHC} for Donald and Donner is,

$$\chi^2_{mhc} = \frac{\left(\displaystyle\sum_{i=1}^{k} \frac{y_{i1}(n_{i2} - y_{i2}) - y_{i2}(n_{i1} - y_{i1})}{n_{i1}B_{i1} + n_{i1}B_{i2}} \right)^2}{\displaystyle\sum_{i=1}^{k} \frac{n_{i1}n_{i2}y_{i.}(n_i - y_{i.})}{(n_{i1}B_{i1} + n_{i2}B_{i2})n_i^2}}$$

The estimated common intraclass correlation is given by:

$$\hat{\rho}_A = \frac{1}{2k}\sum\sum\hat{\rho}_{ij} = \frac{1}{2(2)}(-0.0186 + 0.08654 + 0.5537 + 0.330810) = 0.238 .$$

Now, the correction factor for each cluster is

$$\hat{D}_{ijt} = 1 + (n_{ijt} - 1)\hat{\rho}_A ,$$

and the weighted average of the correction factors is

$$B_{ij} = \frac{\sum n_{ijt}\hat{D}_{ijt}}{\sum n_{ijt}} .$$

So,

B_{11} = Region 1 BLV^- = 1.870026 $\qquad$ B_{12} = Region 1 BLV^+ = 1.768454
B_{21} = Region 2 BLV^- = 1.952450 $\qquad$ B_{22} = Region 2 BLV^+ = 2.047695

and the χ^2_{MHC} equation using Donald and Donner's adjustment is,

$$\chi^2_{MHC} = \frac{\left[\dfrac{18(52-4)-4(44-18)}{44(1.7685)+52(1.820)} + \dfrac{15(19-3)-3(23-15)}{23(2.048)+19(1.952)}\right]^2}{\dfrac{52(44)(22)(96-22)}{(96)^2(44(1.7685)+52(1.870))} + \dfrac{23(19)(18)(42-18)}{(42)^2(23(2.048)+19(1.952))}}$$

$$= 13.33$$

Because 13.33 is larger than the $\chi^2_{(1,0.05)}$ value of 3.84 the null hypothesis is rejected implying that when we adjust for the intracluster correlation, there is a significant association between the BLV and the BIV status.

The value of the variance is now,

$$Var_C\left(\Psi_{MH}\right) = \left(\Psi_{MH}\right)^2 \frac{\sum \hat{w}_i^2 \hat{b}_i^*}{\left(\sum \hat{w}_i\right)^2}$$

where,

$$\hat{b}_i^{*} = \frac{B_{i1}}{n_{i1}\hat{p}_{i1}\hat{q}_{i1}} + \frac{B_{i2}}{n_{i2}\hat{p}_{i2}\hat{q}_{i2}}$$

and

$$\hat{w}_i = \left(\frac{1}{n_{i1}} + \frac{1}{n_{i2}}\right)^{-1} \hat{p}_{i2}\hat{q}_{i1} \ .$$

The results of these computations show that, $\hat{w}_1 = 1.085$, $\hat{w}_2 = 0.572$, $\hat{b}_1^{*} = 0.673$, $\hat{b}_2^{*} = 1.165$ and that the variance is thus equal to 33.791.

Rao and Scott's Adjustment

H_0: no association between BIV and BLV status.
Here the chi square is adjusted by the variance inflation factor, d_{ij} First, we compute the v_{ij},

$$v_{ij} = \frac{k_{ij}}{(k_{ij}-1)n_{ij}^2} \sum_{t=1}^{k_{ij}} (y_{ijt} - n_{ijt}\hat{p}_{ij})^2$$

where, $\hat{p}_{ij} = \dfrac{y_{ij}}{n_{ij}}$. The inflation factor d_{ij} is calculated using

$$d_{ij} = \frac{n_{ij}v_{ij}}{\hat{p}_{ij}\hat{q}_{ij}},$$

which is then used to adjust y_{ij} and n_{ij}, as suggested by Rao and Scott:

$$\tilde{y}_{ij} = \frac{y_{ij}}{d_{ij}} \qquad \tilde{n}_{ij} = \frac{n_{ij}}{d_{ij}} \ .$$

Here, $d_{11}=0.93$, $d_{12}=1.32$, $d_{21}=2.16$, and $d_{22}=1.17$. Calculation of the adjusted chi square, χ_{mhc}^2, is as follows,

$$\chi_{MHC}^2 = \frac{\left(\dfrac{19.35(36.36) - 3.03(27.96)}{(86.70)} + \dfrac{6.94(13.68) - 2.56(3.71)}{(26.89)}\right)^2}{\dfrac{(47.31)(39.39)(26.30)(60.41)}{(86.71)^2(85.71)} + \dfrac{(10.65)(16.24)(5.59)(21.30)}{(26.89)^2(25.89)}}$$

$$= 18.69 \ .$$

84

Therefore, once again we reject the null hypothesis which means that there is a significant association between the BLV and the BIV status.

With the Rao and Scott adjustment, the variance estimate is now

$$Var\left(\Psi_{mhc}\right) = \frac{\left(\Psi_{mhc}\right)^2\left(\sum \tilde{w}^2_i \tilde{b}_i\right)}{\left(\sum \tilde{w}_i\right)^2} = 29.851 \quad .$$

An extensive literature has been developed on the methodologic issues which arise when the risk of disease is compared across treatment groups, when the sampling units are clusters of individuals. The presence of so-called 'cluster effects' invalidates the standard Pearson's chi-square test of homogeneity, and makes the reported confidence interval on the common odds ratio unrealistically narrow. In the previous section we introduced the most commonly used techniques of adjusting clustering effects. It is debatable as to which technique is most appropriate.

Extensive simulation studies conducted by Ahn and Maryon (1995) , and Donner et al. (1994) showed that each approach has its advantages. Ahn and Maryon preferred Rao and Scott's adjustment on the length of the confidence interval of the Mantel-Haenszel common odds ratio. The reason was that the ANOVA estimator of the intraclass correlation ρ, which is needed for Donner's adjustment, is positively biased, and hence the confidence interval would be spuriously wide when $\rho = 0$. On the other hand, Donner's et al. simulations showed that the adjustment to Pearson's chi-square test of homogeneity based on the pooled estimate of the intraclass correlations performs better than methods which separately estimate design effects from each group. The reason was that under randomization, the assumption of common population design effect is more likely to hold, at least under the null hypothesis. They also stressed the need for more research in this area, to investigate the finite-sample properties of methods which do not require the assumption of a common design effect.

VI. MEASURE OF INTERCLINICIAN AGREEMENT FOR CATEGORICAL DATA

Efficient delivery of health care requires proficient and error free assessment of clinical measures taken by physicians, for they serve as the basis for prognostic and therapeutic evaluations. Therefore, researchers have become increasingly aware of the need for reliable measurements of prognostic variables. This means that measurements made during medical evaluation should be as consistent as possible whether recorded by the same clinician on the same patient or by different clinicians performing the same evaluation. Consequently, reliability studies should be conducted in clinical investigations to assess the level of clinician variability in the measurement procedures to be used in medical records. When data arising from such studies are quantitative, tests for interclinician bias and measures of interclinician agreement are obtained from standard ANOVA mixed models or random effects models where the intraclass correlation is the reliability statistic commonly used, as we have shown in Chapter 2.

On the other hand, much of the medical data are categorical whereby the response variable is classified into nominal (or ordinal) multinominal categories. For this type of data the *kappa coefficient* is the measure of agreement statistic and is analogous to the intraclass correlation.

The main objective of this section is to discuss measures of agreement for categorical data. First we introduce the very simple situation of two clinicians classifying n patients into two categories. In this case we discuss two probabilistic models, the first assumes that the two raters are unbiased relative to each other and the second assumes the existence of bias. This is followed by the situation where multiple raters (clinicians) are assigned to each of n patients. Generalization to multiple categories will be discussed at the end of this section.

A. AGREEMENT IN 2X2 TABLES (TWO CLINICIANS AND TWO CATEGORIES)

Most of the early statistical research dealing with analyzing observer or rater agreement has focused on the development of summary measures of agreement. The aim was to develop a statistic that indicated the extent of agreement between two raters. Cohen's kappa was, and still is , the most popular of these types of measures. The value of kappa can fall between 0 and 1, with values near 0 indicating agreement no better than would be expected by chance, and a value of exactly 1 indicating perfect agreement. To clarify this concept let us consider the following example as a prelude to the construction of the kappa coefficient.

Example 3.8

Consider Table 3.15, which shows the examination results of 20 x-rays of the spinal cords of young horses who showed neurological signs of Cervical Vertebral Malformation (CVM). Each of two clinicians categorize the x-ray as yes (which means that the spinal cord is damaged) or no (not damaged).

Table 3.15
CVM Classification Results

X-Ray	Clinician (1)	Clinician (2)
1	Yes	Yes
2	No	No
3	No	Yes
4	No	No
5	Yes	Yes
6	No	No
7	Yes	Yes
8	No	Yes
9	No	Yes
10	Yes	Yes
11	No	No
12	No	No
13	Yes	Yes
14	Yes	No
15	Yes	Yes
16	No	No
17	No	No
18	No	Yes
19	Yes	Yes
20	No	No

The observations in Table 3.15 can be summarized by a simple two-way contingency table as follows:

		clinician (1)		
		Yes	No	Total
clinician (2)	Yes	7 (a)	4 (b)	11
	No	1 (c)	8 (d)	9
	Total	8	12	20 = n

Note that the simple matching or proportion of agreement is given by

$$p_0 = \frac{7+8}{7+4+1+8} = \frac{15}{20} = 0.75$$

which in general is given by:

$$p_0 = \frac{a+d}{a+b+c+d} \quad .$$

Jaccard argued that (see Dunn and Everitt, 1982) if the clinicians are diagnosing a rare condition, the fact that they agree on the absence of the condition (the count "d") should provide no information. He then suggested

$$p_0' = \frac{a}{a+b+c}$$

as a better measure of agreement.

A new dimension to this is provided by realizing that except in the extreme circumstances when $b=0$ or $c=0$, some degree of agreement is to be expected by chance. This may happen if clinician 1 uses a set of criteria for evaluating each patient that is entirely different from the set of criteria used by clinician 2. Different opinions have been expressed on the need to incorporate agreement due to chance while assessing the interclinician reliability. The natural approach to correct for chance agreement is to consider a coefficient that assumes the value 1 when there is complete agreement (that is the maximum agreement is 1). Let p_0 be the observed agreement as defined above and p_e the amount of chance agreement.

Cohen's kappa (1960) is the ratio of p_e subtracted from p_0, divided by p_e subtracted from the maximum agreement (1). Algebraically, this is:

$$\hat{\kappa} = \frac{p_0 - p_e}{1 - p_e} \quad . \tag{3.23}$$

It should be emphasized that estimating kappa by using any of the probabilistic models described in this section involves the following assumptions:

i)   *the patients, slides or specimens to be evaluated are collected independently*
ii)   *patients are blindly assigned to clinicians, that is to say that neither clinician is aware of the result of the other clinician's assessment*
iii) *the rating categories are mutually exclusive and exhaustive*

1. PROBABILISTIC MODELS

Suppose the same two clinicians each evaluate a sample of n patients blindly (independently) with ratings denoted as either yes (1) or no (0). An underlying model proposed by Bahadur (1961) and used by Shoukri et al. (1995) to investigate the properties of estimated kappa is described as follows: let x_m and y_m denote the rating for the mth patient assigned by clinician 1 and clinician 2 respectively. Let π_{ij} ($i,j = 0,1$) denote the probability that $x_m = i$ and $y_m = j$. Furthermore, let $p_1 = \pi_{11} + \pi_{10}$ denote the probability of a positive ("yes") evaluation by rater 1 and $p_2 = \pi_{11} + \pi_{01}$ denote the probability of a positive evaluation by rater 2; $q_1 = 1-p_1$ and $q_2 = 1-p_2$. The following table shows the cell and marginal probabilities of the cross classification.

		x (Clinician 1)		
		1	0	
y (Clinician 2)	1	π_{11}	π_{01}	p_2
	0	π_{10}	π_{00}	q_2
		p_1	q_1	

Since $\pi_{11} = \Pr[x_m = 1, y_m = 1] = E(xy)$, and $E(xy) = E(x)E(y) + \rho(p_1 q_1 p_2 q_2)^{1/2}$, then

$$\pi_{11} = p_1 p_2 + \tau \qquad \pi_{10} = p_1 q_2 - \tau$$

$$\pi_{01} = p_2 q_1 - \tau \qquad \pi_{00} = q_1 q_2 + \tau$$

where,
$$\tau = \rho(p_1 q_1 p_2 q_2)^{1/2}$$

Since the percentage of agreement is $P_o = \pi_{11} + \pi_{00} = p_1 p_2 + q_1 q_2 + 2\tau$, and the percentage of agreement on the basis of chance alone is $P_e = p_1 p_2 + q_1 q_2$, then substituting in

$$\kappa = \frac{P_o - P_e}{1 - P_e},$$

we get

$$\kappa = \frac{2\rho(p_1 q_1 p_2 q_2)^{1/2}}{p_1 q_2 + p_2 q_1}, \tag{3.24}$$

and hence $\tau = \kappa(p_1 q_2 + p_2 q_1)/2$.

Note that when the two clinicians are unbiased relative to each other (i.e. $p_1 = p_2$) the kappa coefficient is ρ (the intraclass coefficient between the two ratings), for if $p_1 = p_2 = p$, then $\tau = \rho pq$, $P_o = p^2 + q^2 + 2\rho pq$ and $P_e = p^2 + q^2$. Then, $P_o - P_e = 2\rho pq$, $1 - P_e = 2pq$ and hence $\kappa = \rho$.

Moreover, when rater's bias is present, a perfect association (i.e. $\rho = 1$) does not imply complete agreement.

The maximum likelihood estimator (MLE) is obtained by maximizing the log-likelihood function L with respect to p_1, p_2, and κ. Shoukri et al. (1995) showed that the log-likelihood function is given by

$$\ln(L) = n\,\bar{s}\,\ln\left[p_1 p_2 + \frac{k}{2}\,(p_1 q_2 + p_2 q_1)\right]$$

$$+ n(\bar{y} - \bar{s})\,\ln\left[p_1 q_2 - \frac{k}{2}\,(p_1 q_2 + p_2 q_1)\right]$$

$$+ n\,(\bar{x} - \bar{s})\,\ln\left[p_2 q_1 - \frac{k}{2}\,(p_1 q_2 + p_2 q_1)\right]$$

$$+ n\,(\bar{s} - \bar{x} - \bar{y} + 1)\,\ln\left[q_1 q_2 + \frac{k}{2}\,(p_1 q_2 + p_2 q_1)\right]$$

where

$$\bar{x} = \frac{1}{n} \sum_{i=1}^{n} x_i \quad, \quad \bar{y} = \frac{1}{n} \sum_{i=1}^{n} y_i \quad,$$

and

$$\bar{s} = \frac{1}{n} \sum_{i=1}^{n} x_i y_i = s/n\,.$$

The MLE's $\hat{p}_1$, $\hat{p}_2$, $\hat{\kappa}$ of p_1, p_2 and κ are $\hat{p}_1 = \bar{x}$, $\hat{p}_2 = \bar{y}$, and

$$\hat{\kappa} = \frac{2(\bar{s} - \bar{x}\,\bar{y})}{\bar{y}(1 - \bar{x}) + \bar{x}(1 - \bar{y})} \tag{3.25}$$

For the case of no rater bias (ie. $p_1 = p_2 = p$), the MLE's are given as

$$\hat{p} = \frac{t}{2n} \quad, \quad \hat{\kappa}^* = \frac{2\left(s - \dfrac{t^2}{4n}\right)}{\left(t - \dfrac{t^2}{2n}\right)} \tag{3.26}$$

where

$$t = \sum_{i=1}^{n} (x_i + y_i) \quad .$$

It is also worthwhile to note that when $p_1 = p_2$, $\hat{\kappa}$ is algebraically equivalent to Scott's (1955) index of agreement, and to the usual estimate of ρ obtained by applying a one-way analysis of variance to the binary ratings.

B. CONFIDENCE INTERVAL ON KAPPA IN 2X2 TABLE

To obtain a large sample confidence interval on κ from a 2x2 table we need an expression for its asymptotic variance. The large sample variance of $\hat{\kappa}$ is given by

$$Var(\hat{\kappa}) = \frac{A+B-C}{n(p_1 q_2 + p_2 q_1)^2} \tag{3.27}$$

where
$$A = (p_1 p_2 + \tau)[1-(p_1+p_2)(1-\kappa)]^2 + (q_1 q_2 + \tau)[1-(q_1+q_2)(1-\kappa)]^2$$
$$B = (1-\kappa)^2[(p_1+q_2)^2(p_2 q_1 - \tau) + (p_2+q_1)^2(p_1 q_2 - \tau)]$$
$$C = [\kappa - P_e(1-\kappa)]^2$$

This is algebraically equivalent to the variance of $\hat{\kappa}$ derived by Fleiss et al. (1969). Assuming $\hat{\kappa}$ is normally distributed with mean κ and variance var($\hat{\kappa}$), the $(1-\alpha)$ 100% asymptotic confidence interval on κ is thus given by

$$\hat{\kappa} \pm Z_{\alpha_{/2}} \sqrt{var(\hat{\kappa})} \quad . \tag{3.28}$$

Shoukri et al. (1995) investigated the adequacy of the above approximation through a Monte-Carlo simulation study. They concluded that the approximation is good for samples of at least 50 individuals.

For the case of no rater bias ($p_1 = p_2$), construction of approximate confidence intervals on κ has been investigated by Bloch and Kraemer (1989) and Donner and Eliasziw (1992).

Bloch and Kraemer showed that the large sample variance of $\hat{\kappa}^*$ (the maximum likelihood of kappa when $p_1 = p_2$) is

$$var(\hat{\kappa}^*) = \frac{1-\kappa}{n}\left[(1-\kappa)(1-2\kappa)+\frac{\kappa(2-\kappa)}{2\,p(1-p)}\right] = \frac{v(\kappa)}{n} \tag{3.29}$$

Assuming $\hat{\kappa}^*$ is normally distributed with mean κ and variance $var(\hat{\kappa}^*)$, the resulting $(1-\alpha)$ 100% limits are given by

$$\hat{\kappa}^* \pm Z_{\alpha_{/2}}\sqrt{var(\hat{\kappa}^*)} \tag{3.30}$$

Bloch and Kraemer also suggested that a variance stabilizing transformation for $\hat{\kappa}^*$ should improve the accuracy for confidence interval estimation, which for $p \neq \frac{1}{2}$ is described as:

$$Z(\hat{\kappa}^*) = \begin{cases} \dfrac{1}{C_u\sqrt{v(\kappa_o)}}\arcsin\left[C_u(\hat{\kappa}^*-\kappa_o)\right], & \hat{\kappa}^*\geq\kappa_o \\[2ex] \dfrac{1}{C_L\sqrt{v(\kappa_o)}}\arcsin\left[C_L(\hat{\kappa}^*-\kappa_o)\right], & \hat{\kappa}^*\leq\kappa_o \end{cases}$$

where,

$$\kappa_o = \frac{(\gamma-\sqrt{\gamma^2-24})}{6}$$

$$\gamma = \frac{2[3-10\,p(1-p)]}{[1-4\,p(1-p)]}$$

$$C_u = \frac{1}{(1-\kappa_o)}$$

$$C_L = \frac{\left(\dfrac{1-v(\kappa_L)}{v(\kappa_o)}\right)^{1/2}}{(\kappa_o-\hat{\kappa}^*)}$$

$$\kappa_L = -\min\left(\frac{p}{1-p},\frac{1-p}{p}\right)\quad.$$

If p=½, then $Z(\hat{\kappa}^*) = arcsin(\hat{\kappa}^*)$.   Since $Z(\hat{\kappa}^*)$ is approximately normal with mean $Z(\kappa)$ and variance 1/n, an (1-α) 100% confidence limit on the transformed scale is obtained using

$$Z(\hat{\kappa}^*) \pm Z_{\alpha/2} \, n^{-1/2} \; . \tag{3.31}$$

An inverse transformation on these limits yields an approximate (1-α) 100% confidence interval on κ.

Remarks

(1)   Under the assumption of no rater bias, that is that each clinician may be characterized by the same underlying success rate p, the estimator $\hat{\kappa}^*$ is quite appropriate if the chief objective of the study is to evaluate the reliability of the measurement process itself rather than detecting potential bias between the clinicians.

(2)   It seems appropriate then that a test for clinician bias should precede estimating kappa. This is achieved by using McNemar's test for the comparison of the marginal probabilities p_1 and p_2 and is given by

$$\chi^2 = \frac{(n_{10} - n_{01})^2}{n_{10} + n_{01}} \tag{3.32}$$

where, n_{10} is the count in the cell (x=1, y=0), and n_{01} is the count in the cell (x=0, y=1). For large samples and under the assumption of marginal homogeneity (no rater bias) the statistic (3.32) is approximately distributed as a chi-square with one degree of freedom. Large values of this statistic are evidence against $H_0:p_1=p_2$.

(3)   If $H_0:p_1=p_2$ is justified by the data, then we recommend constructing confidence limits on kappa from (3.31), and kappa itself should be estimated by $\hat{\kappa}^*$ or $\hat{\kappa}$. If there is no evidence in the data to support the null hypothesis, then kappa is estimated by $\hat{\kappa}$ in (3.25) and the confidence limits are given by (3.28).

(4)   As an answer to the question; "What is good agreement", Landis and Koch (1977) have given six categories that can be used as a benchmark for determining the closeness of the comparisons.

kappa	0.00	.01-.20	.21-.40	.41-.60	.61-.8	.81-1.00
strength of agreement	poor	slight	fair	moderate	substantial	almost perfect

Example 3.9

Cervical Vertebral Malformation (CVM) is a commonly diagnosed cause of progressive spinal cord disease in young horses (foals). Neurological signs due to spinal cord compression result from morphologic changes involving the cervical vertebral and surrounding structures. A quantitative score based on radiograph diagnosis was used to predict CVM in foals predisposed to development signs of spinal cord disease. A cumulative score of 12 or more is considered a positive diagnosis ($=1$), otherwise is negative ($=0$). The results of diagnoses performed by two clinicians are given below.

		Clinician X	
		1	0
Clinician Y	**1**	32	7
	0	10	56

Even though the McNemar's chi-square statistic supports the hypothesis of marginal homogeneity ($P_1 = P_2$), for illustrative purposes we shall construct confidence limits on kappa using (3.28) and (3.31).

Since $P_0 = 0.838$, and $P_e = 0.526$, then $\hat{\kappa} = 0.658$. It can be shown that $\hat{\tau} = 0.16$, $C = 0.29$, $B = 0.019$, $A = 0.344$. Hence from (3.27), $Var(\hat{\kappa}) = 0.0056$, and a 95% confidence interval on kappa is

$$0.658 \pm 1.96 \sqrt{.0056}$$

or

$$0.511 < \kappa < 0.805 \ .$$

To use Bloch and Kraemer's expression, we find $\hat{P} = 0.386$, $\gamma = 24.141$, $\kappa_0 = 0.084$, $v(\kappa_0) = 1.009$, $C_u = 1.091$. Hence

$$0.427 < Z(\hat{\kappa}) < 0.810 \ .$$

Using the inverse transformation on each side we have

$$0.498 < \kappa < 0.795 \ .$$

C. AGREEMENT BETWEEN TWO RATERS WITH MULTIPLE CATEGORIES

A more general situation than has been discussed previously is that of two clinicians being asked to classify n patients into one of m categories. Examples of multiple categories could be levels of severity of a particular condition (slight, moderate, severe), or the classification of tumor sizes from the x-rays of n cancer patients into small, medium, large, and metastasized.

Under the same conditions of independence, blinded assignment and mutual exclusiveness of the categories, the results may be summarized as per the format of Table 3.16.

Table 3.16
Classification by Two Clinicians Into m Categories

		Clinician (2)				
		c_1	c_2	c_j	c_m	Total
	c_1	n_{11}	n_{12}	n_{1j}	n_{1m}	$n_{1.}$
Clinician (1)	c_2	n_{21}	n_{22}	n_{2j}	n_{2m}	$n_{2.}$
	c_i	n_{i1}	n_{i2}	n_{ij}	n_{im}	$n_{i.}$
	c_m	n_{m1}	n_{m2}	n_{mi}	n_{mm}	$n_{m.}$
	Total	$n_{.1}$	$n_{.2}$	$n_{.j}$	$n_{.m}$	n

Here, n_{ij} is the number of patients assigned to category i by Clinician 1 and to category j by Clinician 2.

From the definition of Cohen's kappa,

$$\hat{\kappa} = \frac{P_o - P_e}{1 - P_e} \tag{3.33}$$

the observed proportion of agreement is

$$P_o = \frac{1}{n} \sum_{i=1}^{m} n_{ii}$$

and the chance expected proportion of agreement is

$$P_e = \sum_{i=1}^{m} \left(\frac{n_{i.}}{n} \right) \left(\frac{n_{.i}}{n} \right) .$$

Fleiss, Cohen, and Everitt (1969) showed that the asymptotic variance of $\hat{\kappa}$ is estimated by,

95

$$Var_m(\hat{\kappa}) = \frac{A+B-C}{n(1-P_e)^2} , \qquad (3.34)$$

where

$$A = \sum_{i-1}^{m} \frac{n_{ii}}{n} \left[1-\left(\frac{n_{i.}+n_{.i}}{n}\right)(1-\hat{\kappa})\right]^2$$

$$B = (1-\hat{\kappa})^2 \sum_{i \neq j} \sum \frac{n_{ij}}{n^3}\left(n_{.i}+n_{j.}\right)^2$$

and $\quad C = [\hat{k}-P_e(1-\hat{\kappa})]^2 \quad .$

An approximate (1-α) 100% confidence interval for κ is

$$\hat{\kappa} \pm Z_{\alpha_{/2}}\sqrt{Var_m(\hat{\kappa})} \quad . \qquad (3.35)$$

1. Weighted Kappa

In certain studies some disagreements between the two raters may be considered more critical than others. Therefore, it may be sensible to assign weights to reflect the seriousness of the disagreements between the clinicians. Cohen (1968) suggested using a weighted kappa by assigning weights to the observed agreement and the chance expected agreement. Accordingly the observed agreement is

$$P_{ow} = \sum \sum \frac{w_{ij}n_{ij}}{n}$$

and the chance expected agreement is

$$P_{ew} = \sum_{i=1}^{m} \sum_{j=1}^{m} w_{ij} \left(\frac{n_{i.}}{n}\right) \left(\frac{n_{j}}{n}\right)$$

where w_{ij} are the chosen weights. The weighted kappa (κ_w) is calculated using equation (3.33) on replacing P_o and P_e with P_{ow} and P_{ew} respectively. That is

$$\hat{\kappa}_w = \frac{P_{ow}-P_{ew}}{1-P_{ew}} \quad .$$

96

The weights are defined so as to satisfy the following conditions:
(a) $w_{ij} = 1$ when $i = j$
(b) $0 \leq w_{ij} < 1$ for $i \neq j$, and
(c) $w_{ij} = w_{ji}$.

Fleiss and Cohen (1973) suggested defining the weights as:

$$w_{ij} = 1 - \frac{(i-j)^2}{(m-1)^2} \quad . \tag{3.36}$$

Another set of weights suggested by Ciccetti and Allison (1971) are

$$w_{ij} = 1 - \frac{|i-j|}{m-1} \quad .$$

Note that if $w_{ij} = 0$ for all $i \neq j$, then the weighted kappa reduces to Cohen's unweighted kappa.

Remarks

Let

$$\bar{x}_1 = \frac{1}{n} \sum_{i=1}^{m} i \, n_{i.} \quad \text{and} \quad \bar{x}_2 = \frac{1}{n} \sum_{j=1}^{m} j \, n_{.j}$$

be the mean of the ratings over the n patients for the first and second clinician respectively,

$$s_1^2 = \sum_{i=1}^{m} n_{i.}(i-\bar{x}_1)^2 \, , \quad s_2^2 = \sum_{j=1}^{m} n_{.j}(j-\bar{x}_2)^2$$

are the corresponding sum of squares, and

$$s_{12} = \sum_i \sum_j n_{ij}(i-\bar{x}_1)(j-\bar{x}_2)$$

would be the sum of the cross products of the two sets of ratings.

Since with Fleiss and Cohen's weights (3.26), κ_w can be written

$$\kappa_w = 1 - \frac{\sum_i \sum_j (i-j)^2 n_{ij}}{\frac{1}{n} \sum_i \sum_j n_{i.} n_{.j} (i-j)^2}$$

and, since

$$(i-j)^2 = (i-\bar{x}_1)^2 + (j-\bar{x}_2)^2 + (\bar{x}_1-\bar{x}_2)^2$$

$$+ 2(\bar{x}_1-\bar{x}_2)(i-\bar{x}_1) - 2(\bar{x}_1-\bar{x}_2)(j-\bar{x}_2)$$

$$- 2(i-\bar{x}_1)(j-\bar{x}_2)$$

then,

$$\sum \sum n_{ij}(i-j)^2 = s_1^2 + s_2^2 + n(\bar{x}_1-\bar{x}_2)^2 - 2s_{12}$$

and

$$\sum \sum n_{i.} n_{.j}(i-j)^2 = n s_1^2 + n s_2^2 + n(\bar{x}_1-\bar{x}_2)^2.$$

Hence, the final expression for the weighted kappa is,

$$\kappa_w = \frac{2 s_{12}}{s_1^2 + s_2^2 + n(\bar{x}_1-\bar{x}_2)^2} . \tag{3.37}$$

The expression (3.37) was first given by Kreppendorff (1970) and can be found in Schouten (1985, 1986). Note also that it is the concordance correlation that was given by Lin (1989) as a measure of agreement between two sets of continuous measurements (see Chapter 2).

The large sample variance of weighted kappa with arbitrary weights satisfying conditions (a),(b), and (c) was derived by Fleiss, Cohen, and Everitt (1969) and confirmed by Cicchetti and Fleiss (1977), Landis and Koch (1977). This is given as

$$Var(\hat{\kappa}_w) = \frac{1}{n(1-P_{ew})^2} \left[\sum_i \sum_j \frac{n_{ij}}{n} \left[w_{ij} - (\bar{w}_{i.} + \bar{w}_{.j})(1-\hat{\kappa}_w) \right]^2 - [\hat{\kappa}_w - P_{ew}(1-\hat{\kappa}_w)]^2 \right] \tag{3.38}$$

where

$$\overline{w}_{i.} = \sum_{j=1}^{k} \frac{n_{.j}}{n} w_{ij}$$

$$\overline{w}_{.j} = \sum_{i=1}^{k} \frac{n_{i.}}{n} w_{ij}$$

An approximate $(1-\alpha)$ 100% confidence interval on kappa is

$$\hat{\kappa}_w \pm Z_{\alpha_{/2}} \sqrt{Var(\hat{\kappa}_w)} \ .$$

Example 3.10

The following data are extracted from the VMIMS (Veterinary Medical Information and Management System). Two clinicians were asked to classify 200 dogs into 4 categories of dehydration:

0   ≡ normal
1   <5% dehydration
2   5%-10% dehydration
3   above 10%

The classification was based on subjective physical evaluation.

		Clinician (2)				
		0	1	2	3	Total
	0	119	10	2	0	131
Clinician (1)	1	7	28	1	0	36
	2	2	13	14	2	31
	3	0	0	1	1	2
	Total	128	51	18	3	200

First we construct 95% confidence limits on unweighted kappa. Since $P_0 = .81$, $P_e = .479$, then $\hat{\kappa} = 0.635$ $Var(\hat{\kappa}) = 0.00235$. Hence the 95% confidence limits are $0.54 < \kappa < 0.73$.

Using $w_{ij} = 1-(i-j)^2 / (m-1)^2$, $\hat{\kappa}_w = 0.781$ $Var(\hat{\kappa}_w) = 0.0017$, hence $0.70 < \hat{\kappa}_w < .86$ with 95% confidence. On using $w_{ij} = 1-|i-j| / m-1$ we get $\hat{\kappa}_w = 0.707$, $Var(\hat{\kappa}_w) = 0.00178$, and the 95% confidence limits are, $.62 < \kappa_w < .79$

99

A SAS program which produces the above results is as follows,

```
data kappa;
input clin1 clin2 count;
cards;
0   0   119
0   1    10
0   2     2
0   3     0
1   0     7
1   1    28
...
3   0     0
3   1     0
3   2     1
3   3     1
;
proc freq;
weight count;
tables clin1*clin2 / agree;
run;
```

SAS OUTPUT FROM ABOVE PROGRAM:

STATISTICS FOR TABLE OF CLIN1 BY CLIN2
Kappa Coefficients

Statistic	Value	ASE	95% Confidence Bounds	
Simple Kappa	0.635	0.048	0.541	0.730
Weighted Kappa	0.707	0.042	0.624	0.790

Sample Size = 200

D. MORE THAN TWO CLINICIANS

The generalization of Cohen's kappa to the case of more than two clinicians has been considered by Landis and Koch (1977). In this section we distinguish between two situations; the first is when the set of clinicians classify the patients on a binary scale and the second situation is when the set of clinicians classify the patients into one of several mutually exclusive categories. In both situations the proposed index of agreement will be derived along the lines of work done by Davies and Fleiss (1982).

1. Case I: Two Categories and Multiple Raters

a.   Test for Interclinician Bias

Let us consider the following example.

Example 3.11

As part of the problem based learning approach, senior undergraduate students at the Ontario Veterinary College were asked to identify (from x-rays) foals with Cervical Vertebral Malformation (CVM). Four students took part in the exercise, and were asked to independently classify each of 20 x-rays as affected ("1") or not ("0"). The data are given in Table 3.17.

Table 3.17
Independent Assessments of 20 X-Rays by 4 Students For Identification of CVM in Foals.

X-Ray	Student (clinician) A	B	C	D	Total
1	0	0	0	0	0
2	0	0	1	0	1
3	1	1	1	1	4
4	1	1	1	1	4
5	1	1	0	1	3
6	0	0	0	0	0
7	1	0	0	0	1
8	0	0	0	0	0
9	1	1	1	1	4
10	1	0	1	1	3
11	1	1	1	1	4
12	1	1	0	1	3
13	1	1	0	0	2
14	1	0	1	0	2
15	1	0	0	0	1
16	0	0	1	0	1
17	1	1	1	1	4
18	1	0	0	0	1
19	1	1	1	1	4
20	1	1	1	1	4
proportion	0.75	0.50	0.55	0.50	46
Total	15	10	11	10	

As can be seen, the 4 students clearly differ in their probability of classifying the presence of CVM using the x-rays. Before we construct an index of agreement it is of interest to test whether their differences are statistically significant, that is, testing the equality of the marginal proportions.

101

The appropriate procedure to test for marginal homogeneity of binary data is to use Cochran's Q_A statistic. This is derived using the layout of Table 3.18 and defining y_{ij}, as the assessment of the ith patient by the jth clinician ($i=1,2,...n$, $j=1,2,...k$), where $y_{ij}=1$ if the ith patient is judged by the jth clinician as a case and as 0 otherwise.

Table 3.18
Classification of Outcomes for Multiple Clinicians and Multiple Patients

Patient	Clinician 1	2	....	k	Total
1	y_{11}	y_{12}		y_{1k}	$y_{1.}$
2	y_{21}	y_{22}		y_{2k}	$y_{2.}$
.					
.					
i	y_{i1}	y_{i2}		y_{ik}	$y_{i.}$
.					
.					
n	y_{n1}	y_{n2}		y_{nk}	$y_{n.}$
Total	$y_{.1}$	$y_{.2}$		$y_{.k}$	$y_{..}$

Let $y_{i.}$ be the total number of clinicians who judge the ith patient as a case, and let $y_{.j}$ be the total number of patients the jth clinician judges to be cases. Finally, let $y_{..}$ be the total number of patients classified (judged) as cases.

Cochran's Q_A statistic is then given by:

$$Q_A = \frac{k(k-1) \sum_{j=1}^{k} \left(y_{.j} - \frac{y_{..}}{k} \right)^2}{k y_{..} - \sum_{i=1}^{n} y_{i.}^2} \quad .$$

Under the null hypothesis of marginal homogeneity, Q_A is approximately distributed as chi-square with k-1 degrees of freedom.

Remarks

(i)   In the case of two clinicians (i.e. k=2) Cochran's Q_A test is equivalent to McNemar's test which tests for marginal homogeneity as discussed in section B.

(ii)   For computational purposes, Q_A has the simpler expression:

$$Q_A = k(k-1) \left[\frac{\sum_{j=1}^{k} y_j^2 - \frac{y_{..}^2}{k}}{ky_{..} - \sum_{i=1}^{n} y_{i.}^2} \right] \qquad (3.39)$$

For the CVM assessments given in Table 3.17, $Q_A = 6.375$ with 3 degrees of freedom, which would lead us to accept the hypothesis that the 4 student clinicians have similar classification probabilities, at $\alpha = .05$.

b. Reliability Kappa

In this section we will be concerned with estimating kappa as a measure of reliability. This is achieved by the analysis of variance (ANOVA) procedure for the estimation of reliability coefficients. The ANOVA will be carried out just as if the results were continuous measurements (see Chapter 2) rather than binary ratings.

If the test on the hypothesis of no interclinician bias based on Q_A is justified by the data, then the one-way ANOVA mean squares can provide an estimate of reliability using the expression,

$$r_1 = \frac{MSB - MSW}{MSB + (k-1)MSW},$$

where MSB is the between patients mean square, MSW is the within patients mean square, and k is the number of assessments per patient.

A simple one-way ANOVA is performed on the data in example 3.12 with the following SAS program; an summary of the output is given in Table 3.19.

SAS program:

```
data xrays;
input xray student $ score @@;
1 a 0 1 b 0 1 c 0 1 d 0
2 a 0 2 b 0 2 c 1 2 d 0
3 a 1 3 b 1 3 c 1 3 d 1
...
18 a 1 18 b 0 18 c 0 18 d 0
19 a 1 19 b 1 19 c 1 19 d 1
20 a 1 20 b 1 20 c 1 20 d 1
;
```

```
proc glm;
class xray;
model score=xray;
run;
```

Table 3.19

The Results of the One-Way ANOVA for the Data in Table 3.17

Source	df	Sum of square	Means square
X-ray	19	11.550	0.608
Error	60	8.000	0.133

$$r_1 = \frac{0.608 - .133}{.608 + 3(.133)} = 0.472$$

This value indicates a good reliability score.

If one were not prepared to assume lack of interclinician bias, then we would have two possible models to consider: (i) two-way random effects ANOVA or (ii) a two-way mixed model.

(i) For the two-way random effects ANOVA, the appropriate intraclass correlation was derived by Bartko (1966) and is given in Chapter 2 (equation 2.5). Using the data of Table 3.17, the results of applying the two-way ANOVA using a SAS program as described below, are found in Table 3.20.

SAS PROGRAM:

```
/** two way random effects ANOVA **/

proc glm data=xrays;
class xray clinicn;
model score=xray clinicn;
run;
```

Table 3.20
The Result of the Two-Way ANOVA For the Data in Table 3.16.

Source	df	Sum of squares	Mean square
X-ray (Patient)	19	11.55	.608 (PMS)
Clinician	3	0.85	.283 (CMS)
Error	57	7.15	.125 (MSW)

Now, the estimate of the coefficient of agreement, r_2, is calculated by:

$$r_2 = \frac{\hat{\sigma}_g^2}{\hat{\sigma}_g^2 + \hat{\sigma}_c^2 + \hat{\sigma}_e^2} = 0.475$$

where

$$\hat{\sigma}_g^2 = \frac{PMS - MSW}{k} = .1206$$

$$\hat{\sigma}_c^2 = \frac{CMS - MSW}{n} = 0.008$$

and
$$\hat{\sigma}_e^2 = .125 .$$

Note that $\hat{\sigma}_g^2$ is the variance component estimate of the patient's effect, $\hat{\sigma}_c^2$ is the variance component estimate of the clinician's effect, and $\hat{\sigma}_e^2$ is estimated error variance. These estimates are valid only under the additive model,

$$y_{ij} = \text{patient's effect} + \text{clinician's effect} + \text{error} \qquad (3.40)$$

which assumes no patient-clinician interactions and that the k clinicians involved in the study are a random sample from a larger pool of clinicians.

(ii)   The two-way mixed effects ANOVA, assumes that the assessment scores have the same representation (3.40) except that the clinician's effect is fixed. From Fleiss (1986), the agreement coefficient is r_3 where

$$r_3 = \frac{n(PMS - MSW)}{n(PMS) + (k-1)CMS + (n-1)(k-1)MSW} ,$$

which when applied to the data in Table 3.16 is equal to 0.479.

2. Case II. Multiple Categories and Multiple Raters

Suppose that each of n patients is classified into one of c mutually exclusive and exhaustive categories by each of the same k clinicians; let the vector $y_{ij} = (y_{ij1}, y_{ij2}, \dots y_{ijc})$ represent the classification scores of the ith patient by the jth clinician ($i=1,2,\dots n$, $j=1,2,\dots k$). Hence,

$$y_{ijm} = \begin{cases} 1 & \text{if } i^{th} \text{ patient is classified by } j^{th} \text{ clinician as falling into the } m^{th} \text{ category; } m=1,2,\dots c \\ 0 & \text{else} \end{cases}$$

clearly

$$\sum_{m=1}^{c} y_{ijm} = 1 \quad \text{for all (i,j)}$$

Let y_{im} = number of clinicians who assign the ith patient to the mth category, so,

$$y_{im} = \sum_{j=1}^{k} y_{ijm}$$

Note that;

$$y_i = \sum_{m=1}^{c} y_{im}$$

$$= \sum_{m=1}^{c} \left[\sum_{j=1}^{k} y_{ijm} \right]$$

$$= \sum_{j=1}^{k} \left[\sum_{m=1}^{c} y_{ijm} \right] = k$$

Since we have multiple categories we denote Cohen's kappa by κ_c, where

$$\kappa_c = \frac{P_o - P_e}{1 - P_e} \quad . \tag{3.41}$$

P_o, which is the observed proportion of agreement, is equal to:

$$\frac{\text{Total number of pairs of classification that are in agreement}}{\text{Total number of possible pairs of classification}} \quad .$$

Since the number of pairs of classifications that are in agreement for the ith patient is:

$$\sum_{m=1}^{c} \binom{y_{im}}{2},$$

then the total number of pairs of classifications that are in agreement is

$$\sum_{i=1}^{n} \sum_{m=1}^{c} \binom{y_{im}}{2} \quad . \tag{3.42}$$

For each patient there are $\binom{k}{2}$ possible pairs of classifications, and the total number of possible pairs of classifications is thus:

$$\sum_{i=1}^{n} \binom{k}{2} = n \binom{k}{2} \quad . \tag{3.43}$$

Hence,

$$P_o = \frac{\sum_{i=1}^{n} \sum_{m=1}^{c} \binom{y_{im}}{2}}{n \binom{k}{2}}$$

$$= \frac{1}{n\ k(k-1)} \sum_{i=1}^{n} \sum_{m=1}^{c} y_{im}\ (y_{im}-1)$$

and finally

$$P_o = \frac{1}{n\ k(k-1)} \left[\sum_{i=1}^{n} \sum_{m=1}^{c} y_{im}^2 - nk \right] \tag{3.44}$$

107

(iii)   If we let $y_{jm} = \sum\limits_{i=1}^{m} y_{ijm}$

denote the total number of patients assigned to the mth category by the jth clinician, then

$$P_{jm} = \frac{y_{jm}}{n}$$

is the observed proportion of patients assigned to the mth category by the jth clinician.

If the assignments by clinicians are statistically independent, then the probability that any two raters (j and l) will agree in their classification of a randomly selected patient is

$$\sum\limits_{m=1}^{c} P_{jm} P_{lm}$$

and the average chance-expected probability of pairwise agreement is then

$$P_e = \frac{1}{k(k-1)} \sum\limits_{\substack{j=1 \\ j \neq l}}^{} \sum\limits_{l=1}^{} \left[\sum\limits_{m=1}^{c} P_{jm} P_{lm} \right] \tag{3.45}$$

Substituting 3.44 and 3.45 in 3.41 we get an estimate of κ_c.

Example 3.12 "CVM Data"

Four clinicians are asked to classify 20 x-rays to detect spinal cord damage in young foals believed to have Cervical Vertebral Malformation (CVM). The results are given in Table 3.21.

Classification is based on scores where:

1-5   = N (slight or no damage)
6-11   = I (intermediate damage)
≥ 12 = S (severe damage)

108

Table 3.21

CVM Classification by 4 Clinicians

X-ray	Clinician			
	1	2	3	4
1	S	I	S	I
2	I	I	S	I
3	N	N	I	N
4	N	N	N	N
5	S	S	S	I
6	N	N	N	I
7	N	N	I	N
8	N	N	N	N
9	S	S	I	S
10	S	S	S	S
11	I	I	S	I
12	I	S	S	S
13	S	I	I	I
14	N	I	I	N
15	I	I	N	N
16	I	N	N	I
17	S	S	S	S
18	S	I	S	S
19	I	S	S	S
20	N	N	I	N

The estimation of κ_c proceeds along the following steps:

First we construct the table of y_{im} (Table 3.22) from which P_o is calculated.

Table 3.22
The Number of Clinicians Who Assign the ith X-Ray to the mth Category (y_{im}).

X-ray	S	I	N	$\sum_i \sum_m y_i^2$
1	2	2		8
2	1	3		10
3		1	3	10
4			4	16
5	3	1		10
6		1	3	10
7		1	3	10
8			4	16
9	3	1		10
10	4			16
11	1	3		10
12	3	1		10
13	1	3		10
14		2	2	8
15		2	2	8
16		2	2	8
17	4			16
18	3	1		10
19	3	1		10
20		1	3	10
Total	28	26	26	216

$$P_o = \frac{1}{(20)(4)(3)} \left[216 - (20)(4)\right] = 0.567$$

Next, we collapse the information into the form found in Table 3.23a, from which the cell probabilities (Table 3.23b) and P_e are calculated.

Table 3.23a
Total Number of Patients Assigned to the mth Category by the jth Clinician (y_{im}).

Clinician	Category			Total y_i
	S	I	N	
1	7	6	7	20
2	6	7	7	20
3	9	6	5	20
4	6	7	7	20
Total	28	26	26	80

110

Table 3.23b

$$p_{jm} = \frac{y_{jm}}{n}$$

Clinician	Category		
	S	I	N
1	$\frac{7}{20}(p_{11})$	$\frac{6}{20}(p_{12})$	$\frac{7}{20}(p_{13})$
2	$\frac{6}{20}(p_{21})$	$\frac{7}{20}(p_{22})$	$\frac{7}{20}(p_{23})$
3	$\frac{9}{20}(p_{31})$	$\frac{6}{20}(p_{32})$	$\frac{5}{20}(p_{33})$
4	$\frac{6}{20}(p_{41})$	$\frac{7}{20}(p_{42})$	$\frac{7}{20}(p_{43})$

The last step is to calculate P_e from Table 3.23b.

$$P_e = \frac{1}{12} \left[\sum p_{1m} p_{2m} + \sum p_{1m} p_{3m} + \sum p_{1m} p_{4m} \right.$$

$$+ \sum p_{2m} p_{1m} + \sum p_{2m} p_{3m} + \sum p_{2m} p_{4m}$$

$$+ \sum p_{3m} p_{1m} + \sum p_{3m} p_{2m} + \sum p_{3m} p_{4m}$$

$$+ \left. \sum p_{4m} p_{1m} + \sum p_{4m} p_{2m} + \sum p_{4m} p_{3m} \right]$$

$$= .3298$$

where all the sums are from $m=1$ to $m=4$.
Hence,

$$\kappa_c = \frac{0.567 - 0.3298}{1 - 0.3298} = 0.35$$

According to Landis and Koch's (1977) benchmark, this is indicative of a fair agreement among the clinicians.

E. ASSESSMENT OF BIAS

The assessment of interclinician bias cannot be achieved in a straightforward manner using Cochran's Q_A statistic as in the case of two categories. In the multiple categories situation we assess bias for each category separately. This is done using the chi-square statistic to test the marginal homogeneity of the k clinicians.

For the mth category we evaluate the statistic

$$\chi^2_{(m)} = \sum_{j=1}^{k} \frac{n}{\bar{p}_m(1-\bar{p}_m)} (p_{jm} - \bar{p}_m)^2$$

where

$$\bar{p}_m = \sum_{j=1}^{k} \frac{y_{jm}}{kn} \quad .$$

Under the null hypothesis of no interclinician bias $\chi^2_{(m)}$ has, for large k, a chi-square distribution with k-1 degrees of freedom. In the above example, the statistics $\chi^2(S)$, $\chi^2(I)$, $\chi^2(N)$ are 1.319, 0.228, and 0.68 respectively. This shows that the four clinicians are not biased relative to each other in their classification of the 20 x-rays.

VII. STATISTICAL ANALYSIS OF MEDICAL SCREENING TESTS

A.   INTRODUCTION

Medical screening programmes are frequently used to provide estimates of prevalence where prevalence is defined as the number of diseased individuals at a specified point in time divided by the number of individuals exposed to risk at that point in time. Estimates of disease prevalence are important in assessing disease impact, in delivery of health care, and in the measurement of attributable risk. Since screening tests are less than perfect, estimated prevalence must be adjusted for *sensitivity* and *specificity* which are standard measures used in evaluating the performance of such tests. Before we examine the effect of sensitivity and specificity on the estimate of prevalence it should be recognized that the experimental error associated with the study design needed to evaluate the screening test must be as small as possible. One way to achieve this goal is by providing a precise definition of the condition which the test is intended to detect. This is often a difficult thing to do for the following reasons. First, the presence or absence of a disease is usually not an all or none phenomenon. For a wide range of conditions, there is a gradual transition from healthy to diseased, and thus any "dividing line" between these two states is often somewhat arbitrary. The second reason which is of interest to most biometricians is that repeated examinations also result in experimental errors which are due to the non-zero probability of not selecting the same cases.

This section gives a somewhat detailed treatment of the statistical problems related to medical screening tests. We shall focus on the situation where we have only two outcomes for the diagnostic tests, disease (D) and no-disease ($\bar{D}$). Situations in which the test has more than two outcomes or where the results of the test are measured by a continuous variable will not be considered here.

B.   ESTIMATING PREVALENCE

The purpose of a screening test is to determine whether a person belongs to the class (D) of people who have a specific disease. The test result indicating that a person is a member of this class will be denoted by T, and $\bar{T}$ for those who are non-members.

The sensitivity of the test is $\eta = P(T|D)$ which describes the probability that a person with the disease is correctly diagnosed; the specificity of the test is $\theta = P(\bar{T}|\bar{D})$ which is the probability of a disease free person being correctly diagnosed.

Let $\pi = P(D)$ denote the prevalence of the disease in the population tested. The results of a screening test can be summarized in terms of θ, η, and π as in the following table (Table 3.24):

Table 3.24 Layout of the Results of a Screening Test.

		Test result		
		T	$\bar{T}$	
Disease	D	$\eta\pi$	$\pi(1-\eta)$	π
Status	$\bar{D}$	$(1-\theta)(1-\pi)$	$\theta(1-\pi)$	$1-\pi$
		p	1-p	

If we let $P(T) = p$, denote the apparent prevalence then from the above table we have

$$p = \eta\pi + (1-\theta)(1-\pi) \qquad (3.46)$$

and

$$1-p = \pi(1-\eta) + \theta(1-\pi).$$

In mass screening programs the prevalence π needs to be estimated. If a random sample of n individuals are tested, we estimate p by the proportion ($\hat{p}$) of those who are classified in T.

Solving equation (3.46) for π we get the estimate of prevalence given by Rogan and Gladen (1978) as

$$\hat{\pi} = \frac{\hat{p}-(1-\theta)}{\eta+\theta-1} \tag{3.47}$$

It should be realized that the expression (3.47) may yield an estimate that does not fall between 0 and 1. For example if p is less than $(1-\theta)$ then π is negative. To remedy this problem, one can define a truncated version of $\hat{\pi}$ so that

$$\hat{\pi} = 0 \text{ when } \qquad \hat{\pi} < 0$$
$$\hat{\pi} = 1 \text{ when } \qquad \hat{\pi} > 1 .$$

When θ and η are known, $\hat{\pi}$ is an unbiased estimate with variance

$$var(\hat{\pi}) = \frac{p(1-p)}{n(\eta+\theta-1)^2} . \tag{3.48}$$

If η and θ are not known but instead are estimated from independent experiments that involved n_1 and n_2 individuals respectively, then π is estimated by

$$\hat{\pi}_2 = \frac{\hat{p}+\hat{\theta}-1}{\hat{\eta}+\hat{\theta}-1} \tag{3.49}$$

Rogan and Gladen showed that $\hat{\pi}_2$ is biased and its variance is larger than (3.48). To evaluate the bias and variance of $\hat{\pi}_2$, they employed an asymptotic Taylor's series expansion as:

$$\hat{\pi}_2 = \pi + (\hat{p}-p)\frac{\partial \hat{\pi}_2}{\partial p} + (\hat{\eta}-\eta)\frac{\partial \hat{\pi}_2}{\partial \eta} + (\hat{\theta}-\theta)\frac{\partial \hat{\pi}_2}{\partial \theta}$$

$$+ \frac{1}{2!}\left[(\hat{p}-p)^2\frac{\partial^2 \hat{\pi}_2}{\partial p^2} + (\hat{\eta}-\eta)^2\frac{\partial^2 \hat{\pi}_2}{\partial \eta^2} + (\hat{\theta}-\theta)^2\frac{\partial^2 \hat{\pi}_2}{\partial \theta^2} + \textit{terms with zero expectations}\right] + ...$$

Since $\hat{p}$, $\hat{\eta}$, and are unbiased with $E(\hat{\eta}-\eta)^2 = \frac{\eta(1-\eta)}{n_1}$, and $E(\hat{\theta}-\theta)^2 = \frac{\theta(1-\theta)}{n_2}$, it can be easily seen that

$$E(\hat{\pi}_2-\pi) = \frac{\eta(1-\eta)\pi}{n_1(\eta+\theta-1)^2} - \frac{\theta(1-\theta)(1-\pi)}{n_2(\eta+\theta-1)^2} \tag{3.50}$$

and

$$Var(\hat{\pi}_2) = \frac{p(1-p)}{n(\eta+\theta-1)^2} + \frac{\eta(1-\eta)(p+\theta-1)^2}{n_1(\eta+\theta-1)^4} + \frac{\theta(1-\theta)(\eta-p)^2}{n_2(\eta+\theta-1)^4}.$$

Since $p+\theta-1 = (\eta+\theta-1)\pi$ and $\eta-p = (1-\pi)(\eta+\theta-1)$, then

$$Var(\hat{\pi}_2) = \frac{p(1-p)}{n(\eta+\theta-1)^2} + \frac{\eta(1-\eta)\pi^2}{n_1(\eta+\theta-1)^2} + \frac{\theta(1-\theta)(1-\pi)^2}{n_2(\eta+\theta-1)^2} \tag{3.51}$$

The asymptotic bias (3.50) and variance (3.51) are correct to the first order of approximation.

C. ESTIMATING PREDICTIVE VALUE POSITIVE AND PREDICTIVE VALUE NEGATIVE

An important measure of performance of a screening test, in addition to η and θ, is the predictive value of a positive test, $C=(PV+)$, or $P(D|T)$. From Table (3.24) we have

$$C = \frac{P(D \cap T)}{P(T)} = \frac{\eta\pi}{p}.$$

When η and θ are known, we replace $\hat{\pi}$ by its estimate (3.47) to get an estimate, $\hat{C}$, for $P(D|T)$.

$$\hat{C} = \frac{\hat{\pi}\eta}{\hat{p}} = \frac{\eta}{\eta+\theta-1}\left[1-\frac{1-\theta}{\hat{p}}\right] \tag{3.52}$$

The asymptotic variance of $\hat{C}$ in this case is

$$Var(\hat{C}) = \left[\frac{\eta(1-\theta)}{\eta+\theta-1}\right]^2 \frac{(1-p)}{np^3}. \tag{3.53}$$

When η and θ are not known, but are estimated by $\hat{\eta}$ and $\hat{\theta}$ based on samples of sizes n_1 and n_2 where the screening test is used on persons whose disease status is known, we replace η and θ by these estimated values such that an estimate of C is given by

$$\hat{C}_1 = \frac{\hat{\eta}}{\hat{\eta}+\hat{\theta}-1}\left[1-\frac{1-\hat{\theta}}{\hat{p}}\right]. \tag{3.54}$$

Gastwirth (1987) showed that as n, n_1, and n_2 increase, $\hat{C}_1 \sim N(C, Var(\hat{C}_1))$ where

$$Var(\hat{C}_1) = \left\{\frac{\eta(1-\theta)}{p(\eta+\theta-1)}\right\}^2 \frac{p(1-p)}{np^2} + \left\{\frac{\pi(1-\theta)}{p(\eta+\theta-1)}\right\}^2 \frac{\eta(1-\eta)}{n_1}$$

$$+ \left\{\frac{\eta(1-\pi)}{p(\eta+\theta-1)}\right\}^2 \frac{\theta(1-\theta)}{n_2}$$

(3.55)

From (3.55) it can be seen that in the case when η and θ are near 1, but the prevalence π is low, that the third term can be the dominant term. Moreover, both the first and third terms increase as π decreases which means that using the test on groups that have low prevalence will often yield a low value of C_1 with a large variance.

Another quantity which is of much importance, as described by Gastwirth (1987), is the predictive value negative,

$$F = P(D|\overline{T}) = \frac{\pi(1-\eta)}{1-p}$$

(3.56)

which is estimated by

$$\hat{F} = \frac{\hat{\pi}(1-\hat{\eta})}{\hat{\pi}(1-\hat{\eta})+\hat{\theta}(1-\hat{\pi})}$$

$$= \frac{1-\hat{\eta}}{1-\hat{p}}\left(\frac{\hat{p}+\hat{\theta}-1}{\hat{\eta}+\hat{\theta}-1}\right).$$

(3.57)

Gastwirth (1987) showed that for large samples, $\hat{F}$ has a normal distribution with mean F and variance given by:

$$Var(\hat{F}) = \left(\frac{(1-\eta)^2\theta^2}{(\eta+\theta-1)^2(1-p)^4}\right)\frac{p(1-p)}{n}$$

$$+ \left(\frac{\theta^2\pi^2}{(1-p)^2(\eta+\theta-1)^2}\right)\frac{\eta(1-\eta)}{n_1}$$

(3.58)

$$+ \left(\frac{(1-\eta)^2(1-\pi)^2}{(1-p)^2(\eta+\theta-1)^2}\right)\frac{\theta(1-\theta)}{n_2}$$

Gastwirth recommended against the use of (3.58) for developing confidence intervals unless n, n_1, and n_2 are large enough to ensure that the normal approximation holds. This is important particularly in situations where η, θ, and π are near their boundaries.

D.   ESTIMATION IN DOUBLE SAMPLING

In the previous section we estimated the prevalence using a screening test when the true status of an individual is unknown. If however the true disease status of an individual can be determined by a test or device where it is not subject to misclassification, then we have what is traditionally known as the "gold standard".

In most practical situations though, the screening test (ST) is a relatively inexpensive procedure having less than perfect sensitivity and specificity, thus the ST tend to misclassify individuals. Using only the ST on all the n individuals results in a biased estimate of π; this bias is given by (3.50). Because the use of a gold standard may be very costly due to the requirement for n_1 and n_2 additional individuals, the estimation of η and θ by this means may not be easily obtained.

To compromise between the two extremes, a double sampling scheme (DSS) was proposed by Tenenbein (1970). This DSS requires that:

(i)   a random sample of n individuals is drawn from the target population
(ii)   a subsample of n_1 units is drawn from n and each of these n_1 units is classified by both the ST and the gold standard
(iii)   the remaining $n-n_1$ individuals are classified only by the ST. Let x denote the number of individuals whose ST classification is diseased and let y denote the number of individuals whose ST classification is not disease.

Using Tenenbein's notation, the resulting data can be presented as follows:

		ST		
		S	S	
Gold Standard	D	n_{11}	n_{10}	$n_{1.}$
	$\bar{D}$	n_{01}	n_{00}	$n_{0.}$
		$n_{.1}$	$n_{.0}$	n_1
		X	Y	$n-n_1$

The likelihood function of n_{ij} $(n_{11}, n_{10}, n_{01}, n_{00})'$ is proportional to

$$L(n_{ij}) = (\eta\pi)^{n_{11}} (\pi(1-\eta))^{n_{10}} ((1-\theta)(1-\pi))^{n_{01}} (\theta(1-\pi))^{n_{00}} .$$

Conditional on n_{ij}, x has binomial distribution Bin $(p, n-n_1)$ and the likelihood of the experiment is proportional to

$$L = L(n_{ij}) \; \text{Bin} \; (p, n-n_1)$$

$$\propto (\eta\pi)^{n_{11}} \; (\pi(1-\eta))^{n_{10}} \; ((1-\theta)(1-\pi))^{n_{01}} \qquad\qquad (3.59)$$

$$(\theta(1-\pi))^{n_{00}} \; p^{x}(1-p)^{y}$$

Since

$$C = \frac{\eta\pi}{p} \;, \quad F = \frac{\pi(1-\eta)}{1-p} \;,$$

then from (3.46)

$$1-C = \frac{(1-\theta)(1-\pi)}{p}$$

and

$$1-F = \frac{\theta(1-\pi)}{1-p} \;.$$

Writing (3.59) in terms of C,F, and P we get

$$L \; \alpha \; C^{n_{11}} \; (1-C)^{n_{01}} \; F^{n_{10}} \; (1-F)^{n_{00}} \; P^{x+n_{.1}} \; (1-P)^{y+n_{.0}}$$

Differentiating the logarithm of L with respect to C,F, and P and equating to zero, we get their maximum likelihood estimates (MLE) as

$$\hat{C} = \frac{n_{11}}{n_{.1}} \;, \quad \hat{F} = \frac{n_{10}}{n_{.0}} \;, \quad \text{and} \quad \hat{P} = \frac{x+n_{.1}}{x+y+n_1}$$

therefore, the MLE of π, η and θ are given respectively as

$$\hat{\pi} = \frac{n_{11}}{n_{.1}} \frac{x+n_{.1}}{x+y+n_1} + \frac{n_{10}}{n_{.0}} \frac{y+n_{.0}}{x+y+n_1}$$

118

$$\hat{\eta} = 1 - \frac{\dfrac{n_{10}}{n_{.0}} \; \dfrac{y+n_{.0}}{x+y+n_1}}{\hat{\pi}}$$

and

$$\hat{\theta} = 1 - \frac{\dfrac{n_{01}}{n_{.1}} \; \dfrac{x+n_{.1}}{x+y+n_1}}{1-\hat{\pi}} .$$

Note that $x+y+n_1 = n$.

The variance-covariance matrix of the MLE's is obtained by inverting Fisher's information matrix whose diagonal elements are given by

$$-E\left[\frac{\partial^2 \log L}{\partial C^2}\right], \quad -E\left[\frac{\partial^2 \log L}{\partial F^2}\right], \quad -E\left[\frac{\partial^2 \log L}{\partial p^2}\right],$$

and where,

$$-E\left[\frac{\partial^2 \log L}{\partial C \, \partial F}\right] = -E\left[\frac{\partial^2 \log L}{\partial C \, \partial p}\right] = -E\left[\frac{\partial^2 \log L}{\partial F \, \partial p}\right] = 0$$

$$\frac{\partial^2 \log L}{\partial C^2} = -\frac{n_{11}}{C^2} - \frac{n_{01}}{(1-C)^2}$$

$$-E\left[\frac{\partial^2 \log L}{\partial C^2}\right] = \frac{1}{C^2} E(n_{11}) + \frac{1}{(1-C)^2} E(n_{01})$$

$$= \frac{n_1 \eta \pi}{C^2} + \frac{n_1(1-\theta)(1-\pi)}{(1-C)^2}$$

Since

$$C = \frac{\eta \pi}{p} \quad \text{and} \quad (1-\theta)(1-\pi) = p(1-C)$$

119

then

$$-E\left[\frac{\partial^2 \log L}{\partial C^2}\right] = \frac{n_1 p}{C} + \frac{n_1 p}{1-C} = \frac{n_1 p}{C(1-C)} \; .$$

Since

$$-E\left[\frac{\partial^2 \log L}{\partial F^2}\right] = \frac{1}{F^2} E(n_{10}) + \frac{1}{(1-F)^2} E(n_{00})$$

$$= n_1 \frac{\pi(1-\eta)}{F^2} + n_1 \frac{\theta(1-\pi)}{(1-F)^2}$$

Since

$$F = \frac{\pi(1-\eta)}{1-P} \quad \text{and} \quad (1-P)(1-F) = \theta(1-\pi) \, ,$$

then

$$-E\left[\frac{\partial^2 \log L}{\partial F^2}\right] = n_1 \frac{1-p}{F} + n_1 \frac{1-p}{1-F}$$

$$= \frac{n_1(1-p)}{F(1-F)}$$

Finally;

$$-E\left[\frac{\partial^2 \log L}{\partial p^2}\right] = \frac{1}{p^2} E(x+n_{.1}) + \frac{1}{(1-p)^2} E(y+n_{.0})$$

$$= \frac{1}{p^2}\left[(n-n_1)\, p + n_1\, p\right] + \frac{1}{(1-p)^2}\left[(n-n_1)(1-p) + (1-p)n_1\right]$$

$$= \frac{n}{p(1-p)}$$

As n_1 and n increase, then

$$Var(\hat{C}) = \frac{C(1-C)}{n_1 p}$$

$$Var(\hat{F}) = \frac{F(1-F)}{n_1(1-p)}$$

$$Var(\hat{p}) = \frac{p(1-p)}{n}$$

To find the variance of $\hat{\pi}$, we note that

$$\hat{\pi} = \hat{C}\,\hat{p} + \hat{F}(1-\hat{p}) \quad .$$

Using the delta method, we have to the first order of approximation

$$Var(\hat{\pi}) = Var(\hat{C})\left(\frac{\partial \pi}{\partial C}\right)^2 + Var(\hat{F})\left(\frac{\partial \pi}{\partial F}\right)^2 +$$

$$Var(\hat{p})\left(\frac{\partial \pi}{\partial p}\right)^2$$

where the partial derivatives are evaluated at the true values of the parameters. Hence

$$Var(\hat{\pi}) = \frac{C(1-C)}{n_1}\,p + \frac{F(1-F)}{n_1}\,(1-p) + \frac{p(1-p)}{n}(C-F)^2$$

Using the following identities by Tenenbein (1970)

$$\pi\,\eta(1-\eta) + (1-\pi)\,\theta(1-\theta) \equiv p(1-p) - \pi(1-\pi)(\eta+\theta-1)^2$$

$$C(1-C)\,p + F\,(1-F)(1-p) \equiv \pi(1-\pi)\left[1 - \frac{\pi(1-\pi)}{p}(1-p)\,(\eta+\theta-1)^2\right],$$

and

$$p(1-p)(C-F)^2 \equiv \frac{\pi^2(1-\pi)^2}{p(1-p)}\,(\eta+\theta-1)^2$$

we have

$$Var(\hat{\pi}) = \frac{\pi(1-\pi)}{n_1}\left[1 - \frac{\pi(1-\pi)}{p(1-p)}(\eta+\theta-1)^2\right] + \frac{\pi^2(1-\pi)^2}{n\ p(1-p)}(\eta+\theta-1)^2 \qquad (3.60)$$

For two dichotomous random variables whose joint distribution is given as in Table 3.24, the intraclass correlation is given by

$$\rho = \frac{P(D \cap T) - P(T)P(D)}{\sqrt{P(D)(1-P(D))P(T)(1-P(T))}}$$

$$= \frac{\eta\pi - p\pi}{\sqrt{\pi(1-\pi)p(1-p)}} = \frac{\pi(\eta-p)}{\sqrt{\pi(1-\pi)p(1-p)}} \quad .$$

Since

$$\eta - p = (1-\pi)(\eta+\theta-1)$$

then

$$\rho = \sqrt{\frac{\pi(1-\pi)}{p(1-p)}}\ (\eta+\theta-1) \quad .$$

Tenenbein (1970) defined the coefficient of reliability $K = \rho^2$ as a measure of the strength between the ST and the gold standard. From 3.60 we get:

$$Var(\hat{\pi}) = \frac{\pi(1-\pi)}{n_1}(1-K) + \frac{\pi(1-\pi)}{n}K \qquad (3.61)$$

It turns out that the asymptotic variance of $\hat{\pi}$ is the weighted average of the variance of a binomial estimate of π based on n_1 measurements from the gold standard and the variance of a binomial estimate of π based on n measurements from the gold standard. One should also note that:

(i)   when $\rho = 0$, that is when the ST is useless, then the precision of $\hat{\pi}$ depends entirely on the n_1 measurements from the gold standard

(ii)   where $\rho = 1$, that is when the ST is as good as the gold standard, the precision of $\hat{\pi}$ depends on the n measurements from the gold standard. This is because in using n measurements from the ST we get the same precision as a binomial estimate of π based on n measurements from the gold standard.

Example 3.13

Mastitis is a frequently occurring disease in dairy cows. In most cases it is caused by bacteria. To treat a cow it is quite important to identify the specific bacteria causing the infection. Bacterial culture is the gold standard, but it is a time consuming and expensive procedure. A new test of interest, pro-Staph ELISA procedure, is cheap and easily identifies the *S.aureus* bacteria.

Milk samples were taken from 160 cows. Each sample was split into two halves, one half was tested by bacterial culture and the other half was tested by the ELISA. Further milk samples taken from 240 cows were evaluated by the ELISA test. The data obtained are found in Table 3.25.

Table 3.25
Test Outcomes For Bacterial Identification Using
ELISA and a 'NEW TEST'.

		NEW TEST		
		1	0	Total
ELISA	1	23	13	36
	0	18	106	124
	Total	41	119	160
		62	178	240

Here $n_{11} = 23$, $n_{10} = 13$, $n_{01} = 18$, $n_{00} = 106$, $x = 62$, $y = 178$. The sample sizes are $n_1 = 160$, and $n = 400$.

$$\hat{\pi} = \frac{23}{41}\frac{62+41}{400} + \frac{13}{119}\frac{178+119}{400} = .226$$

$$\hat{\eta} = 1 - \frac{\frac{13}{119}\frac{178+119}{400}}{.226} = 1 - \frac{.0811}{.226} = .641$$

$$\hat{\theta} = 1 - \frac{\frac{18}{41}\frac{62+41}{400}}{.774} = 0.854$$

$$\hat{p} = (.641)(.226)+(1-.854)(1-.226) = .258$$

$$\hat{K} = \frac{(.226)(1-.226)}{.258(1-.258)}(.641+.854-1)^2 = .224$$

123

$$Var(\hat{\pi}) = \frac{.226(1-.226)}{160}(1-.224) + \frac{.226(1-.226)}{400}(.224)$$

$$= 0.00085 + .000095 = 0.00095$$

A 95% confidence interval on π is (0.166, 0.286). These results indicate that the true prevalence lies between 17 and 29%.

VIII. DEPENDENT SCREENING TESTS (MULTIPLE READINGS)

The standard definitions of sensitivity and specificity imply that all true positives have the same probability of a positive test result, and similarly that all true negatives have the same probability of a negative test result. It is realistic to expect that, within each of the two groups - true positives and true negatives, the probability of a positive test result may vary from person to person, or from time to time in the same person.

To improve the performance of a screening test, a recommended device is the use of multiple tests instead of a single test on each person.

Let n be the number of tests obtained from each person and $y_1, \ldots y_n$ denote the n test results where $y_i = 1$ if positive and 0 if negative. Since the tests are repeated on the same individuals, their results are not independent. Indeed, we may expect that any two tests are positively correlated. It is of interest to study the dependence among tests and its effect on sensitivity, specificity and predictive values of the composite test.

Let π denote the probability of a positive outcome of a single test in a given person. The population to which the test is applied is considered as being composed of two subpopulations: true positives and true negatives. In each subpopulation, π has a probability density. Following Meyer (1964) it is proposed that

$$f(\pi \mid \text{true positive}) \equiv f_1(\pi) = \frac{\pi^{a_1-1}(1-\pi)^{b_1-1}}{\beta(a_1,b_1)}$$

and

$$f(\pi \mid \text{true negative}) \equiv f_2(\pi) = \frac{\pi^{a_2-1}(1-\pi)^{b_2-1}}{\beta(a_2,b_2)}$$

Now, let $y = \sum_{i=1}^{n} y_i$; clearly,

$$P(y=k \mid \pi) = \binom{n}{k} \pi^k(1-\pi)^{n-k}$$

$$k=0,1,2,\ldots n$$

hence,

$$P(y=k \mid \text{true positive}) = \binom{n}{k} \int_0^1 \pi^k(1-\pi)^{n-k} f_1(\pi)d\pi$$

and

$$P(y=k \mid \text{true negative}) = \binom{n}{k} \int_0^1 \pi^k(1-\pi)^{n-k} f_2(\pi)d\pi$$

Note that

$$P(y=n \mid \text{true positive}) = E_1(\pi^n)$$
$$P(y=n \mid \text{true negative}) = E_2(\pi^n)$$

i.e.   the nth moment of π has probability interpretation.

Suppose we assign a person with more than k positive responses as screen-positive ($+$) and a person with less than $k+1$ positive results as screen negative (-). The sensitivity and specificity of the composite test can be defined by

$$\eta'_{nk} = Pr(k<y \le n \mid \text{true positive})$$

$$= \sum_{r=k+1}^n \binom{n}{r} \int_0^1 \pi^r (1-\pi)^{n-r} f_1(\pi)d\pi$$

$$\theta'_{nk} = Pr(0<y \le k \mid \text{true negative})$$

$$= \sum_{r=0}^k \binom{n}{r} \int_0^1 \pi^r (1-\pi)^{n-r} f_2(\pi)d\pi$$

Note that

$$\eta = \eta'_{10} = \int_0^1 \pi f_1(\pi)d\pi = E_1(\pi)$$

and

$$\theta = \theta'_{10} = 1 - \int_0^1 \pi \, f_2(\pi) d\pi = 1 - E_2(\pi) \quad .$$

Clearly, given $n \geq 2$, η'_{nk} can be increased by using smaller k, and θ'_{nk} can be increased by using a larger k. Moreover, using a sufficiently larger n and an appropriate k make either η'_{nk} or θ'_{nk} close to unity. Generally, this cannot be achieved with the same k for both parameters regardless of how large n is (see Meyer 1964).

A question of practical importance is, therefore, how much a replacement of a single test by a composite test can improve both η'_{nk} and θ'_{nk} simultaneously, for various choices of the division between observed positives and negatives.

To address this question Meyer (1964) proved that
(i) For any given value of the ratio:

$$\lambda = \frac{k+1}{n+1} \quad ,$$

$$\eta'_{nk} < \int_\lambda^1 f_1(\pi) d\pi \quad \text{and} \quad \theta'_{nk} < \int_0^\lambda f_2(\pi) d\pi$$

Moreover, under fairly general conditions

$$\lim_{n \to \infty} \eta'_{nk} = \int_\lambda^1 f_1(\pi) d\pi = Pr[\pi \geq \lambda | f_1]$$

and

$$\lim_{n \to \infty} \theta'_{nk} = \int_0^\lambda f_2(\pi) d\pi = Pr[\pi \leq \lambda | f_2]$$

The implications of these relationships are as follows. If the distributions of the non-diseased and diseased according to π overlap, then either η'_{nk} or θ'_{nk}, or both, is less than unity no matter what values of n and k we choose. In other words, the screening test can never give a perfect separation between diseased and nondiseased no matter how many tests are given to each individual regardless of how k is chosen. Therefore the usefulness of a multiple readings test is limited by the prior distributions of π.

126

(ii) Under certain conditions, we can always find two integers $n > 1$ and $0 \leq k \leq n$, such that

$$\eta'_{nk} > \eta'_{10} \quad \text{and} \quad \theta'_{nk} > \theta'_{10}$$

are fulfilled simultaneously.

The important problem here is to find a dividing line (i.e. a value of k) which gives the best estimates of η'_{nk} and θ'_{nk}. Two factors should be considered when the decision is made; first, the relative importance of obtaining higher sensitivity or higher specificity and, second the prevalence of true positives in the population.

A. ESTIMATION OF PARAMETERS

From the previous presentation, it is clear that much of the usefulness of estimating sensitivity and specificity using a multiple reading procedure depends on the prior knowledge of the shape of $f_1(\pi)$ and $f_2(\pi)$ which for practical reasons are chosen to follow the already specified beta distributions. The strategies in estimating these parameters are

(i)   for each of N persons in the sample we apply a series of n independent screening tests,
(ii)   for all individuals in the sample (or a fraction of them), we apply a non-faulty (confirmatory) test to identify the true positives.

Here we discuss the situation wherein all screened individuals are followed by a confirmatory test. The results are arranged as in Table 3.26.

Table 3.26
Results of Screened Individuals Followed by a
Confirmatory Test

		Confirmatory		
		D	$\bar{D}$	
Screening Test	T	N_{tp}	N_{fp}	$N_{1.}$
	T	N_{fn}	N_{tn}	$N_{2.}$
		$N_{.1}$	$N_{.2}$	

Let N_k be the number of persons with k positive screening results, $(N = \sum_{k=0}^{n} N_k)$

and let N_k^+ and $N_k^- = N_k - N_k^+$ denote the number of true positives and true negatives among individuals with k positive tests respectively. The likelihood of the sample is written as

127

$$L = \prod_{k=0}^{n} \left(p_k^+ \right)^{N_k^+} \left(p_k^- \right)^{N_k^-}$$

where

$$p_k^+ = Pr \; [y=k \cap true \quad positive]$$

$$p_k^+ = \pi \binom{n}{k} \int_0^1 \pi^k (1-\pi)^{n-k} f_1(\pi) d\pi$$

$$p_k^- = (1-\pi) \binom{n}{k} \int_0^1 \pi^k (1-\pi)^{n-k} f_2(\pi) d\pi$$

$$f_i(\pi) = \frac{\pi^{a_i-1}(1-\pi)^{b_i-1}}{\beta(a_i,b_i)} \qquad i=1,2$$

$$\therefore \; p_k^+ = \pi \binom{n}{k} \int_0^1 \pi^k (1-\pi)^{n-k} \frac{\pi^{a_i-1}(1-\pi)^{b_i-1}}{\beta(a_1,b_1)} \, d\pi$$

$$= \pi \binom{n}{k} \frac{1}{\beta(a_1,b_1)} \int_0^1 \pi^{a_1+k-1} (1-\pi)^{n+b_1-k-1} d\pi$$

$$= \pi \binom{n}{k} \frac{\beta(a_1+k,n+b_1-k)}{\beta(a_1,b_1)}$$

$$= \pi \binom{n}{k} \frac{\Gamma(a_1+k)\Gamma(n+b_1-k)\Gamma(a_1+b_1)}{\Gamma(a_1+b_1+n)\Gamma(a_1)\Gamma(b_1)}$$

Now, since

$$\ell = \log L = \sum_{k=0}^{n} \left\{ N_k^+ \ln p_k^+ + N_k^- \ln p_k^- \right\}$$

then

$$\frac{\partial \ell}{\partial \pi} = \sum_{k=0}^{n} \left[N_k^+ \left\{ \frac{\partial \ln p_k^+}{\partial \pi} \right\} + N_k^- \left\{ \frac{\partial \ln p_k^-}{\partial \pi} \right\} \right]$$

$$= \sum_{k=0}^{n} \left[\frac{N_k^+}{\pi} - \frac{N_k^-}{1-\pi} \right]$$

Equating $\dfrac{\partial \ell}{\partial \pi}$ to zero and solving for π we get:

$$\hat{\pi} = \frac{N^+}{N} \ , \ \text{where} \ N^+ = \sum_{k=0}^{n} N_k^+ \ , \ N^- = \sum_{k=0}^{n} N_k^- \ , \ \text{and} \ N = \sum_{k=0}^{n} (N_k^+ + N_k^-) \ .$$

Since

$$E\left[\frac{-\partial^2 \ell}{\partial \pi^2} \right] = -E \left[\sum_{k=0}^{n} - \frac{N_k^+}{\pi^2} - \frac{N_k^-}{(1-\pi)^2} \right]$$

$$= \sum_{k=0}^{n} \left\{ \frac{E(N_k^+)}{\pi^2} + \frac{E(N_k^-)}{(1-\pi)^2} \right\}$$

and

$$E(N_k^+) = N \, p_k^+ = N \, \pi \binom{n}{k} \frac{\beta(a_1+k, n+b_1-k)}{\beta(a_1, b_1)} \ ,$$

$$E(N_k^-) = N \, p_k^- = N(1-\pi) \binom{n}{k} \frac{\beta(a_2+k, n+b_2-k)}{\beta(a_2, b_2)}$$

$$\therefore i_{\pi\pi} = \sum_{k=0}^{n} \left\{ \frac{N}{\pi} \binom{n}{k} \frac{\beta(a_1+k, n+b_1-k)}{\beta(a_1,b_1)} + \frac{N}{1-\pi} \binom{n}{k} \frac{\beta(a_2+k, n+b_2-k)}{\beta(a_2,b_2)} \right\}$$

$$= N \sum_{k=0}^{n} \left\{ \frac{\binom{n}{k}}{\pi} \frac{\beta(a_1+k, n+b_1-k)}{\beta(a_1,b_1)} + \frac{\binom{n}{k}}{1-\pi} \frac{\beta(a_2+k, n+b_2-k)}{\beta(a_2,b_2)} \right\}$$

Noting that

$$\sum_{k=0}^{n} \frac{\Gamma(a_1+k)\Gamma(n+b_1-k)}{k!(n-k)!} = \frac{\Gamma(a_1)\Gamma(b_1)\Gamma(a_1+b_1+n)}{\Gamma(a_1+b_1) \ n!}$$

then we can show that

$$Var(\hat{\pi}) = (i_{\pi\pi})^{-1} = \frac{\pi(1-\pi)}{N}. \tag{3.62}$$

Example 3.14:

Four teats of 270 healthy cows were infected with a pathogen that causes mastitis. After 21 days each teat was tested for the mastitis infection by a new ELISA test followed by a confirmatory bacterial culture test. The result of the experiment is summarized in Table 3.27.

Table 3.27
Results of Mastitis Testing Using ELISA and Bacterial Culture Test

Number of positive teats (tests) k	Result of follow-up		Total
	N_k^+ True positives	N_k^- True negatives	N_k
0	3	220	223
1	2	10	12
2	5	2	7
3	9	2	11
4	16	1	17
	35	235	270

This gives $\hat{\pi} = \dfrac{35}{270} = 0.13$, and $Var(\hat{\pi}) = 0.00042$.

130

IX. SPECIAL TOPICS

A.   GROUP TESTING

Applications of a screening program to estimate disease prevalence or to identify diseased individuals in a large population is often an expensive and tedious process particularly if individuals are tested one-by-one. In situations where the disease prevalence is low and the cost of testing each individual is high, the goal of proper screening may not be achieved. In such cases, it is often preferable to form pools of units, and then to test all units in a pool simultaneously. This is called group testing. When the test result is positive (disease is present) one concludes that there is at least one diseased individual in the group, but do not know which one(s) or how many. When individuals in the group are considered independent and have the same probability P of being positive, the probability that a group of size k is negative is $(1-P)^k$. Hence, with probability $1-(1-P)^k$ the group will contain at least one diseased individual.

Dorfman (1943) introduced the group testing approach to the statistical literature under the assumption of a simple binomial model. He applied this to the problem of blood testing for syphilis. The purpose of this section is to outline some procedures for estimating prevalence using group testing.

1. Estimating Prevalence by Group Testing

Suppose that we have n groups tested, each of size k, and let x be the number of nonconformant or defective groups. This means than $\frac{x}{n}$ can be used as the moment estimator of $1-(1-P)^k$.

That is

$$\frac{x}{n} = 1-(1-\hat{P})^k$$

from which

$$\hat{P} = 1-(1-\frac{x}{n})^{\frac{1}{k}}.$$

Clearly, the smaller the value of P, the larger the advantage of group testing over one-by-one unit testing. Moreover, the smaller the value of P the larger the chance that a group testing will be shown to contain zero defectives. The question is how large can P be before group testing is ineffective. Chen and Swallow (1990) showed that if r is the cost of a unit, then for

$$P < 1 - \frac{(1+2r)}{3+2r}$$

group testing can have lower cost per unit information than will single - unit testing, a result reported earlier by Sobel and Elashoff (1975).

In general it is often the case that data arise from unequal group sizes. Suppose there are m different group sizes $k_1, k_2, \ldots k_m$ and we denote y_i as the number of defective groups out of n_i groups of size k_i.

Assume that y_i follows a binomial distribution with parameters $(n_i, 1-q^{ki})$, where $q = 1-p$.

$$E(y_i) = n_i(1-q^{k_i}) \quad var(y_i) = n_i q^{k_i}(1-q^{k_i})$$

The likelihood of the sample is given by

$$L = \prod_{i=1}^{m} \binom{n_i}{y_i} (1-q^{k_i})^{y_i} (q^{k_i})^{n_i-y_i}$$

The log-likelihood is

$$\ell = \sum_{i=1}^{m} \{ y_i \ln(1-q^{k_i}) + k_i(n_i-y_i)\ln q \}$$

Differentiating ℓ with respect to q and equating to zero we get

$$\frac{\partial \ell}{\partial q} = \sum_{i=1}^{m} \left\{ \frac{-k_i y_i q^{k_i-1}}{1-q^{k_i}} + \frac{k_i(n_i-y_i)}{q} \right\}$$

Solving $\dfrac{\partial \ell}{\partial q} = 0$, we have

$$0 = \sum_{i=1}^{m} \left\{ \frac{k_i(n_i-y_i)(1-q^{k_i}) - q^{k_i}k_i y_i}{1-q^{k_i}} \right\},$$

or

$$\sum_{i=1}^{m} k_i n_i = \sum_{i=1}^{m} k_i y_i(1-q^{k_i})^{-1}$$

$$\frac{\partial^2 \ell}{\partial q^2} = \sum_{i=1}^{m} \left[\frac{-k_i(n_i - y_i)}{q^2} - \frac{k_i y_i}{(1-q^{k_i})^2} \left\{ (1-q^{k_i})(k_i-1)q^{k_i-2} + q^{k_i-1}k_i\ q^{k_i-1} \right\} \right] .$$

Hence,

$$E\left[-\frac{\partial^2 \ell}{\partial q^2} \right] = \sum_{i=1}^{m} \frac{n_i\ k_i^2\ q^{k_i-2}}{(1-q^{k_i})} ,$$

and,

$$Var(\hat{P}) = \left\{ \sum_{i=1}^{m} \frac{n_i\ k_i^2\ q^{k_i-2}}{1-q^{k_i}} \right\}^{-1} ,$$

which is a result obtained by Walter et al. (1980).

If all the groups are of equal size, the variance of $\hat{p}$ reduces to

$$Var(\hat{p}) = \frac{1-q^k}{m\ n\ k^2\ q^{k-2}}$$

Dorfman's (1943) work on group testing aimed at classifying the individuals into diseased and non-diseased while reducing the expected number of tests. His approach requires that if a group is classified as defective, then each individual must be retested to identify its status. In general, Dorfman's approach possesses the following features:

a) an approximate solution for the optimal group size can be easily obtained in closed form
b) every individual is subjected to at most two tests, and
c) the entire testing protocol needs only two time periods given that the resources are available to test the initial groups, and then all individuals in the group identified as diseased are tested.

Let T be the total number of tests required by a group testing procedure. Since N/k is the number of groups of size k, and $N/k[1-(1-p)^k]$ is the expected number of infected groups of k in a population of size N, then the expected number of tests required by the group testing procedure is

$$E(T) = \frac{N}{k} + \frac{N}{k}\ [1-(1-p)^k](k)$$

133

and

$$Var(T) = kN[1-(1-p)^k](1-p)^k$$

$$= kN[(1-p)^k-(1-p)^{2k}]$$

The relative cost r of the group testing as compared to the one-by-one testing is the ratio of E(T) to N. That is,

$$r = \frac{E(T)}{N} = \frac{1}{k} + 1-(1-p)^k$$

$$= \frac{1+k}{k} - (1-p)^k$$

In many biomedical experiments, the tested material is always of a unit portion which is regulated by the testing equipment. Thus a test on a group of size k means that the test portion is a mixture of k items each contributing 1/k units. Therefore the contribution of a diseased item to the test portion will be diluted as the group gets larger. When the screening test is sensitive to the portion of defective units in the group, this "dilution effect" may cause the test to fail in identifying the defective status of the group and hence the binomial model is no longer appropriate. Chen and Swallow, based on Hwang's (1976) dilution-effect model, suggested that for a fixed d (dilution factor) and k_i, the parameter q is estimated by fitting a non-linear regression model:

$$E\left(\frac{n_i-y_i}{n_i}\right) = \frac{q^{k_i}+q-q^{k_i+1}-q^{k_i^d}}{1-q^{k_i^d}} \qquad 0 \le d \le 1$$

Here, d=0 corresponds to no dilution effect, while d=1 indicates complete dilution.

B. EVALUATING THE PERFORMANCE OF MEDICAL SCREENING TESTS IN THE ABSENCE OF A GOLD STANDARD (Latent class models)

We have already seen how sensitivity η and specificity θ are used as measures of evaluating the performance of a screening test. Equally important are their complement $\beta=1-\eta$ and $\alpha=1-\theta$ or respectively, the false negative and the false positive error rates associated with the test.

The estimation of α and β is a simple matter when the true diagnostic status can be determined by using a gold standard. Unfortunately, this is often impractical or impossible, and α and β are estimated by comparing a new test with a reference test, which also has error rates associated with it. Ignoring the error rates of the reference test would result in biased estimates of the new test. If the error rates of the reference test are known, Gart and Buck (*GB*) (1966) derived estimates of the error rates of the new test.

In situations when the error rates of the new test and the reference test are unknown, Hui and Walter (*HW*) (1980) showed how to estimate them. They used the method of maximum likelihood when the two tests (the new and the reference) are simultaneously applied to individuals from two populations with different disease prevalences.

The estimators obtained by *GB* and *HW* are derived under the assumption that the two tests are independent, conditional on the true diagnostic status of the subjects. That is, the misclassification errors of the two tests are assumed to be unrelated. Vacek (1985) showed that there are situations when this assumption of conditional independence cannot be justified.

In the following section we will first introduce the error rates estimation procedure discussed by *HW* and examine the effect of conditional dependence on the estimates as was shown by Vacek.

1. Estimation of Error Rates Under the Assumption of Conditional Independence

Following *HW*, let the standard test (Test 1) and the new test (Test 2) be applied simultaneously to each individual in samples from S populations. Let $N_g(g=1,2,\ldots S)$ be a fixed sample size, π_g the probability of a diseased individual ($Pr(D) = \pi$) and α_{gh} and β_{gh} the false positive and false negative rates of test h (h=1,2) in population g. A crucial assumption for this procedure is that conditional on the true disease status of an individual, the two tests have independent error rates. This assumption may be reasonable if the tests have unrelated bases, e.g. x-ray versus blood test. The results of the two tests in the gth sample can be summarized in the following 2x2 table (Table 3.28).

Table 3.28
The Result of the gth Sample
g=1,2,...S

		Test (1)	
		$T_1=1$ (positive)	$T_1=0$ (negative)
Test (2)	$T_2=1$	P_{g11}	P_{g01}
	$T_2=0$	P_{g10}	P_{g00}

We shall drop the subscript *g* for the time being.

$$P_{11} = Pr[T_1=1, \ T_2=1]$$

$$= Pr[T_1=1, \ T_2=1 \mid D]Pr(D) + Pr[T_1=1, \ T_2=1 \mid \overline{D}]Pr(\overline{D})$$

Because of the assumption of conditional independence we write

$$Pr[T_1=1, \ T_2=1 \mid D] = Pr[T_1 \mid D]Pr[T_2 \mid D]$$

$$= \eta_1\eta_2 = (1-\beta_1)(1-\beta_2)$$

and

$$Pr[T_1=1, \ T_2=1 \mid \overline{D}] = Pr[T_1=1 \mid \overline{D}]Pr[T_2=1 \mid \overline{D}]$$

$$= (1-\theta_1)(1-\theta_2) = \alpha_1\alpha_2 \quad .$$

Hence

$$P_{11} = (1-\beta_1)(1-\beta_2)\pi + \alpha_1\alpha_2(1-\pi)$$

which means that for the gth sample we write

$$P_{g11} = (1-\beta_{g1})(1-\beta_{g2})\pi_g + \alpha_{g1}\alpha_{g2}(1-\pi_g). \quad (g=1,2) \tag{3.63}$$

In a similar fashion we describe the probabilities P_{01} and P_{10}, followed by the corresponding derivations for the gth sample (P_{g01}, P_{g10} and P_{g11}).

Hence,

$$P_{01} = Pr[T_1=0, \ T_2=1]$$

$$= Pr[T_1=0, \ T_2=1 \mid D]Pr(D) + Pr[T_1=0, \ T_2=1 \mid \overline{D}]Pr(\overline{D})$$

$$= Pr[T_1=0 \mid D]Pr[T_2=1 \mid D]Pr(D) + Pr[T_1=0 \mid \overline{D}]Pr(T_2=1 \mid \overline{D})Pr(\overline{D})$$

$$= (1-\eta_1)\eta_2\pi + \theta_1(1-\theta_2)(1-\pi)$$

$$= \beta_1(1-\beta_2)\pi + (1-\alpha_1)\alpha_2(1-\pi)$$

and for the gth sample,

$$P_{g01} = \beta_{g1}(1-\beta_{g2})\pi_g + (1-\alpha_{g1})\alpha_{g2}(1-\pi_g) \tag{3.64}$$

Then,

$$P_{10} = Pr[T_1=1, \ T_2=0]$$

$$= Pr[T_1=1, \ T_2=0 \mid D]Pr(D) + Pr[T_1=1, \ T_2=0 \mid \overline{D}]Pr(\overline{D})$$

$$= Pr[T_1 \mid D]Pr[T_2=0 \mid D]Pr(D) + Pr[T_1=1 \mid \overline{D}]Pr[T_2=0 \mid \overline{D}]Pr(\overline{D})$$

$$= \eta_1(1-\eta_2)\pi + (1-\theta_1)\theta_2(1-\pi)$$

$$= (1-\beta_1)\beta_2\pi + \alpha_1(1-\alpha_2)(1-\pi)$$

which for the gth sample is,

$$P_{g10} = \beta_{g2}(1-\beta_{g1})\pi_g + \alpha_{g1}(1-\alpha_{g2})(1-\pi_g) \tag{3.65}$$

and finally

$$P_{g00} = 1 - P_{g11} - P_{g01} - P_{g10}$$

$$= \beta_{g1}\beta_{g2}\pi_g + (1-\alpha_{g1})(1-\alpha_{g2})(1-\pi_g) \ . \tag{3.66}$$

The likelihood function of the two samples is given by

$$L = \prod_{g=1}^{2} \left(P_{g11}\right)^{N_{g11}}\left(P_{g01}\right)^{N_{g01}}\left(P_{g10}\right)^{N_{g10}}\left(P_{g00}\right)^{N_{g00}}$$

where

$$N_{gij} = (N_g)\, P^*_{gij} \qquad g=1,2; \quad i,j=0,1$$

are the observed frequencies associated with the cells and P^*_{gij} are the observed proportions of sample g with test outcome i,j in tests 1 and 2 respectively (i,j = 1,0).

Parameter estimates can now be obtained by maximizing the loglikelihood function. However, it should be realized that in general, for k tests applied to S populations, there are $(2^k-1)S$ degrees of freedom for estimating $(2k+1)S$ parameters. In the present example there are 6 degrees of freedom and ten parameters (β_{11}, β_{12}, β_{21}, β_{22}, α_{11}, α_{12}, α_{21}, α_{22}, π_1, π_2). This model is therefore overparameterized. The necessary condition for parameter estimability is

$$(2^k - 1)S \geq (2k+1)S$$

$$\text{or} \qquad 2^k \geq 2(k+1)$$

$$k \geq 3$$

To study the case of most practical importance, that is when k=2, *HW* assumed that $\alpha_{gh} = \alpha_h$ and $\beta_{gh} = \beta_h$ (h=1,2) for all g=1...S. The number of parameters to be estimated is now 2k+S and the number of degrees of freedom is larger than the number of parameters if

$$S \geq k / (2^{k-1} - 1) \quad .$$

For k=2 the model can now be applied to data from two populations. The model is still unidentified, and *HW* introduced the constraint

$$\alpha_1 + \beta_1 < 1$$

which is a reasonable assumption for the standard test if it is of any practical value.

Hence,

$$L = \prod_{g=1}^{2} \ [\ [(1-\beta_1)(1-\beta_2)\pi_g + \alpha_1\alpha_2(1-\pi_g)]^{N_{g11}}$$

$$[\beta_1(1-\beta_2)\pi_g + (1-\alpha_1)\alpha_2(1-\pi_g)]^{N_{g01}}$$

$$[\beta_2(1-\beta_1)\pi_g + (1-\alpha_2)\alpha_1(1-\pi_g)]^{N_{g10}}$$

$$[\beta_1\beta_2\pi_g + (1-\alpha_1)(1-\alpha_2)(1-\pi_g)]^{N_{g00}} \ ] \ .$$

2. The Effect of Conditional Dependence (Vacek 1985)

Often the assumption that the misclassification errors of the two tests are assumed to be unrelated cannot be justified, particularly if the tests are based on physiologic phenomena. Thibodeau (1981) has investigated the effect of a positive correlation between diagnostic tests when the error rates of a new test are estimated by comparison with a reference test having known

138

error rates. He demonstrated that the assumption of test independence will result in an underestimation of the error rates of the new test if it is positively correlated with the reference test.

Vacek (1985) examined the effect of test dependence on the error rate and prevalence estimators for situations in which both tests have unknown error rates. She denoted the covariance between the two tests (when the true diagnosis is positive) by e_b, and e_a as the covariance when the true diagnosis is negative. Consequently, if a subject is a true positive, the probability that both tests will be positive is $(1-\beta_1)(1-\beta_2)+e_b$, and if the true diagnosis is negative, the probability that both tests will be negative is $(1-\alpha_1)(1-\alpha_2)+e_a$. Eliminating unnecessary algebra, Vacek showed that

$$\text{Bias}(\hat{\alpha}_1) = \frac{[(1-\beta_1-\alpha_1)(1-\beta_2-\alpha_2)+(e_b-e_a)-D]}{[2(1-\beta_2-\alpha_2)]} \tag{3.67}$$

$$\text{Bias}(\hat{\beta}_1) = \frac{[(1-\beta_1-\alpha_1)(1-\beta_2-\alpha_2)+(e_a-e_b)-D]}{[2(1-\beta_2-\alpha_2)]} \tag{3.68}$$

$$\text{Bias}(\hat{\pi}_g) = \{(2D)^{-1}[(2\pi_g-1)(1-\beta_1-\alpha_1)(1-\beta_2-\alpha_2)-(e_b-e_a)+D]-\pi_g\} \tag{3.69}$$

where

$$D = \pm \left[(1-\beta_1-\alpha_1)^2(1-\beta_2-\alpha_2)^2+2(e_b+e_a)(1-\beta_1-\alpha_1)(1-\beta_2-\alpha_2)+(e_b-e_a)^2\right]^{1/2}$$

The expressions for Bias $(\hat{\alpha}_2)$, Bias $(\hat{\beta}_2)$ are identical to the above equations except for a change of subscripts in the denominator of the bias term.

Since the prevalence parameters π_1 and π_2 do not appear in equations (3.67) and (3.68), they have no effect on the bias of the error rate estimators. Moreover, the bias in $\hat{\pi}_i$ does not depend on π_j ($i \neq j = 1,2$).

One should also notice that the likelihood function of *HW* can be rewritten to include the covariance terms e_a and e_b, but this results in an overparameterization. However, if e_a and e_b are specified as fixed ratios of their maximum value:

$$e_a \leq \beta_1(1-\beta_2) \quad \text{and} \quad e_b \leq \beta_2(1-b_1) \quad ,$$

139

no new parameters are introduced. The likelihood can then be maximized numerically to obtain parameter estimates which are based on the assumption that the two tests are conditionally dependent. Differences between these estimates and those obtained under the assumption of independence indicate the size of bias in the latter when the specified degree of dependence exists.

Chapter 4

LOGISTIC REGRESSION

I. INTRODUCTION

In many research problems, it is of interest to study the effects that some variables exert on others. One sensible way to describe this relationship is to relate the variables by some sort of mathematical equation.

In most applications statistical models are mathematical equations constructed to approximately relate the response (dependent) variables to a group of extraneous (explanatory) variables.

When the response variable, denoted by y, is continuous and believed to depend linearly on k variables $x_1, x_2,...x_k$ through unknown parameters $\beta_0, \beta_1,...\beta_k$, then this linear (where "linear" is used to indicate linearity in the unknown parameters) relationship is give as

$$y_i = \sum_{j=0}^{k} \beta_j \, x_{ji} + \epsilon_i \qquad\qquad (4.1)$$

where $x_{0i} = 1$ for all i=1,2,...n.

The term ϵ_i is unobservable random error representing the residual variation and is assumed to be independent of the systematic component $\sum_{j=0}^{k} \beta_j \, x_{ji}$ It is also assumed that $E(\epsilon_i) = 0$ and Var $(\epsilon_i) = \sigma^2$; hence,

$$E(y_i) = \sum_{j=0}^{k} \beta_j x_{ji} \quad \text{and} \quad Var(y_i) = \sigma^2 \, .$$

To fit the model (4.1) to the data (y_i, x_i) one has to estimate the parameters $\beta_0...\beta_k$. The most commonly used methods of estimation are (i) the method of least squares and (ii) the method of maximum likelihood.

Applications of those methods of estimation to the linear regression model (4.1) are extensively discussed in Draper & Smith (1981), Mosteller & Tukey (1977) and many other sources. It should be noted that no assumptions on the distribution of the response variable y are needed (except the independence of $y_1...y_n$) to estimate the parameters by the method of least squares. However, the maximum likelihood requires that the sample $y = (y_1...y_n)$ is randomly drawn from a distribution where the specified structure of that distribution in most applications is,

$$N\left(\sum_{j=0}^{k} \beta_j x_{ji} \, , \, \sigma^2\right) \quad .$$

The least squares estimates of the regression parameters will then coincide with those obtained by the method of maximum likelihood. Another remark that should be made here is that there is nothing in the theory of least squares that restricts the distribution of the response variable to be continuous, discrete or of bounded range. For example suppose that we would like to model the proportion $\hat{p}_i = \dfrac{y_i}{n_i}$ (i=1,2,...m) of individuals suffering from some respiratory illness, observed over several geographical regions, as a function of k covariates, where

$$y_i = \sum_{j=1}^{n_i} y_{ij}$$

$$y_{ij} = \left\{ \begin{array}{ll} 1 & \text{if diseased} \\ 0 & \text{else} \end{array} \right.$$

That is, we assume the relationship between p_i and the covariates to be

$$p_i = \sum_{j=0}^{k} \beta_j x_{ji} \tag{4.2}$$

The least squares estimates are obtained by minimizing

$$s = \sum_{i=1}^{n} \left(\hat{p}_i - \sum_{j=0}^{k} \beta_j x_{ji}\right)^2$$

Several problems are encountered when the least squares method is used to fit model (4.2);

(i) One of the assumptions of the method of least squares is variance homogeneity; that is, $Var(y_i) = \sigma^2$ does not vary from one observation to another. Since for binary data, y_i follows a binomial distribution with mean $n_i p_i$ and variance $n_i p_i (1-p_i)$, then $Var(\hat{p}_i) = \dfrac{p_i(1-p_i)}{n_i}$.

As was proposed by Cox and Snell (1989), one can deal with the variance heterogeneity by applying the method of weighted least squares using the reciprocal variance as a weight. The weighted least squares estimates are thus obtained by minimizing

$$s_w = \sum_{i=1}^{n} [Var(\hat{p}_i)]^{-1} \left(\hat{p}_i - \sum_{j=0}^{k} \beta_j x_{ji}\right)^2 \tag{4.3}$$

Note that $w_i^{-1} = Var(\hat{p}_i)$ depends on p_i which in turn depends on the unknown parameters $\beta_0, \beta_1, ... \beta_k$ through the relationship (4.2).

Fitting the model (4.2) is done quite easily using PROC REG in SAS together with the WEIGHT statement, where $\hat{w}_i$ are the specified weights.

(ii)   Note that $0 \le p_i \le 1$, and the estimates $\beta_0, ... \beta_k$ of the regression parameters are not constrained. That is, they are permitted to attain any value in the interval $(-\infty, \infty)$. Since the fitted values are obtained by substituting $\beta_0, ... \beta_k$ in (4.2),

$$\hat{p}_i = \sum_{j=0}^{k} \hat{\beta}_j x_{ji}$$

thus, there is no guarantee that the fitted values should fall in the interval [0,1].

To overcome the difficulties of using the method of least squares to fit a model where the response variable has a restricted range, it is suggested that a suitable transformation be employed so that the fitted values of the transformed parameter vary over the interval $(-\infty, \infty)$. This chapter is devoted to the analysis of binomially distributed data where the binary responses are independent. Attention will be paid to situations where the binary responses are obtained from clusters and hence cannot be assumed to be independent.

II. THE LOGISTIC TRANSFORMATION

Let us consider the following example which may help the reader understand the motives of a logistic transformation.

Suppose that we have a binary variable y that takes the value 1 if a sampled individual is diseased and takes the value 0 if not diseased.

Let $P(D) = \Pr[y=1] = \pi$ and $P(\bar{D}) = \Pr[y=0] = 1-\pi$. Moreover, suppose that X is a risk factor that has normal distribution with mean μ_1 and variance σ^2 in the population of diseased, that is

$$X|D \sim N(\mu_1, \sigma^2)$$

which reads, given the information that the individual is diseased, the conditional distribution of x is $N(\mu_1, \sigma^2)$. Hence

$$f(X|D) = \frac{1}{\sigma\sqrt{2\pi}} \exp\ [-(x-\mu_1)^2/2\sigma^2]\ .$$

In a similar manner, we assume that the risk factor X, in the population of non-diseased has a mean μ_2 and variance σ^2. That is

$$f(X|\bar{D}) = \frac{1}{\sigma\sqrt{2\pi}} \exp\ [-(x-\mu_2)^2/2\sigma^2]$$

Since

$$p = Pr[y=1|X=x] = \frac{p(y=1\ ,\ X=x)}{f(x)}$$

then from Bayes' theorem,

$$p = \frac{f(X|D)\,P(D)}{f(X|D)\,P(D) + f(X|\bar{D})P(\bar{D})}$$

$$= \frac{\left(\dfrac{\pi}{\sigma\sqrt{2\pi}}\right)\exp\left[\dfrac{-(x-\mu_1)^2}{2\sigma^2}\right]}{\left(\dfrac{\pi}{\sigma\sqrt{2\pi}}\right)\exp\left[\dfrac{-(x-\mu_1)^2}{2\sigma^2}\right] + \dfrac{(1-\pi)}{(\sigma\sqrt{2\pi})}\exp\left[\dfrac{-(x-\mu_2)^2}{2\sigma^2}\right]}$$

Simple manipulation shows that

$$p = Pr[y=1|X=x] = \frac{e^{\beta_0+\beta_1 x}}{1+e^{\beta_0+\beta_1 x}} \tag{4.4}$$

where

$$\beta_0 = -\ln\frac{1-\pi}{\pi} - \frac{1}{2\sigma^2}(\mu_1-\mu_2)(\mu_1+\mu_2)$$

and

$$\beta_1 = \frac{\mu_1-\mu_2}{\sigma^2}\ .$$

Note that

$$\ln\left(\frac{p}{1-p}\right) = \beta_0 + \beta_1 x. \qquad (4.5)$$

So, the log-odds is a linear function of the explanatory variable (here, the risk factor) X. The logarithmic transformation on the odds is called "logit".

Remarks

The regression parameter β_1 has log-odds ratio interpretation in epidemiologic studies. To show this, suppose that the exposure variable has two levels (exposed, not exposed). Let us define a dummy variable X that takes the value 1 if the individual is exposed to the risk factor, and 0 if not exposed.

Since from equation (4.4) we have

$$P_{11} = Pr(y=1 \mid X=1) = \frac{e^{\beta_0+\beta_1}}{1+e^{\beta_0+\beta_1}}$$

$$P_{01} = Pr(y=0 \mid X=1) = \frac{1}{1+e^{\beta_0+\beta_1}}$$

$$P_{10} = Pr(y=1 \mid X=0) = \frac{e^{\beta_0}}{1+e^{\beta_0}}$$

$$P_{00} = Pr(y=0 \mid X=0) = \frac{1}{1+e^{\beta_0}}$$

and that the odds ratio is

$$\psi = \frac{P_{11}P_{00}}{P_{10}P_{01}} = e^{\beta_1},$$

it follows that $\ln \psi = \beta_1$.

The representation (4.5) can be extended such that logit (p) is a function of more than just one explanatory variable.

Let $y_1, y_2, \ldots y_n$ be a random sample of n successes out of $n_1, n_2, \ldots n_n$ trials, and let the corresponding probabilities of success be $p_1, p_2, \ldots p_n$. If we wish to express the probability p_i as a function of the explanatory variables $x_{1i}, \ldots x_{ki}$, then the generalization of (4.5) is

$$\text{logit } (p_i) = \log \left(\frac{p_i}{1-p_i} \right) = \sum_{j=0}^{k} \beta_j \, x_{ji} \tag{4.6}$$

$x_{0i}=1$ for all $i=1,2,\dots n$.

We shall denote the linear function $\sum_{j=0}^{k} \beta_j \, x_{ji}$ by η_i, which is usually known as the *link function*.

Hence,

$$p_i = e^{\eta_i} / (1+e^{\eta_i}) \tag{4.7}$$

The binomially distributed random variables y_i ($i=1\dots n$) have mean $\mu_i = n_i p_i$ and variance $n_i p_i q_i$. Since we can write $y_i=\mu_i+\epsilon_i$, then the residuals $\epsilon_i=y_i-\mu_i$ have zero mean. Note that, in contrast to the normal linear regression theory, ϵ_i do not have a distribution of a recognized form.

Fitting the model to the data is achieved after the model parameters $\beta_0,\dots\beta_k$ have been estimated.

The maximum likelihood method is used to estimate the parameters where the likelihood function is given by:

$$L(\beta) = \prod_{i=1}^{n} \binom{n_i}{y_i} p_i^{y_i} \, q_i^{n_i-y_i}$$

$$= \prod_{i=1}^{n} \binom{n_i}{y_i} \left(e^{\eta_i}\right)^{y_i} \left(\frac{1}{1+e^{\eta_i}} \right)^{n_i} \tag{4.8}$$

The loglikelihood is given by:

$$\ell(\beta) = \sum_{i=1}^{n} \left\{ y_i \eta_i - n_i \ln(1+e^{\eta_i}) \right\} \tag{4.9}$$

Differentiating $\ell(\beta)$ with respect to β_r, we have

$$\ell_r = \frac{\partial l(\beta)}{\partial \beta_r} = \sum_{i=1}^{n} \left\{ y_j \, x_{ri} - n_i \, x_{ri} \, e^{\eta_i}(1+e^{\eta_i})^{-1} \right\} \tag{4.10}$$

$$= \sum_{i=1}^{n} x_{ri} \, (y_i - n_i \, p_i) \qquad r=0,1,2,\dots k$$

The (k+1) equations in (4.10) can be solved numerically.
Since

$$I = -E[\ell_{rs}] = -E\left[\frac{\partial^2 \ell(\beta)}{\partial \beta_r \partial \beta_s}\right] = \sum_{i=1}^{n} n_i \, x_{ri} \, x_{si} \, p_i(1-p_i) \qquad (4.11)$$

the large sample variance-covariance matrix is I^{-1}.
Once the parameters have been estimated, the predicted probability of success is given by

$$\hat{p}_i = e^{\hat{\eta}_i} / 1 + e^{\hat{\eta}_i} \quad , \text{ where}$$

$$\hat{\eta}_i = \sum_{j=0}^{k} \hat{\beta}_j \, x_{ji}$$

with $\hat{Var}(\hat{\beta}_r) = v_{rr}$, where v_{rr} is the rth diagonal element of $\hat{I}^{-1}$.

Using PROC LOGISTIC in SAS, we obtain the maximum likelihood estimates, their standard errors $SE(\hat{\beta}_r)=v_{rr}^{1/2}$ and the Wald chi-square values, $(\hat{\beta}_r^2)/v_{rr}$, r=0,1,...k which can be used to test the hypothesis that the corresponding coefficient in η_i is zero.

III. CODING CATEGORICAL EXPLANATORY VARIABLES AND INTERPRETATION OF COEFFICIENTS

Recall that when (4.4) is applied to the simple case of one independent variable which has two levels (exposed, not exposed) we defined a dummy variable X such that $X_i=1$ if the ith individual is exposed, and $X_i=0$ if the ith individual is not exposed.

Now, suppose that the exposure variable X has m>2 categories. For example X may be the strain or the breed of the animal, or it may be the ecological zone from which the sample is collected. Each of these variables (strain, breed, zone,...etc.) is *qualitative*, or a factor variable which can take a finite number of values known as the levels of the factor. To see how qualitative independent variables or factors are included in a logistic regression model, suppose that F is a factor with m distinct levels. There are various methods in which the indicator variables or the dummy variables can be defined. The choice of a particular method will depend on the goals of the analysis. One way to represent the m levels of the factor variable F is to define m-1 dummy variables f_1, f_2...f_{m-1} such that the portion of the design matrix corresponding to those variables looks like Table 4.1.

147

Table 4.1
Dummy Variables for the Factor Variable F Within m Levels

	Dummy Variables		
Factor level	f_1	f_2	f_{m-1}
1	0	0	0
2	1	0	0
3	0	1	0
.	.	.	.
.	.	.	.
m	0	0	1

Example 4.1

The following data are the results of a carcinogenic experiment. Different strains of rats have been injected with carcinogen in their foot pad. They were then put on a high fat diet and at the end of week 21 the number with tumours (y) were counted.

Table 4.2
Hypothetical Results of Carcinogenic Experiment. Counts of Rats with Tumours

	Strain (1)	Strain (2)	Strain (3)	Strain (4)
y_i	10	20	5	2
n_i-y_i	45	20	20	43
n_i	55	40	25	45
ψ	4.78	21.50	5.375	

Using strain (4) as a reference group we define the three dummy variables

$$X_j = \begin{cases} 1 & \text{if rat is from strain j} \qquad j=1,2,3 \\ 0 & \text{otherwise} \end{cases}$$

The results of fitting the logistic regression model to the data in Table 4.2 are given in Table 4.3.

Table 4.3
Maximum Likelihood Analysis

Variable	Estimates		SE	Wald Chi-square	ψ
Intercept	-3.068	β_0	.723	17.990	
Strain (1)	1.564	β_1	.803	3.790	4.780
Strain (2)	3.068	β_2	.790	15.103	21.500
Strain (3)	1.682	β_3	.879	3.658	5.375

148

Note that the last row of Table 4.2 gives the odds ratio for each strain using strain (4) as the reference level. For example, for strain (1) the estimated odds ratio is

$$\psi \ (\text{strain}(1) \ ; \ \text{strain}(4)) \ = \ \frac{(10)(43)}{(45)(2)} \ = \ 4.78$$

Moreover, $\ln \psi$ (strain (1); strain (4)) $= \ln (4.78) = 1.56 = \hat{\beta}_1$. Hence, as we mentioned earlier, the estimated parameters maintain their log-odds ratio interpretation even if we have more than one explanatory variable in the logistic regression function. One should also note that the estimated standard error of the odds ratio estimate from a univariate analysis is identical to the standard error of the corresponding parameter estimate obtained from the logistic regression analysis.

For example,

$$SE \ [\ln \ \psi \ (\text{strain}(3) \ ; \ \text{strain}(4)] \ = \ \left(\frac{1}{5} + \frac{1}{20} + \frac{1}{2} + \frac{1}{43} \right)^{1/2} \ = \ .879 \ ,$$

which is identical to SE $(\hat{\beta}_3)$ as shown in Table 4.3.

We can construct approximate confidence limits using the same approach used in Chapter 3. The $(1-\alpha)$ 100% confidence limits for the parameter are

$$\hat{\beta}_j \ \pm \ Z_{\alpha_{/2}} \ \hat{SE}(\hat{\beta}_j) \ ,$$

and the limits for the odds ratio are

$$\exp \ [\hat{\beta}_j \ \pm \ Z_{\alpha_{/2}} \ \hat{SE}(\hat{\beta}_j)] \ .$$

IV. INTERACTION AND CONFOUNDING

In Example 4.2 we showed how the logistic regression model can be used to model the relationship between the proportion or the probability of developing tumours and the strain. Other variables could also be included in the model such as sex, the initial weight of each mouse, age or other relevant variables. The main goal of what then becomes a rather comprehensive model is to adjust for the effect of all other variables and the effect of the differences in their distributions on the estimated odds ratios. It should be pointed out (Hosmer and Lemeshow, 1989) that the effectiveness of the adjustment (measured by the reduction in the bias of the estimated coefficient) depends on the appropriateness of the logit transformation and the assumption of constant slopes across the levels of a factor variable. Departure from the constancy of slopes is explained by the existence of interaction. For example, if we have two factor variables, F_1 and F_2, where the response at a particular level of F_1 varies over the levels of F_2, then F_1 and F_2 are said to interact. To model the interaction effect, one should include terms representing the main effects of the two factors as well as terms representing the two-factor interaction which is represented by a product of two dummy variables. Interaction is known to epidemiologists as "effect modifications". Thus, a variable that interacts with a risk factor of exposure variable is termed an "effect modifier". To illustrate, suppose that factor F_1 has three levels (a_1, a_2, a_3), and factor F_2 has two levels (b_1, b_2). To model the main effects and their interactions we first define the dummy variables.

$$x_i = \begin{cases} 1 & \text{if an observation belongs to the } i^{th} \text{ level of factor } F_1 \ (i=1,2) \\ 0 & \text{otherwise} \end{cases}$$

which means that a_3 is the referent group. Similarly, we define a dummy variable $b=1$ if the individual belongs to the second level of factor F_2. Suppose that we have the following 5 data points (Table 4.4):

Table 4.4
Hypothetical Data Points of a Two Factor Experiment

observation	y	F_1	F_2
1	1	a_1	b_1
2	1	a_1	b_2
3	0	a_2	b_2
4	0	a_3	b_2
5	1	a_3	b_1

To model the interaction effect of F_1 and F_2, the data layout (Table 4.5) becomes

Table 4.5
Dummy Variables for Modelling Interaction of Two Factors

observation	y	x_1	x_2	b	bx_1	bx_2
1	1	1	0	0	0	0
2	1	1	0	1	1	0
3	0	0	1	1	0	1
4	0	0	0	1	0	0
5	1	0	0	0	0	0

Another important concept to epidemiologists is confounding. A confounder is a covariate that is associated with both the outcome variable (the risk of disease, say) and a risk factor. When both associations are detected then the relationship between the risk factor and the outcome variable is said to be confounded.

Kleinbaum et al. (1988) recommended the following approach (which we extend to logistic regression) to detect confounding in the context of multiple linear regression. Suppose that we would like to describe the relationship between the outcome variable y and a risk factor x, adjusting for the effect of other covariates $x_1, \ldots x_{k-1}$ (assuming no interactions are involved) so that

$$\eta = \beta_0 + \beta_1 x_1 + \ldots \beta_k x_{k-1} \qquad (4.12)$$

Kleinbaum et al. argued that confounding is present if the estimated regression coefficient β_1 of the risk factor x, ignoring the effects of $x_1 \ldots x_{k-1}$ is meaningfully different from the estimate of β_1 based on the linear combination (4.12) which adjusts for the effect of $x_1 \ldots x_{k-1}$.

Example 4.2

To determine the disease status of a herd with respect to listeriosis (a disease caused by bacterial infection), fecal samples are collected from selected animals, and a 'group testing' technique is adopted to detect the agent responsible for the occurrence. One positive sample means that there is at least one affected animal in the herd, hence the herd is identified as a 'case'.

In this example, we will consider possible risk factors, such as *herd size, type of feed* and *level of mycotoxin in the feed*.

Herd sizes are defined as: *small* (25-50), *medium* (50-150), and *large* (150-300).
The two types of feed are *dry* and *non dry*; the three levels of mycotoxin are *low, medium* and *high*. Table 4.6 indicates the levels of mycotoxin found in different farms and the associated disease occurrences.

Table 4.6
Occurrence of Listeriosis in Herds of Different Sizes with Different Levels of Mycotoxin

Level of Mycotoxin	Type of Feed	Herd Size					
		Small		Medium		Large	
		Case	Ctrl	Case	Ctrl	Case	Ctrl
Low	Dry	2	200	15	405	40	300
	Non Dry	1	45	5	60	5	4
Medium	Dry	5	160	15	280	40	100
	Non Dry	4	55	3	32	0	2
High	Dry	2	75	60	155	40	58
	Non Dry	8	10	10	20	4	4

Before we analyze this data using multiple logistic regression, we should calculate some basic results. The study looks at the type of feed as the risk factor of primary interest. Suppose that we would like to look at the crude association between the occurrence of the disease (case/ctrl) and the level of mycotoxin (low, medium, high). Table 4.7 summarizes the relationship between the disease and levels of mycotoxin (LM) ignoring the effect of herd size (HS) and the type of feed (TF).

Table 4.7
Disease Occurrence at Different Levels
of Mycotoxin

	Disease			
LM	**D⁺**	**D⁻**	**Total**	ψ
low	68	1014	1082	1
medium	67	629	696	1.59
high	124	322	446	5.74

('low' is the referent group)

The estimated odds of disease for a farm with medium level of mycotoxin relative to a farm with low level is

$$\hat{\psi} = \frac{(1014)(67)}{(68)(629)} = 1.59 \quad .$$

This suggests that herds with medium levels are about 1.5 times more likely to have diseased animals than herds with low levels. On the other hand, the likelihood of disease on a farm with high levels is about 6 times relative to a farm with low levels

$$\hat{\psi} = \frac{(1014)(124)}{(322)(68)} = 5.74$$

Now we reorganize the data (Table 4.8) so as to look at the association between disease and the type of feed while adjusting for the herd size.

Table 4.8
Listeriosis Occurrence when Considering Herd Size
and Type of Feed

Type of Feed	Herd Size					
	Small		Medium		Large	
	case	ctrl	case	ctrl	case	ctrl
dry	9	435	90	840	120	458
non-dry	13	110	18	112	9	10
N_i	567		1060		597	
ψ	0.175		0.667		0.291	

When considering each herd size separately we have the following results.

Small

	case	control	
dry	9	435	
not dry	13	110	
			567

$$\psi_1 = \frac{990}{5655} = 0.175$$

$$\ln \psi_1 = -1.743$$

$$SE(\ln \psi_1) = (\frac{1}{9} + \frac{1}{435} + \frac{1}{13} + \frac{1}{110})^{\frac{1}{2}}$$

$$= 0.446$$

Medium

	case	control	
dry	90	840	
not dry	18	112	
			1060

153

$$\psi_2 = 0.667$$

$$\ln\psi_2 = -0.405$$

$$SE(\ln\psi_2) = 0.277$$

Large

	case	control	
dry	120	458	
not dry	9	10	
			597

$$\psi_3 = 0.291$$

$$\ln\psi_3 = -1.234$$

$$SE(\ln\psi_3) = 0.471$$

Testing for interaction between herd size and type of feed is done using Woolf's test.

$$\ln\psi = \frac{\sum \frac{\ln(\psi_j)}{var(\ln\psi_j)}}{\sum \frac{1}{var(\ln\psi_j)}}$$

$$= \frac{(\frac{-1.743}{0.1991})+(\frac{-0.405}{0.0767})+(\frac{-1.234}{0.2216})}{(\frac{1}{0.1991})+(\frac{1}{0.0767})+(\frac{1}{0.2216})}$$

$$= -0.867 \ .$$

Thus, $\psi = e^{-0.867} = 0.420$

The χ^2 test for interaction using the Woolf χ^2 is obtained as follows:

$$Var(\ln \psi_W) = \cfrac{1}{\displaystyle\sum \cfrac{1}{var(\ln \psi_j)}}$$

$$= \frac{1}{22.5732}$$

$$= 0.0443$$

$$\chi^2_W = \sum \frac{1}{var(\ln \psi_j)}[\ln\psi_j - \ln\psi]^2$$

$$= \frac{(-1.74+0.867)^2}{0.1991} + \frac{(-0.405+0.867)^2}{0.0767} + \frac{(-1.234+0.867)^2}{0.2216}$$

$$= 7.22$$

Since χ^2_w is greater than $\chi^2_{(.05,2)}$ there is evidence of an interaction between the herd size and the type of feed used on the farm. We investigate this in Example 4.3.

V. THE GOODNESS OF FIT AND MODEL COMPARISONS

Measures of goodness of fit are statistical tools used to explore the extent to which the fitted responses obtained from the postulated model compare with the observed data. Clearly, the fit is good if there is a good agreement between the fitted and the observed data. The Pearson's chi-square (χ^2) and the likelihood ratio test (LRT) are the most commonly used measures of goodness of fit for categorical data. The following sections will give a brief discussion on how each of the chi-square and the LRT criteria can be used as measures of goodness of fit of a logistic model.

A.   PEARSON'S X² - STATISTIC

This statistic is defined by

$$\chi^2 = \sum_{i=1}^{n} \frac{(y_i - n_i \hat{p}_i)^2}{n_i \hat{p}_i \hat{q}_i}$$

155

where

$$\hat{p}_i = \frac{e^{\hat{\eta}_i}}{1 + e^{\hat{\eta}_i}} \quad ,$$

and

$$\hat{\eta}_i = \sum_{j=1}^{k} \hat{\beta}_j x_{ji}$$

the linear predictor, obtained by substituting the MLE of the β_j in η_i. The distribution of χ^2 is asymptotically that of a chi-square with (n-k-1) degrees of freedom. Large values of χ^2 can be taken as evidence that the model does not adequately fit the data. Because the model parameters are estimated by the method of maximum likelihood, it is recommended that one uses the LRT statistic as a criterion for goodness of fit of the logistic regression model.

B.   THE LIKELIHOOD RATIO CRITERION (Deviance)

Suppose that the model we would like to fit (called current model) has $k+1$ parameters, and that the loglikelihood of this model given by (4.8) is denoted by ℓ_c. That is

$$\ell_c = \sum_{i=1}^{n} \left[y_i \eta_i - n_i \ln(1 + e^{\eta_i}) \right]$$
$$= \sum_{i=1}^{n} \left[y_i \ln p_i + (n_i - y_i) \ln(1 - p_i) \right]$$

Let $\hat{p}_i$ be the maximum likelihood estimator of p_i under the current model. Therefore, the maximized loglikelihood function under the current model is given by

$$\hat{\ell}_c = \sum_{i=1}^{n} \left[y_i \ln \hat{p}_i + (n_i - y_i) \ln(1 - \hat{p}_i) \right] \quad .$$

McCullagh and Nelder (1983) indicated that, in order to assess the goodness of fit of the current model, $\hat{\ell}_c$ should be compared with another log likelihood of a model where the fitted responses coincide with the observed responses. Such a model has as many parameters as the number of data points and is thus called the *full* or *saturated* model and is denoted by $\tilde{\ell}_s$. Since under the saturated model the fitted are the same as the observed proportions

$$\tilde{p} = y_i / n_i \quad ,$$

the maximized loglikelihood function under the saturated model is

$$\tilde{\ell}_s = \sum_{i=1}^{n} \left[y_i \ln \tilde{p}_i + (n_i - y_i) \ln (1 - \tilde{p}_i) \right]$$

The metric $D = -2[\hat{\ell}_c - \tilde{\ell}_s]$ which is called the *Deviance*, was suggested by McCullagh and Nelder (1983) as a measure of goodness of fit of the current model.

As can be seen, the Deviance is in fact the likelihood ratio criterion for comparing the current model with the saturated model. Now, since the two models are trivially nested, it is tempting to conclude from the large sample of the likelihood theory that the deviance is distributed as a chi-square with (n-k-1) degrees of freedom if the current model holds. However, from the standard theory leading to the chi-square approximation for the null distribution of the likelihood ratio statistic we find that: if model A has p_A parameters and model B (nested in model A) has p_B parameters with $p_B < p_A$, then the likelihood ratio statistic that compares the two models has chi-square distribution with degrees of freedom $p_A - p_B$ as $n \to \infty$ (with p_A and p_B both fixed). If A is the saturated model, $p_A = n$ so the standard theory does not hold. In contrast to what has been reported in the literature, Firth (1990) pointed out that the deviance does not, in general, have an asymptotic chi-square distribution in the limit as the number of data points increase. Consequently, the distribution of the deviance may be far from chi-square, even if n is large.

There are situations however when the distribution of the deviance can be reasonably approximated by a chi-square. The binomial model with large n_i is an example. In this situation a binomial observation y_i may be considered a sufficient statistic for a sample of n_i independent binary observations each with the same mean, so that $n_i \to \infty$ (i=1,2,...n) plays the role in asymptotic computations as the usual assumption $n \to \infty$. In other words, the validity of the large-sample approximation to the distribution of the deviance, in logistic regression model fit, depends on the total number of individual binary observations,

$$\sum n_i$$

rather than on n, the actual number of data points y_i. Therefore, even if the number of binomial observations is small, the chi-square approximation to the distribution of the deviance can be used so long as $\sum n_i$ is reasonably large.

More fundamental problems with the use of the deviance for measuring goodness of fit arise in the important special case of binary data, where $n_i = 1$ (i=1,2,...n). The likelihood function for this case is

$$L(\beta) = \prod_{i=1}^{n} p_i^{y_i} (1 - p_i)^{1 - y_i}$$

and

$$\hat{\ell}_c = \sum_{i=1}^{n} y_i \ln \hat{p}_i + (1-y_i) \ln(1-\hat{p}_i)$$

for the saturated model $\hat{p}_i = y_i$. Now since $y_i = 0$ or 1, $\ln y_i = (1-y_i) \ln (1-y_i) = 0$; $\hat{\ell}_c$; and the deviance is

$$D = -2 \sum_i \left\{ y_i \ln \left(\frac{\hat{p}_i}{1-\hat{p}_i} \right) + \ln (1-\hat{p}_i) \right\} \tag{4.13}$$

We now show that D depends only on the fitted values $\hat{p}_i$ and so is uninformative about the goodness of fit of the model. To see this, using (4.10) where $n_i = 1$, we have

$$\frac{\partial l(\beta)}{\partial \beta_r} = \sum_i x_{ri}(y_i - \hat{p}_i) = 0 \tag{4.14}$$

Multiplying both sides of (4.14) by β_r and summing over r:

$$0 = \sum (y_i - \hat{p}_i) \sum_r \beta_r x_{ri}$$

$$= \sum_i (y_i - \hat{p}_i) \hat{\eta}_i$$

$$= \sum (y_i - \hat{p}_i) \ln \left(\frac{\hat{p}_i}{1-\hat{p}_i} \right)$$

from which

$$\sum y_i \ln \left(\frac{\hat{p}_i}{1-\hat{p}_i} \right) = \sum \hat{p}_i \ln \left(\frac{\hat{p}_i}{1-\hat{p}_i} \right) \tag{4.15}$$

Substituting (4.15) into (4.13) we get

$$D = -2 \sum_i \left\{ \hat{p}_i \ln \left(\frac{\hat{p}_i}{1-\hat{p}_i} \right) + \ln (1-\hat{p}_i) \right\} .$$

Therefore D is completely determined by $\hat{p}_i$ and hence useless as a measure of goodness of fit of the current model. However, in situations where model (A) is nested in another model (B), the difference in deviance of the two models can be used to test the importance of those additional parameters in model (B).

The Pearson's χ^2:

$$\chi^2 = \sum_{i=1}^{n} \frac{(y_i - \hat{p}_i)^2}{\hat{p}_i \hat{q}_i}$$

encounters similar difficulties. It can be verified that, for binary data ($n_i = 1$ for all i), χ^2 always takes the value n and is therefore completely uninformative. Moreover, the χ^2 statistic, unlike the difference in deviance, cannot be used to judge the importance of additional parameters in a model that contains the parameters of another model.

We now explain how the difference in deviance can be used to compare models. Suppose that we have two models with the following link functions

Model	Link (η)
A	$\displaystyle\sum_{j=0}^{p_1} \beta_j x_{ji}$
B	$\displaystyle\sum_{j=0}^{p_1 + p_2} \beta_j x_{ji}$

Model (A) contains $p_1 + 1$ parameters and is therefore nested in model (B) which contains $p_1 + p_2 + 1$ parameters. The deviance under model (A) denoted by, D_A carries ($n-p_1-1$) degrees of freedom while that of model (B) denoted by D_B carries ($n-p_1-p_2-1$) degrees of freedom. The difference D_A-D_B has an approximate chi-square distribution with ($n-p_1-1$)-($n-p_1-p_2-1$)=p_2 degrees of freedom. This chi-square approximation to the difference between two deviances can be used to assess the combined effect of the p_2 covariates in model (B) on the response variable y.

The model comparisons based on the difference between deviances is equivalent to the analysis based on the likelihood ratio test. In the following example, we illustrate how to compare between models using the likelihood ratio test (LRT).

Example 4.3

Here we apply the concept of difference between deviances, or LRT, to test for the significance of added variables. First, we fitted a logistic regression model with main effects, where the variables were coded as:

$rf=1$ if feed is dry,   *else* $rf=0$
$h1=1$ if herd size if large,   *else* $h1=0$
$h2=1$ if herd size is medium,   *else* $h2=0$
$m1=1$ if mycotoxin level is high,   *else* $m1=0$
$m2=1$ if mycotoxin level is medium,   *else* $m2=0$

The results of fitting a model using the following SAS program are given below.

```
data myco;
input rf herdsize $ lmyco $ dis nodis;

n=dis+nodis;
if herdsiz='lg' then h1=1;
else h1=0;
if herdsize='m' then h2=1;
else h2=0;
if lmyco='h' then m1=1;
else m1=0;
if lmyco='m' then m2=1;
else m2=0;
rfm1=rf*m1;
rfm2=rf*m2;

cards;
1 s l    2  200
0 s l    1   45
1 s m    5  160
0 s m    4   55
1 s h    2   75
0 s h    8   10
1 m l   15  405
0 m l    5   60
1 m m   15  280
0 m m    3   32
1 m h   60  155
0 m h   10   20
1 lg l   40  300
0 lg l    5    4
1 lg m  40  100
0 lg m   0    2
1 lg h   40   58
0 lg h    4    4
;
run;

/*  Fitting the logistic model with main effects */

proc logistic;
model dis/n = rf h1 h2 m1 m2  / covb;
run;
```

/* Fitting the logistic model with interaction between type of feed and level of mycotoxin */

```
proc logistic;
model dis/n = rf h1 h2 m1 m2 rfm1 rfm2 / covb;
run;
```

An excerpt from the output that results from running the model without interaction is as follows:

Model Fitting Information and Testing Global Null Hypothesis BETA=0

Criterion	Intercept Only	Intercept and Covariates	Chi-Square for Covariates
AIC	1602.415	1362.475	.
SC	1608.122	1396.718	.
-2 LOG L	1600.415	1350.475	249.940 with 5 DF (p=0.0001)
Score	.	.	261.269 with 5 DF (p=0.0001)

Analysis of Maximum Likelihood Estimates

Variable	DF	Parameter Estimate	Standard Error	Wald Chi-Square	Pr > Chi-Square	Standardized Estimate	Odds Ratio
INTERCPT	1	-3.4712	0.2870	146.3261	0.0001	.	.
RF	1	-0.8637	0.2131	16.4326	0.0001	-0.156055	0.422
H1	1	2.3401	0.2581	82.2196	0.0001	0.571864	10.383
H2	1	1.0819	0.2487	18.9246	0.0001	0.297994	2.950
M1	1	1.9898	0.1751	129.2011	0.0001	0.439363	7.314
M2	1	0.7288	0.1860	15.3544	0.0001	0.186369	2.073

All the covariates have a significant effect on the disease. We shall call this model M_a. When the interaction between type of feed and levels of mycotoxin (rfm1 = rf*m1, rfm2 = rf*m2) are included we get the following results (we shall refer to this model as M_b:

Model Fitting Information and Testing Global Null Hypothesis BETA=0

Criterion	Intercept Only	Intercept and Covariates	Chi-Square for Covariates
AIC	1602.415	1364.527	.
SC	1608.122	1410.184	.
-2 LOG L	1600.415	1348.527	251.888 with 7 DF (p=0.0001)
Score	.	.	266.485 with 7 DF (p=0.0001)

Analysis of Maximum Likelihood Estimates

Variable	DF	Parameter Estimate	Standard Error	Wald Chi-Square	Pr > Chi-Square	Standardized Estimate	Odds Ratio
INTERCPT	1	-3.2236	0.3804	71.8042	0.0001	.	.
RF	1	-1.1221	0.3604	9.6937	0.0018	-0.202738	0.326
H1	1	2.3129	0.2578	80.4693	0.0001	0.565209	10.104
H2	1	1.0530	0.2491	17.8745	0.0001	0.290025	2.866
M1	1	1.8497	0.4359	18.0091	0.0001	0.408416	6.358
M2	1	0.0966	0.5167	0.0350	0.8517	0.024702	1.101
RFM1	1	0.1617	0.4749	0.1160	0.7334	0.033918	1.176
RFM2	1	0.7283	0.5527	1.7364	0.1876	0.178256	2.072

One should realize that the parameter estimates of the main effects in M_b are not much different from what they are under M_a. Now, to test for the significance of the interaction effects using the deviance, we use the following notation:

Let, $\lambda_a = -2\,Log\,L\;\;under\;\;M_a$ and $\lambda_b = -2\,Log\,L\;\;under\;\;M_b$, and hence,
$$Deviance = (\lambda_a - \lambda_b) = 1350 - 1348.527 = 1.948\,.$$

Asymptotically, the deviance has a chi-square distribution with degrees of freedom equal to 7-5=2. Therefore, we do not have sufficient evidence to reject the null hypothesis of no interaction effect.

General Comments:

1. The main risk factor of interest in this study was the type of feed (rf). Since the sign of t estimated coefficient of rf is negative, then according to the way we coded that variable, dry feed seems to have sparing effect on the risk of listeriosis.

2. From M_a, adjusting for herd size, the odds of a farm being a case with dry feed and high level of mycotoxin, relative to a farm with low level of mycotoxin is $exp[-0.864 + 1.9898] = 3.08$.

3. From M_b, the same odds ratio estimate is,

$$exp[-1.122(1-0) + 1.8497(1-0) + 0.1617(1-0)] = 2.43\,.$$

The difference, 3.08-2.43=0.65, is the bias in the estimated odds ratio from M_a, if the interaction effects were ignored.

VI. LOGISTIC REGRESSION OF CLUSTERED BINARY DATA

A. INTRODUCTION

Clustered samples of binary data arise frequently in many statistical and epidemiologic investigations. This clustering may be as a result of observations that are repeatedly collected on the experimental units as in cohort studies, or may be due to sampling blocks of experimental units such as families, herds, litters ... etc. The data of interest consist of a binary outcome variable y_{ij}. The data are collected in clusters or blocks, and i=1,2,...k indexes the clusters, while j=1,2...n_i indexes units within clusters. A distinguishing feature of such clustered data is that they tend to exhibit intracluster correlation. To obtain valid inference, the analysis of clustered data must account for the effect of this correlation. The following section illustrates the diversity of the problems in which intracluster effect may be present through a number of examples.

B. INTRACLUSTER EFFECTS

1.   Toxicological experiments that are designed to study the teratogenic effects of chemical compounds on lifeborn fetuses are common situations in which clustering arises. Here, a compound is given to pregnant females with a resultant binary outcome, such as fetal death or presence of abnormality in the individual fetuses. There is a tendency for fetuses within the same litter to respond more similarly than fetuses from different litters. This tendency has been termed a "litter effect".

2.   McDermott et al. (1992) carried out an investigation designed to describe culling practices of Ontario Cow-Calf producers. Since culled cows are subject to the same management decisions or because they share the same herd risk factors, there is a clustering effect created when the herd is selected as the experimental unit. This clustering effect is known as the "herd effect".

3.   Rosenblatt et al. (1985) designed a study to assess the safety of obstetrics in small hospitals. Data on births and perinatal deaths for the years 1978-1981 were obtained for all public maternity hospitals in New Zealand. Hospitals were classified on the basis of neonatal care facilities and the annual number of deliveries. Higher percentages of low birth weights were found to cluster among large hospitals with sophisticated technologies.

4.   Studying the familial aggregation of chronic obstructive pulmonary disease was the focus of investigation by Connolly and Liang (1988). Here, clusters of families of diseased and non-diseased individuals formed the sampling units. For a variety of environmental and genetic backgrounds it is expected that members of the same family may respond in a like manner as opposed to members from different families.

5.   Correlated binary data arise in longitudinal studies where repeated measures of a binary outcome are gathered from independent samples of individuals. There are usually two types of covariates measured; the first set is measured at the baseline and does not change during the course of the study. The second set is measured at each time point. For longitudinal studies we may be interested in how an individual's response changes over time, or more generally in the effect of change in covariates on the binary outcome.

Although different in their specific objectives and their scientific significance, the above studies share some common characteristics that allowed statisticians to devise a unified approach in the analysis of clustered correlated binary outcomes.

It should be noted that if there are no covariates measured on the individuals within the cluster, rather that the covariates are measured only at the cluster level, we can summarize the binary outcomes for a cluster as a binomial proportion. In this case, positive correlation between y_{ij} and y_{il} is then manifest as over-dispersion, or extra-binomial variation. Less often, the observations within a cluster are negatively correlated and hence there is under-dispersion. This may happen if for example the clusters are pens, and animals within a pen are competing for a common food source. Another example for under-dispersion may occur in family studies where children may be competing for maternal care.

As we have already seen in the previous section of this chapter, when studying the association between binary outcomes and sets of covariates it is standard practice to use logistic regression. However, if the data are correlated this technique may not be appropriate. While the estimates of the regression parameters are essentially correct, as we demonstrate later, their variances will be wrongly estimated. There is a vast literature on methods for over-dispersed binomial data (regression for correlated binomial data). However, it is only recently that statistical methods have been developed for situations where covariates are collected at the individual level (regression for correlated binary outcomes). In this case the vector of individual binary outcomes constitutes a vector of multivariate observations gathered on each cluster.

In the next section we investigate the effect of covariates on over-dispersed binomial data, and on correlated binary data. Our investigation will include semi-parametric, full likelihood methods and the recently developed "Generalized Estimating Equations" approach of Liang and Zeger (1986) and Prentice (1988). To understand the technique several examples will be presented.

C. ANALYSIS OF OVERDISPERSED BINOMIAL DATA

1. Modelling Sources of Variation

Suppose that we have k clusters and the data consists of the observed counts $y_1, y_2,...y_k$ where

$$y_i = \sum_{j=1}^{n_i} y_{ij} \ ,$$

and y_{ij} is the binary response from the jth individual within the ith cluster. Suppose also that corresponding to the ith cluster, the response probability depends on p explanatory variables $x_1 ... x_p$, through a linear logistic model. One way to introduce the variability in the response probability p_i is to assume that they vary about a mean π_i. This means that $E(p_i)=\pi_i$. Moreover we assume that p_i has a variance, Var $(p_i) = \sigma_i^2\pi_i(1-\pi_i)$, where $\sigma_i^2 > 0$ is an unknown scale parameter.

Since

$$E\ (y_i|p_i)\ =\ n_i p_i \quad \text{and} \quad Var(y_i|p_i)\ =\ n_i p_i(1-p_i)$$

then from the standard results of conditional probability we have

$$E\ (y_i)\ =\ E[E(y_i|p_i)]\ =\ n_i E(p_i)\ =\ n_i\pi_i$$

and

$$
\begin{aligned}
Var(y_i)\ &=\ E[\,Var\,(y_i|p_i)]\ +\ Var[E(y_i|p_i)] \\[4pt]
&=\ n_i E[p_i(1-p_i)]\ +\ n_i^2 Var(p_i) \\[4pt]
&=\ n_i[E(p_i)-E(p_i^2)]\ +\ n_i^2 Var(p_i) \\[4pt]
&=\ n_i\{E\,(p_i)-Var(p_i)-[E(p_i)]^2\}\ +\ n_i^2 Var(p_i) \\[4pt]
&=\ n_i[\pi_i(1-\pi_i)]\ +\ n_i(n_i-1)\ Var(p_i) \\[4pt]
&=\ n_i\pi_i(1-\pi_i)[1+(n_i-1)\sigma_i^2]
\end{aligned}
$$

(4.16)

In most applications the dispersion parameter is assumed to be constant $\sigma_i^2=\sigma^2$.

It should be noted that the dispersion factor $1+(n_i-1)\sigma^2$ can be derived under the assumption that p_i has a beta distribution with parameters α and β. The pdf of this beta distribution is given by

$$f(Z)\ =\ Z^{\alpha-1}\ (1-Z)^{\beta-1}/B(\alpha,\beta)$$

165

$$E(p_i) = \frac{\alpha}{\alpha+\beta} = \pi_i$$

$$Var(p_i) = \sigma^2 \pi_i(1-\pi_i) \quad \text{where} \quad \sigma^2 = (1+\alpha+\beta)^{-1}$$

The marginal distribution of y_i is

$$p(y_i) = \int_0^1 \frac{Z^{\alpha-1}(1-Z)^{\beta-1}}{B(\alpha,\beta)} \binom{n_i}{y_i} Z^{y_i} (1-Z)^{n_i-y_i} \, dZ$$

$$= \binom{n_i}{y_i} \frac{\Gamma(\alpha+y_i)\Gamma(\beta+n_i-y_i)}{\Gamma(\alpha+\beta+n_i)} \frac{\Gamma(\alpha+\beta)}{\Gamma(\alpha)\Gamma(\beta)} .$$

The unconditional variance of y_i is thus identical to (4.16).

When there is no variability among the response probabilities, that is, when $\sigma^2=0$, equation (4.16) reduces to Var $(y_i) = n_i \, \pi_i(1-\pi_i)$ which is the variance of a binomially distributed random variable. If on the other hand we assume that $\sigma^2>0$ then Var (y_i) will be larger than $n_i\pi_i(1-\pi_i)$ by an amount equal to the dispersion factor $1+(n_i-1)\sigma^2$.

In the special case of binary data with $n_i=1$, $i=1,2,...k$ equation (4.16) becomes Var$(y_i)=\pi_i(1-\pi_i)$. Consequently, no information about the dispersion parameter σ^2 can be obtained from the data, and hence we cannot model overdispersion under such circumstances.

2. Correlated Binary Responses

Since $y_i = \sum_{j=1}^{n_i} y_{ij}$ where y_{ij} are Bernoulli random variables, then E $(y_i) = n_i \, p_i$ and

$$Var(y_i) = \sum Var(y_{ij}) + \sum\sum^i Cov(y_{ij},y_{il}) . \tag{4.17}$$

Assuming that y_{ij} and y_{il} (the jth and l^{th} binary responses within the ith cluster) are correlated so that

$$\rho = Corr\,(y_{ij},y_{il}) = \frac{Cov(y_{ij},y_{il})}{\sqrt{Var(y_{ij})\,Var(y_{il})}}$$

then substituting in (4.17) we get

$$Var(y_i) = \sum_{j=1}^{n_i} p_i(1-p_i) + \sum_{j \neq l}^{n_i} \rho \, p_i(1-p_i)$$

$$= n p_i(1-p_i) + n_i(n_i-1)\rho \, p_i(1-p_i)$$

$$= n_i \, p_i(1-p_i)\left[1 + (n_i-1)\rho\right]$$

(4.18)

If the correlation ρ is zero, then Var $(y_i) = n_i \, p_i \, (1-p_i)$ which is the variance under the binomial assumption. On the other hand if $\rho > 0$ then the Var (y_i) as given in (4.18) will be larger than $n_i p_i (1-p_i)$. Thus, positive correlation between pairs of binary responses leads to extra variation in the number of positive responses than would be expected if they were independent.

Although equation (4.16) is essentially the same as (4.18) there is one fundamental difference. While the model that produced equation (4.16) is used to model overdispersion (since $\sigma^2 \geq 0$), the correlation parameter in (4.18) can be negative. In fact, since $1 + (n_i - 1)\rho > 0$, this implies $-1/(n_i - 1) \leq \rho \leq 1$. However, if $\rho > 0$, then equations (4.16) and (4.18) become identical and one cannot differentiate between the effects of correlation between binary responses and the variation in response probabilities on the variance of y_i.

Alternatively, and perhaps conveniently, we might prefer to assume a constant dispersion factor relative to the binomial, namely

$$Var(y_i) = \phi \, n_i \, \pi_i \, (1-\pi_i)$$

(4.19)

Overdispersion is then modelled by $\phi > 1$. Certain cluster sampling procedures can give rise to variances approximately of the form (4.19). While the use of a non-full likelihood approach will in general result in loss of efficiency, its loss is known to be modest in many practical situations (Firth, 1987). We shall elaborate on this remark in a subsequent section.

Example 4.4   Shell's data

This example will compare the outcomes of adjusting for extra binomial using the methods of Rao and Scott, and Donner on data taken from Paul (1982). The data are given in Table 4.9; they describe experimental data on the number of live foetuses affected by treatment (control, low dose and medium dose). The high dose was removed from the original data set.

Table 4.9
Data from Shell Toxicology Laboratory
i) Number of Live Foetuses Affected by Treatment
ii) Total Number of Live Foetuses
(Paul, 1982 *Biometrics*)

Group

1 CONTROL	i)	1	1	4	0	0	0	0	0	1	0	2	0	5	2	1	2	0	0	1	0	0	0	0	3	2	4	0
	ii)	12	7	6	6	7	8	10	7	8	6	11	7	8	9	2	7	9	7	11	10	4	8	10	12	8	7	8

2 LOW	i)	0	1	1	0	2	0	1	0	1	0	0	3	0	0	1	5	0	0	3
	ii)	5	11	7	9	12	8	6	7	6	4	6	9	6	7	5	9	1	6	9

3 MEDIUM	i)	2	3	2	1	2	3	0	4	0	0	4	0	0	6	6	5	4	1	0	3	6
	ii)	4	4	9	8	9	7	8	9	6	4	6	7	3	13	6	8	11	7	6	10	6

To show how RS adjustment is used to correct for the effect within litter correlation within the framework of logistic regression, we first derive the inflation factor of each group. Direct computations show that:

control: $\hat{\rho}_1 = 0.135$, $v_1 = 0.00127$, $d_1 = 2.34$

low $\hat{\rho}_2 = 0.135$, $v_2 = 0.0017$, $d_2 = 1.94$

medium $\hat{\rho}_3 = 0.344$, $v_3 = 0.0036$, $d_3 = 2.41$

The ANOVA estimator of the intralitter correlation needed to adjust Donner's adjustment is $\hat{\rho} = 0.261$. The following is the SAS program that can be used to analyze the data:

```
data read;
input group $ y n litter;
if group='c' then x = -1 and d = 2.34;
if group='l' then x = 0 and d = 1.94;
if group='m' then x = 1 and d = 2.41;
/*if group = 'c' then d = 2.34;  if group = 'm' then d = 1.94;  if group = 'l' then d = 2.41; */
raoscott = 1/d;
donner = 1/(1+(n-1)*0.261);
 cards;
c        1        12       1
c        1        7        2
c        4        6        3
c        0        6        4
....    ;
```

```
proc logistic;
model y/n=x;
run;

/*using the weight statement to adjust for correlation using Rao-Scott */
proc logistic;
model y/n=x;
weight raoscott;
run;

/*using the weight statement to adjust for correlation using Donner's */
proc logistic;
model y/n=x;
weight donner;
run;
```

Note that we have coded the dose groups as a continuous variable $x=-1$ (control), $x=0$ (low) and $x=1$ (medium). The SAS program runs logistic regression under independence and then using the two weighting adjustments by Rao and Scott and Donner (*weight* statements in the program). The *raoscott* adjustment is the reciprocal of the inflation factor d; the *donner* adjustment depends on the estimated intralitter correlation. The results are summarized in Table 4.10.

Table 4.10
Analysis of Shell Toxicology Data

parameter	independence	Rao-Scott	Donner
$\hat{\beta}_o$	-1.4012 (0.1165)	-1.4247	-1.4058 (0.1934)
$\hat{\beta}_1$	0.6334 (0.1372)	0.6354 (0.2126)	0.6180 (0.2287)
p-value	(0.0001)	(0.0028)	(0.0069)

The bracketed numbers are the standard errors (SE) of the estimated parameter The p-values are related to the null hypothesis $H_0: \beta_1 = 0$. Note that there are no marked differences among the estimates obtained from the three procedures. However, the standard errors are inflated when we adjust for the intralitter overdispersion using either Rao-Scott or Donner's approach.

3. Test for Trend in the 2 x I Table

The data in example 4.4 is an illustration of cluster randomization of animals to test for mortality or morbidity pattern in various groups. For example, suppose that animals have been randomized into k experimental groups, and that the animals in the ith group are exposed to a dose level x_i of a chemical compound, with $x_1 < x_2 < \ldots < x_3$. One is usually interested in testing whether the proportion of dead or (affected) subjects increases or decreases monotonically with dose. A conveniently monotonic function is the logistic function (4.7) with $\eta_i = \alpha + \beta x_i$. As we have indicated, this implies that the log odds (logit) is a linear function of the dose. Thus, if the doses are equally spaced (e.g. at unit intervals) the logistic model implies that the odds ratios between adjacent doses are equal.

The data are usually presented in a $2 \times I$ table as in Table 4.11.

Table 4.11

$2 \times I$ Data from a Dose-Response Experiment

	Dose			Total
	x_1	x_2	x_I	
Affected	y_1	y_2	y_I	$y_.$
Not affected	$n_1 - y_1$	$n_2 - y_2$	$n_I - y_I$	$n_. - y_.$
Total	n_1	n_2	n_I	$n_.$

Example 4.5

The data (Table 4.12) summarize the number of fish with liver tumor in three size groups (small, medium and large). It is believed that size is a function of age, so the purpose of the cross-sectional study was to test for trend in the percentage of tumours as the fish get older.

Table 4.12

Data on Liver Tumours for Three Groups

	size (x_i)			
	small (1)	medium (2)	large (3)	Total
With tumour	2	7	60	69
Without tumour	28	43	15	86
Total	30	50	75	155

Under $H_0 : \beta = 0$, the Cochran-Armitage test for trend is

$$\chi_t^2 = \frac{\left[\sum_{i=1}^{I} x_i \left(y_i - n_i \bar{p} \right) \right]^2}{\bar{p}\bar{q} \sum_{i=1}^{I} n_i \left(x_i - \bar{x} \right)^2}$$

where $\bar{x} = \sum_{i=1}^{I} \dfrac{n_i x_i}{n_{..}}$.

The statistic χ_t^2 is an approximate chi-square variate with one degree of freedom. For the data in Table (4.12), $\chi_t^2 = 62.69$. Note that the χ_h^2 given by (3.22) is used to test for heterogeneity among groups (the subscript h is for heterogeneity). For the dose-response case, the approximate chi-square statistic with $I - 2$ degrees of freedom for departure from linear trend is

$$\chi_Q^2 = \chi_h^2 - \chi_t^2 \ . \ \text{The} \ \chi_h^2 = 73.57, \text{ and } \chi_Q^2 = 10.88$$

which is significant at the 0.05 level.

The test statistics χ_h^2 , χ_t^2 are valid when subjects are individually randomized into the groups as in the example in Table 4.12. Under cluster randomization (data of example 4.4), the test for heterogeneity of proportion is done using either χ_D^2 or $\tilde{\chi}^2$. Recall that $\tilde{\chi}^2$ is robust against misspecification of the within cluster correlation. We therefore should use a similar χ^2 test for trend that accounts for the within cluster correlation. The required test statistic of the hypothesis $H_0 : \beta = 0$ was given by Boos (1992) as

$$\chi_B^2 = \frac{\left[\sum_{i=1}^{I} x_i \left(y_i - n_i \bar{p} \right) \right]^2}{\sum_{i=1}^{k} \left(x_i - \bar{x} \right)^2 \sum_{j=1}^{n_i} \left(y_{ji} - n_{ij} \bar{p} \right)} \ .$$

The only difference between χ_t^2 and χ_B^2 is that the binomial variance $n_i \bar{p}\bar{q}$ has been replaced by the empirical variance estimate $\sum_{j=1}^{n_i} \left(y_{ij} - n_{ij} \bar{p} \right)^2$.

For the data in Table 4.9 we have

$$\chi_D^2 = 10.82 \quad \text{with 2 degrees of freedom}$$

$$\chi_B^2 = 8.34 \quad \text{with 1 degree of freedom}$$

Recall that the weighted logistic regression under Donner's adjustment gave $\hat{\beta} = 0.618$ with SE = 0.229. The one degree of freedom chi-square test of the hypothesis $H_0: \beta = 0$, is

$$\left(\frac{\hat{\beta}}{SE}\right)^2 = 7.28, \quad \text{which is quite close to} \quad \chi_B^2.$$

D. FULL LIKELIHOOD MODELS

1. Random Effects Models

In this section we shall discuss models that can be used in the analysis of grouped binomial data as well as ungrouped binary data. In other words, the models we intend to discuss are quite appropriate for situations in which covariates are measured at both the individual and the cluster level. This class of models is know as the "Random Effects Models" (REM).

Suppose that the response probability of the jth individual within the ith cluster depends on the values of p explanatory variables $x_1 \ldots x_p$ and on q unknown explanatory variables $b_1 \ldots b_q$ so that

$$\text{logit } p_{ij} = X_{ij}\beta + b_1 + b_2 + \ldots b_q . \tag{4.20}$$

The inclusion of such unmeasured explanatory variables $b_1, b_2 \ldots b_q$ is another way to account for the extra variation in the data. For example b_i's may represent the genotype of an individual which may be impossible or difficult to measure. Since numerical values cannot be assigned to these quantities, they are taken as random variables with common mean and common variance σ^2. The model in (4.20) is known as a mixed effects model since it consists of two parts: the fixed part $X_{ij}\beta$ and the random part $b_1 + \ldots + b_k$. The mixed effects model can be written in a more compact form as

$$\text{logit}(p_{ij}) = X_{ij}\beta + Z_{ij}b_i \tag{4.21}$$

where β is a px1 vector of fixed effects, b_i is a qx1 vector of random effects, X_{ij} and Z_{ij} are covariate vectors, corresponding to the fixed effects and the random effects respectively. In most applications it is assumed that b_i has a multivariate normal distribution with mean 0 and covariance matrix D, that is, $b_i \sim$ MVN (0,D).

To estimate the model parameters, the method of maximum likelihood can be used. The likelihood function is given by:

$$L\ (\beta,b_i)\ =\ \prod_{i=1}^{k}\ \prod_{j=1}^{n_i}\ p_{ij}^{y_{ij}}\ (1-p_{ij})^{1-y_{ij}}$$

$$=\ \prod_{i=1}^{k}\ \prod_{j=1}^{n_i}\ \frac{[\exp\ (x_{ij}\beta\ +\ Z_{ij}b_i)]^{y_{ij}}}{\exp(x_{ij}\beta\ +\ Z_{ij}b_i)} \tag{4.22}$$

$$=\ \prod_{i=1}^{k}\ l(l_i)\ \ . \tag{4.23}$$

The standard approach to dealing with a likelihood function that contains random variables is to integrate the likelihood function with respect to the distribution of these variables. After integrating out the b_i's, the resulting function is called a "marginal likelihood", which depends on β_0, β_1,... β_p and the parameters of the covariance matrix D. The maximum likelihood estimates of these parameters are those values that maximize the marginal likelihood function given as

$$L\ (\beta,D)\ =\ \prod_{i=1}^{k}\ \int\ l_i\ \frac{\exp\left[(-1/2)b_iD^{-1}b_i^T\right]}{(2\pi)^{k/2}|D|^{1/2}}\ db_i \tag{4.24}$$

The two problems associated with the directed maximization of the marginal likelihood are:
(i)   Closed form expression for the integrals (4.24) is not available, so we cannot find exact maximum likelihood estimates.

(ii)   The maximum likelihood estimator of the variance components (the parameters of the matrix D) does not take into account the loss in the degrees of freedom resulting from estimating fixed effects. This means that the ML of the variance components are biased in small samples.

To find approximation to the ML estimates one has to evaluate the integral in (4.24) numerically. The package EGRET evaluates the integral in (4.24) using the Gauss-quadrature formula for numerical integration, an approach that was followed by Anderson and Aitken (1985). The Gauss-quadrature approximation of an integral is given as

$$\int_{-\infty}^{\infty}\ f(t)e^{-t^2}dt\ \approx\ \sum_{r=1}^{m}\ c_r\ f(u_r) \tag{4.25}$$

173

where the values of c_r and u_r can be obtained from Abromowitz and Stegun (1972). It was reported by many researchers that m need not exceed 20. However, in a recent article Crouch and Spiegelman (1990) showed that m=20 may not be sufficient in some situations.

In most applications we may assume that $D=I\sigma_b^2$ meaning that the random effects are not correlated. This assumption greatly simplifies the integrand of (4.24) and one can obtain the numerical integration much faster. Since, under this assumption, the joint pdf of b_i's is given by

$$g(b) = \frac{\exp\left[-\frac{1}{2}b_i\, D^{-1}\, b_i^T\right]}{(2\pi)^{k/2}|D|^{1/2}} = \prod_{t=1}^{k} \frac{e^{-\frac{1}{2}b_t^2/\sigma_b^2}}{\sqrt{2\pi\sigma_b^2}}$$

then the product

$$l_i\, g(b) = \prod_{j=1}^{n_i} \frac{\left[\exp\,(x_{ij}\beta+Z_{ij}b_i)\right]^{y_{ij}}}{\exp\,(x_{ij}\beta+Z_{ij}b_i)} \prod_{t=1}^{k} \frac{e^{-b_t^2/2\sigma_b^2}}{\sqrt{2\pi\sigma_b^2}} \quad ,$$

under the transformation,

$$U_t = \frac{b_t}{\sqrt{2}\sigma_b} \quad ,$$

results in the likelihood equation (4.24) being written as,

$$L(\beta,D) = \prod_{i=1}^{k} \int \prod_{j=1}^{n_i} \frac{\left[\exp(x_{ij}\beta+Z_{ij}\,\sigma_b\,U_i\sqrt{2})\right]^{y_{ij}}}{\exp(x_{ij}\beta+Z_{ij}\,\sigma_b U_i\sqrt{2})} \frac{e^{-U_i^2}}{\sqrt{\pi}}\, dU_i \quad .$$

This is the required form for the application of the Gauss-quadrature approximation; this is given as,

$$L(\beta,D) \approx \pi^{-k/2} \prod_{i=1}^{k} \sum_{t=1}^{m} \prod_{j=1}^{n_i} C_j \left\{ \frac{\left[\exp(x_{ij}\beta+Z_{ij}\,\sigma_b\,U_i\,\sqrt{2})\right]^{y_{ij}}}{\exp(x_{ij}\beta+Z_{ij}\,\sigma_b\,U_i\sqrt{2})} \right\} \tag{4.26}$$

The maximization of log $L(\beta,D)$ can be achieved numerically.

Williams (1982) proposed an approximate method to estimate β and σ_b^2 when covariates are measured at the cluster level. More generally, an approximation similar to William's may be suggested for the case when covariates are measured at the unit level. Using Taylor's expansion of $E(y_{ij} \mid b_i)$ about $b_i=0$,

$$E(y_i|b_i) \simeq \frac{e^{x_{ij}\beta}}{1+e^{x_{ij}}} + \frac{e^{x_{ij}\beta}}{(1+e^{x_{ij}\beta})^2} Z_{ij}b_i$$

$$= p_{ij} + p_{ij}(1-p_{ij})Z_{ij}b_i \quad .$$

Since

$$Var(y_{ij}) = Var(E(y_{ij}|b_i)) + E(Var(y_{ij}|b_i))$$

then

$$Var(E(y_{ij}|b_i)) = p_{ij}^2 (1-p_{ij})^2 Z_{ij} D Z_{ij}^T \quad . \tag{4.27}$$

Moreover

$$Cov[E(y_{ij}|b_i), E(y_{ik}|b_i)] = p_{ij}p_{ik}(1-p_{ij})(1-p_{ik})Z_{ij} D Z_{ij}^T \quad .$$

When conditional on b_i, y_{ij} is independent of y_{ik}, thus

$$Var(y_{ij}|b_i) = diag\{E(y_{ij}|b_i)(1-E(y_{ij}|b_i))\}$$

and

$$E[Var(y_{ij}|b_i)]$$

$$= E[E(y_{ij}|b_i) - E(E(y_{ij}|b_i))]^2$$

$$= p_{ij} - E[p_{ij}^2 + p_{ij}^2(1-p_{ij})^2 Z_{ij}b_i \; b_i^T Z_{ij}^T + 2p_{ij}^2(1-p_{ij})Z_{ij}b_i^T] \tag{4.28}$$

$$= p_{ij} - p_{ij}^2 - p_{ij}^2(1-p_{ij})^2 Z_{ij}D Z_{ij}^T$$

$$= p_{ij}(1-p_{ij}) - p_{ij}^2(1-p_{ij})^2 Z_{ij}D Z_{ij}^T$$

Therefore, from (4.27) and (4.28)

$$
\text{Cov } (y_{ij}, y_{ij}) = \left\{ \begin{array}{ll} p_{ij} (1-p_{ij}) & j=k \\ p_{ij} (1-p_{ij}) p_{ik} (1-p_{ik}) Z_{ij} D Z_{ik}^{T} & j \neq k \end{array} \right. \tag{4.29}
$$

One should note that if $Z_{ij} = Z_{ik} = 1$, $D = I\sigma_b^2$ and covariates are measured only at the group (cluster) level, then

$$
p_{ij} = p_i, \quad \text{and} \quad y_i = \sum_{j=1}^{n_i} y_{ij}
$$

has approximate variance given by

$$
\begin{aligned}
Var(y_i) &= \sum_{j=1}^{n_i} \left[p_i(1-p_i) + \sum_{j \neq l} \sum p_i^2 (1-p_i)^2 \sigma_b^2 \right] \\
&= n_i p_i (1-p_i) + n_i (n_i-1) p_i^2 (1-p_i)^2 \sigma_b^2 \\
&= n_i p_i (1-p_i) \left[1 + (n_i-1) p_i (1-p_i) \sigma_b^2 \right]
\end{aligned}
$$

This is the same result obtained by Williams (1982).

2. Bahadur's Model

Bahadur (1961) suggested that the joint probability distribution of the vector of binary responses $y_i = (y_{i1}, y_{i2}, \ldots y_{in_i})$ can be written as

$$
f(y_i) = \prod_{j=1}^{n_i} p_{ij}^{y_{ij}} (1-p_{ij})^{1-y_{ij}} \left[1 + \sum_{j<l} \delta_{ijl} W_{ij} W_{il} \right] \tag{4.30}
$$

where

$$
W_{ir} = \frac{y_{ir} - p_{ir}}{\sqrt{p_{ir}(1-p_{ir})}} \tag{4.31}
$$

and

$$\delta_{ijl} = E(W_{ij}W_{il}) , \quad i=1,2,...k \tag{4.32}$$

The likelihood function of the sample of k clusters is then given by

$$L = \prod_{i=1}^{k} f(y_i) . \tag{4.33}$$

Assuming an exchangeable correlation structure (i.e. $\delta_{ijk}=\rho$) the log-likelihood function is given as:

$$l(\beta,\rho) = \sum_{i=1}^{k} \left\{ \sum_{j=1}^{n_i} y_{ij} \log p_{ij} + (1-y_{ij}) \log (1-p_{ij}) \right.$$

$$\left. + \log \left(1+\rho \sum_{j<l} w_{ijl} \right) \right\} \tag{4.34}$$

where

$$w_{ijl} \frac{(y_{ij}-p_{ij})(y_{il}-p_{il})}{\sqrt{p_{ij}q_{ij} \ p_{il}q_{il}}} .$$

Now, suppose also that the logit of p_{ij} is linearly related to X_{ij} through the parameter vector β

$$logit (p_{ij}) = X_{ij}\beta .$$

The derivative of $l(\beta,\rho)$ with respect to β_r and ρ is given by

$$\frac{\partial l}{\partial \beta_r} = \sum_{i=1}^{k} \left\{ \sum_{j=1}^{n_i} (y_{ij}-p_{ij}) x_{ijr} + \rho \frac{\sum_{j<l} \dot{w}_{ijlr}}{1+\rho \sum_{j<l} w_{ijl}} \right\} \tag{4.35}$$

$$\frac{\partial l}{\partial \rho} = \sum_{i=1}^{k} \left\{ \frac{\sum_{j<l} w_{ijl}}{1+\rho \sum_{j<l} w_{ijl}} \right\}$$

where

$$\dot{w}_{ijl} = \frac{\partial\, w_{ijl}}{\partial\, \beta_r} = -\frac{1}{2}(p_{ij}q_{ij}p_{il}q_{il})^{-\frac{1}{2}}\, \{x_{ijr}(y_{il}-p_{il})(y_{ij}-2p_{ij}y_{ij}+p_{ij})$$

$$+ x_{ilr}(y_{ij}-p_{ij})(y_{il}-2p_{il}y_{il}+p_{il})\} \quad .$$

The maximum likelihood estimates can be obtained by solving

$$\frac{\partial\, l}{\partial\, \beta_r} = 0 \quad \text{and} \quad \frac{\partial\, l}{\partial\, \rho} = 0 \quad .$$

One should notice that when $\rho=0$, then equation (4.35) reduces to equation (4.10), which is the log-likelihood equation for estimating β_r under independence for clusters of size $n_i=1$. Consistent estimates of the variances and covariance of the maximum likelihood estimates (β,ρ) are obtained by inverting the matrix,

$$\hat{I} = \begin{bmatrix} -\dfrac{\partial^2\hat{I}}{\partial\beta_r\partial\beta_s} & -\dfrac{\partial^2\hat{I}}{\partial\beta_r\partial\rho} \\[3ex] -\dfrac{\partial^2\hat{I}}{\partial\beta_r\partial\rho} & -\dfrac{\partial^2\hat{I}}{\partial\rho^2} \end{bmatrix}$$

where

$$-\frac{\partial^2\hat{I}}{\partial\beta_r\partial\beta_s} = \sum_{i=1}^{k}\left\{\sum_{j=1}^{n_i} x_{ijr}x_{ijs}\hat{p}_{ij}(1-\hat{p}_{ij}) + \frac{\rho\,\hat{A}_{rs}-\rho^2\,\hat{B}_{rs}}{\left(1+\rho\sum_{j<l}\hat{w}_{ijl}\right)^2}\right\}$$

$$\frac{\partial^2\hat{I}}{\partial\beta_r\partial\rho} = -\sum_{i=1}^{k}\left\{\frac{\sum_{j<l}\dot{w}_{ijlr}}{\left(1+\rho\sum_{j<l}\hat{w}_{ijl}\right)^2}\right\} \tag{4.36}$$

$$-\frac{\partial^2\hat{I}}{\partial\rho^2} = \sum_{i=1}^{k}\left[\frac{\sum_{j<l}\hat{w}_{ijl}}{1+\rho\sum_{j<l}\hat{w}_{ijl}}\right]^2$$

178

$$\hat{A}_{rs} = \left(1+\rho \sum_{j<l} \hat{w}_{ijl}\right)\left(\sum_{j<l} \ddot{w}_{ijlrs}\right)$$

$$\hat{B}_{rs} = \left(\sum_{j<l} \dot{w}_{ijlr}\right)\left(\sum_{j<l} \dot{w}_{ijls}\right)$$

and

$$\ddot{w}_{ijlrs} = \frac{\partial^2 \hat{w}_{ijl}}{\partial \beta_r \partial \beta_s} = \frac{1}{4}(\hat{p}_{ij}\hat{q}_{ij}\hat{p}_{il}\hat{q}_{il})^{-\frac{1}{2}} \{(x_{ijr}\ x_{ijs}$$

$$+x_{ilr}\ x_{ils})(y_{ij}-\hat{p}_{ij})(y_{il}-\hat{p}_{il}) + (x_{ijr}\ x_{ils}$$

$$+x_{ilr}\ x_{ijs})(y_{ij}-2\hat{p}_{ij}y_{ij}+\hat{p}_{ij})(y_{il}-2\hat{p}_{il}y_{il}+\hat{p}_{il})\}$$

The cap symbol (ˆ) indicates that the parameters are to be replaced by their ML estimates.

Remark:

Prentice (1988) introduced Bahadur's representation as a model that is potentially useful for the analysis of correlated binary data when covariates are measured at both the individual and the group levels. One of the advantages of this model is that the estimates of the correlation and regression parameters are obtained from the direct maximization of the full likelihood function. Unfortunately, the requirement that the right hand side of (4.30) be nonnegative places severe restrictions on the range of possible values for the δ_{ijl} (see; Prentice, 1988 for more details). Such restrictions on expression (4.30) make this model attractive for the analysis of data sets consisting of clusters of size two. A full likelihood approach to the estimation of regression effects however, will probably not be as tractable in situations where clusters are of arbitrary sizes. The generalized estimating equation approach which we introduce in the next section is expected to be more applicable in these situations.

3. Models with More Than One Random Effect : "Hierarchical or Multilevel Modelling of Binary Data"

Models with random effects are expected to be more efficient, provided that we have correctly identified the correct distribution for the random component in the model. The random effects model can be extended to include more than one random component. For example, the health status (presence or absence of disease) of individuals in the same household, same counties and in the same geographical area, is often more alike than that of individuals from different households, counties or regions. This may be due to common socio-economic, environmental and/or behavioral factors. The ability to quantify sources of unobserved heterogeneity in the health outcome of individuals is important for several reasons (see Katz et al., 1993). The within household, or counties or regional clustering of disease, alters the effective sample size needed to provide accurate estimates of disease prevalence. Estimates of disease prevalence at each level of and organization (household, county, region, etc.) can provide insight into the dynamics among the risk factors operating at each level. The ability to obtain separate estimates of the variance component for the random effect at each level of clustering may guide the policy makers regarding the level of organization to which we should direct our health management dollars.

Estimating models with random effects for binary data has been the subject of investigation by many authors as a special case of a more general class known as "Generalized Linear Mixed Models" (GLMM). Schall (1991) and McGilchrist (1994) demonstrated that the parameters of the GLMM, with the logit link

$$\eta(p) = \text{logit}(p) = x\beta + z_1 b_1 + ... + z_c b_c = x\beta + zb$$

can be estimated by solving Henderson's (1975) mixed model equations. This approach avoids specifying a particular distribution or class of distributions for random effects. Another advantage of this approach is that we can avoid the computationally intensive numerical multiple integration needed to construct the marginal likelihood as for single random effect in (4.24). Schall's (1991) algorithm is summarized as follows:

First we write $Y = P + e$, apply the logit transformation to the data $\eta(y)$ and then use the first order Taylor's expansion of $\eta(y)$. That is

$$\xi = \eta(y) = \eta(p) + (y-p)\eta'(p)$$

$$= x\beta + zb + e\eta'(p) \ .$$

Here , $E(\xi) = x\beta$, $Cov(b) = D$, $Cov(e) = \sigma^2 I$ and $Cov(e\eta'(p)) = \left(\dfrac{\partial \eta}{\partial p}\right)^2 = W^{-1}$, so that

$Cov(\xi) = W^{-1} + ZDZ'$. Parameter estimates are obtained on iterating between

$$c \begin{bmatrix} \hat{\beta} \\ \hat{b} \end{bmatrix} = \begin{bmatrix} W^{-1}x & W^{1/2}z \\ 0 & D^{-1} \end{bmatrix} \begin{bmatrix} \hat{\beta} \\ \hat{b} \end{bmatrix} = \begin{bmatrix} W^{1/2}\xi \\ 0 \end{bmatrix}$$

and

$$\hat{\sigma}^2_i = \frac{\hat{b}_i' \hat{b}_i}{\left(q_i - v_i\right)}$$

where q_i is the number of levels of the ith random effect, $v_i = \frac{\text{trace}\left(T_{ii}\right)}{\hat{\sigma}^2_i}$. The matrices T_{ii} are

obtained by partitioning a matrix T formed by the last q rows and columns of the $(C'C)^{-1}$, partitioned conformably with D as

$$T = \begin{bmatrix} T_{11} & \cdots & T_{1c} \\ \vdots & \ddots & \vdots \\ T_{c1} & \cdots & T_{cc} \end{bmatrix}.$$

After convergence, σ^2 (extra-binomial variation parameter) is estimated as:

$$\hat{\sigma}^2 = \frac{\left(\xi - x\hat{\beta} - z\hat{b}\right)' W \left(\xi - x\hat{\beta} - z\hat{b}\right)}{df}$$

where $df = n - p - \sum_{i=1}^{c}\left(q_i - v_i\right)$. Values of $\hat{\sigma}^2$ close to 1 indicate an adequately fitted model.

Schall implemented his algorithm in a SAS macro. Another program that implements Schall's algorithm is "GLIMMIX", which is a SAS macro that can be downloaded. Interested readers may contact "sasrdw@unx.sas.com".

Example 4.6 :

The following data are taken from Schall's paper. Four hundred cells were placed on a dish and three dishes were irradiated at a time. After the cells were irradiated, the surviving cells were counted. Since cells would also die naturally, dishes with cells were put in the radiation chamber without being irradiated, to establish the natural mortality. For the purpose of this example, only these zero-dose data are analyzed. Twenty-seven dishes on nine time points, or three per time point, were available. The resulting 27 binomial observations are given in Table 4.13.

Table 4.13
Cell Irradiation Data

Occasion	Dish	No. cells surviving out of 400 placed	Occasion	Dish	No. cells surviving out of 400 placed
1	1	178	6	16	115
1	2	193	6	17	130
1	3	217	6	18	133
2	4	109	7	19	200
2	5	112	7	20	189
2	6	115	7	21	173
3	7	66	8	22	88
3	8	75	8	23	76
3	9	80	8	24	90
4	10	118	9	25	121
4	11	125	9	26	124
4	12	137	9	27	136
5	13	123			
5	14	146			
5	15	170			

Schall fitted two models, the first is with one random effect due to time effect (with 9 levels)

$$\text{logit}\left(p_{ij}\right) = \beta_0 + b_{1i} , \qquad i = 1,\ldots,9 \ , \quad j = 1,\ldots,3$$

and the second is with two random effects, due to time effect (9 levels) and due to dish effect (27 levels). For the purpose of comparion, we fitted the data with "GLIMMIX". The SAS program is

```
Data dish;
        input time dish y;
not y = 400-y;
cards;
1  1   178
1  2   193
 :
9  27  136
;
proc print data=dish;
run;

Data cell;
        set dish;
        do i=1 to y;
```

182

```
        r=1;
          output;
      end;
       do i=1 to noty;
         r=0;
           output;
       end;
       run;

    % include 'glmm612.sas'/nosource;
    run;
    % glimmix (data = cell,
    procopt = emthod = reml,
    stmts = % str (
    class  time  dish;
    model r = = / solution;
    random   time dish(time);
    ),
    error = binomial,
    link = logit
    )
    run;
```

To fit the model with only time as a random effect, we remove "dish(time)" from the "random" statement. The results are summarized in Table 4.14.

Table 4.14
Fitting Cell Data Using Schall's Algorithm and GLIMMIX SAS Macro

Model	Schall	GLIMMIX
one random effect :		
time $\hat{\sigma}_1^2$	0.2250	0.2450
error $\hat{\sigma}^2$	1.8100	1.9700
two random effects :		
time $\hat{\sigma}_1^2$	0.2220	0.2217
dish $\hat{\sigma}_2^2$	0.0100	0.0099
error $\hat{\sigma}^2$	0.9370	0.9977

Note that in the model with two random effects, $\hat{\sigma}^2$ is very close to 1, and the fitting algorithms produce similar results.

E. ESTIMATING EQUATIONS APPROACH

In the previous sections we discussed methods for the analysis of correlated binary response data that require full specifications of the probability distributions of the vector $(y_{i1},...y_{in_i})'$ in order to obtain fully efficient maximum likelihood estimates of the model parameters.

In this section we discuss a semiparametric approach using the so-called estimating equations. This approach requires assumptions only about the mean and the variance of the distribution of the response variables, rather than full specification of the likelihood. For correlated binary response variable Liang and Zeger (1986), and Zeger and Liang (1986) suggested an estimating equations approach that provided estimates of the regression parameters which are consistent and asymptotically normal even if the covariance structure of the response is misspecified. Their approach is considered a generalization of the quasi-likelihood proposed by Weddenburn (1974).

The interested reader will find the article by Godambe and Kale (1991) quite lucid in its explanation of the theory of estimating functions and their applications in parameter estimation. Here we provide a brief summary.

Let us consider the simple case when $y_1,...,y_n$ are independent random variables such that

$$E(y_i) = \mu_i(\beta) \ , \ Var(y_i) = \sigma^2 \ , \ i=1,2,...n \tag{4.37}$$

We assume that μ_i is twice differentiable function of β with unique inverse function μ_i^{-1}. Our main objective here is to estimate β based on the mean and variance structure given in (4.37). We define the linear estimating function,

$$g_n = \sum_{i=1}^{n} b_i(\beta) \ (y_i-\mu_i(\beta)) \tag{4.38}$$

Where $b_i(\beta)$ are differentiable functions of β. If the vector β is p-dimensional, the $b_i(\beta)$ is a px1 vector which yields a set of p estimating equations. The estimate of β is obtained by solving $g_n=0$ for β. Since $g_n(\beta)$ are linear in y_i with $E(g_n)=0$, the functions g_n are called linear unbiased estimating functions.

As an illustration to this concept, recall that the least square (LS) estimate of the regression parameters is obtained by minimizing

$$\sum_i (y_i-\mu_i)^2$$

which when differentiating and equating to zero we get

$$\sum (y_i - \mu_i) \frac{\partial \mu_i}{\partial \beta} = 0 \qquad\qquad (4.39)$$

In the special case when $\mu_i = X_i\beta$, we estimate β by solving

$$\sum (y_i - \mu_i)X_i = 0$$

The Gauss-Markov (GM) theorem states that the estimates of β obtained by solving (4.39) have a minimum variance in the class of unbiased estimators.

Wedderburn's (1974) introduction of the concept of quasi-likelihood (QL) allows the construction of estimating equations when the variance of y_i is a known function of the mean. That is

$$E(y_i) = \mu_i(\beta), \; Var(y_i) = \phi \, V(\mu_i) \qquad\qquad (4.40)$$

where $V(\mu_i)$ is a positive semi-definite matrix whose elements are known functions of the mean μ_i.

The maximum quasi-likelihood equations used to estimate β are

$$\sum D_i^T \, V_i^{-1}(\mu_i)(y_i - \mu_i(\beta)) = 0 \qquad\qquad (4.41)$$

where $D_i = \dfrac{\partial \mu_i}{\partial \beta}$.

White (1982) showed that even though the QL does not necessarily correspond to a specific distribution, maximizing the QL gives estimates of β which are consistent and asymptotically normal. Moreover, the equations (4.41) are optional in the sense that they maximize the asymptotic efficiency among all linear unbiased estimating equations.

Within the framework of longitudinal data Liang and Zeger (1986) and Zeger and Liang (1986) proposed a generalization of the QL in order to account for the correlations between the subunits or the repeated observations on the ith individual (cluster). Their approach requires correct specifications of the mean and postulates a "working" correlation matrix for $(y_{i1}, \dots y_{in_i})'$ which is not assumed to be the correct correlation. This approach is quite important because only in rare situations are we able to specify the correlation structure. We now outline two generalized estimating equations approaches; the first is due to Liang and Zeger (1986) whereby the correlation parameter is considered nuisance; the other is due to Zhao and Prentice (1990) where additional estimating equations are needed to allow for the joint estimation of the regression and correlation parameters.

Let $y_i = (y_{i1}, y_{i2},...,y_{in_i})$ be a random sample of k clusters of correlated binary responses and let $X_i = (x_{i1},... x_{in_i})$ be a matrix of covariates for the ith cluster.
Suppose that

$$E(y_{ij}; x_{ij}, \beta) = p_r[y_{ij}=1|X_{ij}, \beta] = p_{ij} \quad \text{and} \quad \text{logit}(p_{ij}) = x_{ij}\beta$$

A GEE estimator of β based on k clusters is a solution to

$$\sum_{i=1}^{k} D_i^T V_i^{-1}(y_i - p_i) = 0 \tag{4.42}$$

where

$$p_i = (p_{i1}, p_{i2},...p_{in_i})' ,$$

$$D_i = \frac{\partial \mu_i}{\partial \beta}$$

$$B_i = \text{diag} (p_{i1}q_{i1}, p_{i2}q_{i2},...p_{ik}q_{ik})' ,$$

$$V_i = B_i^{\frac{1}{2}} R_i(\alpha) B_i^{\frac{1}{2}} ,$$

and $R_i(\alpha)$ is a working correlation matrix for y_i with parameter vector α. Note that the equations (4.42) are similar to (4.41), the quasi-likelihood equations, except that V_i is a function of β as well as α. For a given α the solution $\tilde{\beta}$ to (4.42) can be obtained by an iteratively reweighed least squares calculation. The solution to these equations is a consistent estimate of β provided that the relationship between p_i and β is correctly specified. The consistency property follows because $D_i^T V_i^{-1}$ does not depend on the y_i's so, equation (4.42) coverges to zero and has consistent roots so long as $E(y_i - p_i) = 0$. If an $\sqrt{k}$ consistent estimate of α is available, $\tilde{\beta}$ are asymptotically normal, even if the correlation structure is misspecified. Correct specification of the correlation gives more efficient estimates of β. Liang and Zeger (1986) proposed a "robust" estimate of the variance of $\tilde{\beta}$ as

$$V(\tilde{\beta}) = A_{11}^{-1} M A_{11}^{-1} \tag{4.43}$$

where

$$A_{11} = \sum_{i=1}^{k} \tilde{D}_i^T \tilde{V}_i^{-1} \tilde{D}_i$$

$$M = \sum_{i=1}^{k} \tilde{D}_i^T \tilde{V}_i^{-1} \tilde{Cov}(y_i) \tilde{V}_i^{-1} \tilde{D}_i$$

(a tilde (~) denotes evaluation at $\tilde{\beta}$ and $\tilde{\alpha}(\tilde{\beta})$) and

$$\tilde{Cov}(y_i) = (y_i - \tilde{p}_i)(y_i - \tilde{p}_i)' \quad .$$

As we have previously mentioned, in order to solve the equation (4.42) for β, we need a consistent estimate of the correlation parameters α. Liang and Zeger (1986) proposed a simple estimator for α based on Pearson's residuals

$$\hat{r}_{ij} = \frac{y_{ij} - \hat{p}_{ij}}{\sqrt{\hat{p}_{ij}\,\hat{q}_{ij}}} \quad .$$

For example under exchangeable (common) correlation structure, that is, corr $(y_{ij}, y_{il}) = \alpha$ for all i,j,l, an estimate of α is

$$\hat{\alpha} = \sum_{i=1}^{k} \sum_{j=1}^{n_i} \sum_{l=j+1}^{n_i-1} \hat{r}_{ij}\,\hat{r}_{il} \Big/ \left\{ \sum_{i=1}^{k} \binom{n_i}{2} - p \right\} \tag{4.44}$$

where the p in the denominator of (4.44) is the number of regression parameters.

A SAS macro for the analysis of correlated binary data was provided by Karim and Zeger (1988). More description will be given during the discussion of a practical example at the end of this chapter.

Prentice (1988) proposed an extension of the GEE approach to allow joint estimation of the regression parameters and the pairwise correlations. To be more specific, a GEE estimator of the correlation parameter α may be derived noting that the sample correlation

$$U_{ijl} = \frac{(y_{ij} - p_{ij})(y_{il} - p_{il})}{(p_{ij}p_{il}\,q_{ij}q_{il})^{1/2}} \tag{4.45}$$

has

$$E(U_{ijl}) = Corr(y_{ij}, y_{il}) = \delta_{ijl}(\alpha)$$

$$Var(U_{ijl}) = W_{ijl} = 1 + (1 - 2p_{ij})(1 - 2p_{jl})(p_{ij}p_{il}q_{ij}q_{il})^{1/2}\delta_{ijl} - \delta_{ijl}^2.$$

Hence a generalized estimating equations estimator $(\tilde{\beta}, \tilde{\alpha})$ for β and α may be defined as a solution to

$$\sum_{i=1}^{k} D_i^T V_i^{-1}(y_i - p_i) = 0$$

(4.46)

$$\sum_{i=1}^{k} E_i^T W_i^{-1}(U_i - \delta_i) = 0$$

where

$$\delta_i = (\delta_{i12}, ...\delta_{i1n_i}, \delta_{i23}, ...,...,\delta_{in_i-1,n_i})'$$
$$W_i = \text{diag}\,(W_{i12}, ...W_{i1n_i}, W_{i23}, ...,...,W_{in_i-1,n_i})'$$

and

$$E_i = \frac{\partial \delta_i}{\partial \alpha}.$$

The above definition of W as proposed by Prentice (1988) uses an $\dfrac{n_i(n_i-1)}{2}$ -dimensional matrix as a working correlation matrix for

$$U_i = (U_{i12}, ...U_{i1n_i}, U_{i23}, ...,....,U_{in_i-1n_i})'$$

Note that the first of equation (4.46) is the same as equation (4.42). Note also that if $\delta_{ijl} = \alpha$ for all $i = 1, 2, ...k$, $j < l \leq n_i$, and W_i is the identity matrix then the second set of estimating equations in 4.46 reduces to

$$\sum_{i=1}^{k} \left\{ \sum_{i=1}^{n_i} \sum_{l=j+1}^{n_i-1} U_{ijl} \right\} \bigg/ \sum_{i=1}^{k} \binom{n_i}{2} = \alpha$$

(4.47)

For practical purposes we take $U_{ijl} = \hat{r}_{ij}\,\hat{r}_{il}$ and hence the estimate of α given in (4.47) differs from (4.44) by a factor of

$$\frac{\Sigma \binom{n_i}{2}}{\left\{ \Sigma \binom{n_i}{2} - p \right\}}$$

The divisor in (4.44), when used instead of that in (4.47), allows for the adjustment for the loss of degrees of freedom due to the estimation of β. Thus the estimate from (4.44) will be less biased than that from (4.47). Under mild regularity conditions, Prentice (1988) showed that the asymptotic distribution of $(\hat{\beta}, \hat{\alpha})$ is multivariate normal with mean zero and covariance matrix consistently estimated by

$$A \begin{pmatrix} \Sigma_{11} & \Sigma_{12} \\ \Sigma_{21} & \Sigma_{22} \end{pmatrix} A^T \tag{4.48}$$

where

$$A = \begin{pmatrix} A_{11} & 0 \\ A_{21} & A_{22} \end{pmatrix} \quad ;$$

A_{11} is defined in (4.43)

$$A_{22} = \sum_{i=1}^{k} \left(\tilde{E}_i^T \tilde{W}_i^{-1} \right)^{-1}$$

$$A_{12} = A_{22} \left(\sum_{i=1}^{k} \tilde{E}_i^T \tilde{W}_i^{-1} \frac{\partial \tilde{U}_i}{\partial \beta} \right) A_{11}$$

$$\Sigma_{11} = M \ (where \ M \ is \ defined \ in \ (4.43))$$

$$\Sigma_{12} = \Sigma_{21}^T = \sum_{i=1}^{k} \tilde{D}_i^T \tilde{V}_i^{-1} \tilde{Cov}(y_i, \tilde{U}_i) \tilde{W}_i^{-1} \tilde{E}_i$$

$$\Sigma_{22} = \sum_{i=1}^{k} \tilde{E}_i^T \tilde{W}_i^{-1} \tilde{Cov}\left(\tilde{U}_i \right) \tilde{W}_i^{-1} \tilde{E}_i$$

$$\tilde{Cov}(y_i) = \left(y_i - \tilde{p}_i \right)\left(y_i - \tilde{p}_i \right)$$

$$\tilde{Cov}(y_i, U_i) = \left(y_i - \tilde{p}_i \right)\left(\tilde{U}_i - \tilde{\delta}_i \right)$$

$$\tilde{Cov}(U_i) = \left(\tilde{U}_i - \tilde{\delta}_i \right)\left(\tilde{U}_i - \tilde{\delta}_i \right)$$

Also,

$$\frac{\partial \tilde{U}_i}{\partial \beta} = -\left\{ \tilde{D}_{ij}\left(y_{ij}-\tilde{p}_{ij}\right) + \tilde{D}_{il}\left(y_{il}-\tilde{p}_{il}\right) + \frac{1}{2}(y_{ij}-p_{ij})\left(y_{il}-\tilde{p}_{il}\right) \right.$$

$$\left. \left[\left(1-2\tilde{p}_{ij}\right)\tilde{p}_{ij}^{-1}\tilde{q}_{ij}^{-1}\tilde{D}_{ij} + (1-2p_{il})\tilde{p}_{il}^{-1}\tilde{q}_{il}^{-1}\tilde{D}_{il}\right]\right\}\tilde{\tau}_{ijl}^{-1/2}$$

where

$$\tau_{ijl} = \tilde{p}_{ij}\,\tilde{q}_{ij}\,\tilde{p}_{il}\,\tilde{q}_{il}\;.$$

Prentice argued that careful modelling of the correlations would increase the efficiency of the GEE estimator of β. In such a case we may replace $\tilde{C}ov(y_i)$ in M by $\tilde{V}_i$ and hence obtain

$$\left(\sum_{i=1}^{k} \tilde{D}_i^T \tilde{V}_i^T \tilde{D}_i \right) \tag{4.49}$$

as the covariance matrix of $\tilde{\beta}$. This estimate of the covariance matrix is called "naive" estimate since some loss of robustness is expected under a misspecified correlation form.

Example 4.7

Mastitis is one of the costliest diseases of the dairy industry. In a 1976 survey, the annual monetary loss resulting from bovine mastitis in the U.S.A. was estimated to be $1.3 billion. Approximately 69% of the loss ($81.32 per cow) was attributed to reduced milk production resulting from subclinical mastitis; 18% ($20.99 per cow) as a result of money spent on treatment of clinical cases; and the remaining 13% represented the cost of replacing cattle. This is strong economic evidence that the disease must be monitored and controlled. The occurrence of subclinical mastitis in herds or individual cows can be monitored by determining the somatic cell count of milk samples. In a cross-sectional study that involved 122 cows from 6 herds, the mastitis status of a cow for each of its four teats was determined by bacterial culturing of the milk samples. The somatic cell counts were also recorded. Since the age of the cow is believed to be a confounding factor it was included in the logistic regression model. Note that the somatic cell counts are measured at the subunit (teat) level, while age is measured at the cluster (cow) level. Let $y_{ij} = 1$ if the jth teat of the ith cow is tested positive to mastitis and zero otherwise ($i=1,2,...122$; $j=1,2,3,4$) clearly y_{ij} and y_{il} ($j \neq l$) are correlated. To account for such correlation we assume an exchangeable (common) correlation structure. That is

$$\text{Corr } (y_{ij}, y_{il}) = \rho \qquad\qquad \text{for all } i=1,2,...n$$

$$j \neq l = 1,2,3,4$$

The scientific question we are posing is whether the somatic cell count is related to the disease status (when age is controlled for). The postulated model is

$$\text{logit } (p_{ij}) = \beta_0^* + \beta_1^* (LSCC)_{ij} + \beta_2^* (Age)_i ,$$

where LSCC is the natural logarithm of somatic cell count and $p_{ij}=p_r[y_{ij}=1|LSCC,Age]$. To account for the higher order effects of LSCC and age, we decided to use the following dummy variable structure,

A_1 $\quad\begin{cases} 1 \text{ if age} \geq 3 \text{ years} \\ 0 \text{ if age} < 3 \text{ years} \end{cases}$

A_2 $\quad\begin{cases} 1 \text{ if age} \geq 9 \text{ years} \\ 0 \text{ if age} < 9 \text{ years} \end{cases}$

L_1 $\quad\begin{cases} 1 \text{ if LSCC} \geq 4.75 \\ 0 \text{ if LSCC} < 4.75 \end{cases}$

L_2 $\quad\begin{cases} 1 \text{ if LSCC} \geq 5.75 \\ 0 \text{ if LSCC} < 5.75 \end{cases}$

The data (see Table 4.15) were analyzed using the ordinary logistic regression (assuming independence of responses), using the GEE approach of Liang and Zeger (1986) and the maximum likelihood method under Bahadur's representation. The results are summarized in Table 4.15. The logistic regression model is

$$\text{logit } (p_{ij}) = \beta_0 + \beta_1 A_1 + \beta_2 A_2 + \beta_3 L_1 + \beta_4 L_2 .$$

Table 4.15
Analysis of the Mastitis Data Provided by Dr. Y. Schukken

Coefficient	Logistic regression	GEE 1	MLE (Bahadur's model)
β_0	-.876 (.533)	-.780 n(.588),r(.486)	-.824 (.539)
β_1	-.804 (.449)	-.849 n(.543),r(.451)	-.853 (.472)
β_2	.605 (.412)	.651 n(.497),r(.426)	.661 (.430)
β_3	-1.547 (.560)	-1.638 n(.578),r(.511)	-1.583 (.556)
β_4	.587 (.426)	.597 n(.459),r(.327)	.560 (.436)
$\hat{\rho}$	0	0.137 (0.122)	.050 (.039)

The GEE provides two estimates of the standard errors; the first is based on the "robust" formula of the variance covariance matrix (4.43) which we denote by r(.); the other is based on the naive formula given in (4.49) which we denote by n(.). It appears that young age has a sparing effect on the disease prevalence, while older cows are more likely to get the disease. The same argument holds for the somatic cell counts; that is teats with higher somatic cell counts are likely to be tested positive and vice versa.

The three approaches tend to produce very similar estimates for the parameters and very little differences among the estimated standard error. It has been confirmed through extensive simulations (McDonald 1993, Lipsitz et al. 1990) that the three methods may produce similar results when the cluster sizes are equal and small, and when the estimated p_{ij} are far from their boundary values. Since the mastitis data have clusters of size $n_i = 4$ for all ($i = 1, 2, \ldots 122$) which is relatively small, this may explain the similarity of the estimates under the three models.

Example 4.8

For the shell toxicology data of example 4.4, we illustrate the use of the GEE. The SAS program given below follows what has already been given in example 4.4.

```
data shell;
 set read;
inter = 1;
```

```
%include 'a:\gee1.mac';
%GEE (data = shell,
        yvar = y,
        xvar=inter x,
        id = litter,
        link = 3,
        vari = 3,
        n = n,
        corr = 4);
run;
```

The results are

$$\hat{B}_o = -1.4012 \,, n(0.180) \,, r(0.173)$$
$$\hat{B}_1 = 0.6334 \,, n(0.212) \,, r(0.217)$$

Note that the robust standard errors are quite close to those resulting from the Rao-Scott adjustment.

VII. LOGISTIC REGRESSION FOR CASE-CONTROL STUDIES

A. COHORT VERSUS CASE-CONTROL MODELS

Although initial applications of the logistic regression model (4.6) were specific to cohort studies this model can be applied to the analysis of data from case-control studies. The specification of the logistic model for case-control studies in which the presence or absence of exposure is taken to be the dependent variable was given by Prentice (1976). His approach assumes that we are interested in the effect of one factor. Suppose that the exposure factor which is the focus of interest is dichotomous, say x_1, where $x_1 = 1$ (exposed) and $x_1 = 0$ (unexposed) and that x_2 is another potential risk factor or a confounder. Hence, the prospective logistic model corresponding to the retrospective study is such that

$$\text{logit}\left[pr(X_1 = 1 | y, X_2) \right] = \beta_0 + \beta_1 y + \beta_2 X_2 \quad .$$

The relative odds of exposure among diseased as compared to the nondiseased may be given as

$$\text{OR} = e^{\beta_0 + \beta_1(1) + \beta_2 X_2} / e^{\beta_0 + \beta_1(0) + \beta_2 X_2}$$

$$= e^{\beta_1}$$

which is mathematically equivalent to the relative odds of disease among the exposed subjects as compared to the unexposed, as we have already shown in Chapter 3.

The rationale behind this argument was provided by Mantel (1973) as follows: let f_1 denote the sampling fraction of cases; that is if n_1 cases were drawn out of a population of N_1, then $f_1 = n_1/N_1$. Similarly we define f_0 as the sampling fraction of control. It is assumed that neither f_1 nor f_0 depends on the covariate vector X. Now consider the following 2x2 table (4.16).

Table 4.16
Description of Sampling and Disease Occurrence in a 2x2 Layout

	Case $y=1$	Control $y=0$	Total
Sampled (S)	$f_1 p_x$	$f_0 q_x$	$f_1 p_x + f_0 q_x = p(S)$
Not Sampled (S)	$(1-f_1)p_x$	$(1-f_0)q_x$	$(1-f_1)p_x + (1-f_0)q_x = p(S)$
	p_x	$q_x = 1 - p_x$	1

where $p_x = \text{pr}[y=1 \mid X]$

In a case-control study we would like to model $\text{logit}[\text{pr}(y=1 \mid x, \text{sampled})]$ as a linear function of the covariate vector X.

Since

$$Pr[y=1 \mid X, \text{ sampled}] = \frac{Pr[y=1, S \mid X]}{p(S)}$$

$$= \frac{f_1 p_x}{f_1 p_x + f_0 q_x}$$

and

$$Pr[y=0 \mid X, \text{ sampled}] = \frac{f_0 q_x}{f_1 p_x + f_0 q_x}$$

then

194

$$\text{logit}[Pr(y=1|X, \text{ sampled})] = \log \left[\frac{f_1}{f_0} \frac{p_x}{q_x} \right] \tag{4.50}$$

$$= \log \frac{f_1}{f_0} + \text{logit } Pr[y=1|X]$$

$$= \log \frac{f_1}{f_0} + \beta_0 + \beta_1 x_1 + ... \beta_p x_k$$

$$= \beta_0^* + \sum_{j=1}^{k} \beta_j x_j \tag{4.51}$$

where

$$\beta_0^* = \log \frac{f_1}{f_0} + \beta_0$$

As can be seen from equation (4.51), the logistic model for the case-control study has the same form as that of the logistic model for cohort study (4.6). This means that the regression parameters, which measure the joint effects of groups of covariates on the risk of disease, can be estimated from the case-control study. The following remarks are emphasized:

i) If β_0^*, β_1,...β_k are estimated from a case-control study, and since β_0^* depends on the ratio f_1/f_0, the risk of disease p_x (which depends on β_0) cannot be estimated unless f_1/f_0 is known. The situations in which f_1/f_0 is known are quite uncommon.

ii) For a given x, equation (4.50) represents the log-odds of disease in the sample of cases and controls which is related to the log-odds of disease in the target population by the factor ($\log f_1/f_0$). However, if we estimate the log-odds of disease in the sample of cases for a subject with covariate pattern X^*, relative to the sampled control whose covariate pattern is $\hat{X}$, then

$$\psi(X^*; \hat{X}) = \log \left[\frac{e^{\beta_0' + \sum\limits_{j=1}^{k} \beta_j X_j^*}}{e^{\beta_0' + \sum\limits_{j=1}^{k} \beta_j \hat{X}_j}} \right]$$

(4.52)

$$= e^{\sum\limits_{j=1}^{k} \beta_j \left(X_j^* - \hat{X}_j \right)}$$

This means that the estimate of β_0^* is irrelevant to the estimation of the odds ratio.

B. MATCHED ANALYSIS

We saw how data from a case-control study can be analyzed using the logistic regression model to measure the effect of a group of covariates on the risk of disease, after adjusting for potential confounders. We have also indicated (Chapter 3) that the primary objective of matching is the elimination of the biased comparison between cases and controls that results when confounding factors are not properly accounted for. A design which enables us to achieve this control over such factors is the "matched case-control study". Unless the logistic regression model properly accounts for the matching used in the selection of cases and controls, the estimated odds ratios can be biased. Thus, matching is only the first step in controlling for confounding. To analyze matched case-control study data using logistic regression we will discuss two situations. The first is called 1:1 matching (which means that each case is matched with one control), and the other is 1:M matching (which means that each case is matched with M controls). Before we show how the data analysis is performed we describe the general set-up of the likelihood function.

Suppose that controls are matched to cases on the basis of a set of variables x_1, x_2,...x_k. These variables may represent risk factors and those potential confounders that have not been used in the matching procedure. Moreover, the risk of disease for any subject may depend on the 'matching variable' that defines a 'matched set' to which an individual belongs. The values of these matching variables will generally differ between each of n matched sets of individuals.

Let $p_j(x_{ij})$ denote the probability that the ith person in the jth matched set is a case (or diseased), $i=0,1,...M$, $j=1,2,...n$. The vector of explanatory variables for the case is x_{0j}, while the vector x_{ij} ($i=1,2,...M$) denotes the explanatory variables for the Mth control in the jth matched set. The disease risk $p_j(x_{ij})$ will be modelled as

$$p_j(x_{ij}) = \frac{e^{\alpha_j + \sum\limits_{l=1}^{k} \beta_l x_{lij}}}{1 + e^{\alpha_j + \sum\limits_{l=1}^{k} \beta_l x_{lij}}} \qquad (4.53)$$

Where x_{lij} is the value of the lth explanatory variable $l=1,2,\ldots k$ for the ith individual, $i=0,1,\ldots M$ in the jth matched set. The term α_j represents the effects of a particular configuration of matching variables for the jth matched set on the risk of disease. It can be seen from (4.53) that the relationship between each explanatory variable x_l and the risk of disease is the same for all matched sets.

From (4.53) the odds of the disease is given by

$$\frac{p_j(X_{ij})}{1 - p_j(X_{ij})} = \exp\left[\alpha_j + \sum_{l=1}^{k} \beta_l x_{lij}\right] . \qquad (4.54)$$

In particular, for two individuals from the same matched set, the odds of disease for a subject with explanatory variable X_{1j} relative to one with explanatory variable X_{2j} is

$$\left\{\frac{p_j(X_{1j})}{1 - p_j(X_{1j})}\right\} \left\{\frac{p_j(X_{2j})}{1 - p_j(X_{2j})}\right\}$$

$$= \exp\left[\beta_1(X_{11j} - X_{12j}) + \ldots + \beta_k(X_{k1j} - X_{k2j})\right] , \qquad (4.55)$$

which is independent of α_j. This means that the odds of disease for two matched individuals with different explanatory variables does not depend on the actual values of the matching variables.

C. CONDITIONAL LIKELIHOOD

The likelihood function based on matched case-control studies is known as "conditional likelihood". Here we construct the likelihood function under the 1:M matched case-control study, where the jth matched set contains M controls ($j=1,2,\ldots n$). Let X_{0j} denote the vector of the explanatory variables for the case and $X_{ij},\ldots X_{Mj}$ denote the vector of explanatory variables for the M controls in the jth matched set. Following Breslow et al. (1978) let $p(X_{ij} \mid y=1)$ be the probability that a diseased individual in the matched set has explanatory variables X_{ij} for $i=0,1,\ldots M$ and $j=1,2,\ldots n$; let $p(X_{ij} \mid y=0)$ be the probability of a disease-free individual with explanatory variable X_{ij}. Therefore, the joint probability that X_{0j} corresponds to the case ($y=1$) and X_{ij} ($i \geq 1$) corresponds to the control ($y=0$) is

$$p(X_{0j}|y=1) \prod_{i=1}^{M} p(X_{ij}|y=0) \tag{4.56}$$

Note that from Bayes theorem we can write

$$p(X_{0j}|y=1) = \frac{p(y=1|X_{0j})\ p(X_{0j})}{p(y=1)} \tag{4.57}$$

and similarly

$$p(X_{ij}|y=0) = \frac{p(y=0|X_{ij})\ p(X_{ij})}{p(y=0)} \tag{4.58}$$

Now, the probability that one of the $M+1$ subjects in the jth matched set is the case and the remainder are controls, is the sum of the probability that the subject with explanatory variable X_{0j} is diseased and the rest are disease free, plus the probability that the subject with explanatory variable X_{ij} is diseased and the rest are disease free and so on. That is

$$p(X_{0j}|y=1) \prod_{i=1}^{M} p(X_{ij}|y=0) + p(X_{1j}|y=1) \prod_{i \neq 1}^{M} p(X_{ij}|y=0)$$

$$+ \ldots + p(X_{Mj}|y=1) \prod_{i \neq M}^{M} p(X_{ij}|y=0) \tag{4.59}$$

$$= \sum_{i=0}^{M} p(X_{ij}|y=1) \prod_{r \neq i}^{M} p(X_{rj}|y=0)$$

Therefore, the conditional probability that the case in the jth matched set has explanatory variable X_{0j}, conditional on X_{0j}, X_{ij},$\ldots X_{Mj}$ being the explanatory variables for the subjects in that matched set is

$$F_j(X_j) = \frac{p(X_{0j}|y=1) \prod_{i=1}^{M} p(X_{ij}|y=0)}{\sum_{i=0}^{M} p(X_{ij}|y=1) \prod_{r \neq i}^{M} p(X_{rj}|y=0)} \tag{4.60}$$

Substituting (4.57) and (4.58) into (4.60) we get

$$\frac{\dfrac{p(y=1|X_{oj})p(X_{oj})}{p(y=1)}\displaystyle\prod_{i=1}^{M}\dfrac{p(y=0|X_{ij})p(X_{ij})}{p_{(}y=0)}}{\displaystyle\sum_{i=0}^{M}\dfrac{p(y=1|X_{ij})p(X_{ij})}{p(y=1)}\displaystyle\prod_{r\neq i}^{M}\dfrac{p(y=0|X_{rj})p(X_{rj})}{p(y=0)}}$$

$$=\frac{p(y=1|X_{0j})\displaystyle\prod_{i=1}^{M}p(y=0|X_{ij})}{\displaystyle\sum_{i=0}^{M}p(y=1|X_{ji})\displaystyle\prod_{r\neq i}^{M}p(y=0|X_{rj})}$$

$$=\frac{p(y=1|X_{0j})\displaystyle\prod_{i=1}^{M}p(y=0|X_{ij})}{p(y=1|X_{0j})\displaystyle\prod_{r=1}^{M}p(y=0|X_{rj})+\displaystyle\sum_{i=1}^{M}p(y=1|X_{ij})\displaystyle\prod_{r\neq i}^{M}p(y=0|X_{rj})}$$

$$=\left[1+\frac{\displaystyle\sum_{i=1}^{M}p(y=1|X_{ij})\displaystyle\prod_{r\neq i}^{M}p(y=0|X_{rj})}{p(y=1|X_{0j})\displaystyle\prod_{i=1}^{M}p(y=0|X_{ij})}\right]^{-1}.$$

Hence

$$F_{j}(X_{j})=\left\{1+\sum_{i=1}^{M}\left[\frac{p(y=1|X_{ij})\ p(y=0|X_{0j})}{p(y=1|X_{0j})\ p(y=0|X_{ij})}\right]\right\}^{-1} \tag{4.61}$$

Similar to equation (4.53), assume that

$$\frac{p(y=1|X_{ij})}{p(y=0|X_{ij})}=\exp\left[\alpha_{j}+\beta_{1}x_{1ij}+\beta_{2}x_{2ij}+\ldots+\beta_{k}x_{kij}\right]$$

which when substituted in (4.61) results in

$$F_{j}(X_{j})=\left\{1+\sum_{i=1}^{M}\exp\left[\beta_{1}(X_{1ij}-X_{10j})+\ldots+\beta_{k}(X_{kij}-X_{k0j})\right]\right\}$$

Hence, the conditional likelihood function of a sample of n matched sets is given by

$$L = \prod_{j=1}^{n} F_j(X_j) \ . \tag{4.62}$$

D. FITTING MATCHED CASE-CONTROL STUDY DATA IN SAS

The main purpose of this section is to discuss, through an example, how the parameters of the conditional likelihood (4.62) are estimated. For the matched pair design, $M=1$, the conditional likelihood (4.62) reduces to

$$L = \prod_{i=1}^{n} \left\{ 1 + \exp\left[\sum_{r=1}^{k} \beta_r (X_{rij} - X_{r0j}) \right] \right\}^{-1}$$

$$= \prod_{i=1}^{n} \left\{ 1 + \exp\left[-\sum_{r=1}^{k} \beta_r Z_{rj} \right] \right\}^{-1} \tag{4.63}$$

where $Z_{rj} = X_{r0j} - X_{r1j}$.

This likelihood (4.63) is identical to the likelihood function of a logistic regression for n binary observations y_i, such that $y_i = 1$ for $i = 1, 2, \ldots n$. Note that the explanatory variables here are $Z_{ij}, \ldots Z_{kj}$, and there is no intercept. Therefore, using SAS, *PROC LOGISTIC* fits the 1:1 matched data by following these steps:

(i)   the number of matched sets n is the number of observations

(ii)   the response variable $y_j = 1$ for all $j = 1, 2, \ldots n$

(iii)   the explanatory variable Z_{rj} ($r = 1, 2 \ldots k$ and $j = 1, 2, \ldots n$) is the difference between the value of the rth explanatory variable for a control and the rth explanatory variable for a case, within the same matched set. Note that if qualitative or factor variables are used, the explanatory variables in (4.63) will correspond to dummy variables. Consequently, variables are the differences between the dummy variables of the case and control in the matched pair. Note also that the interaction terms can be included in the model by representing them as products of the corresponding main effects. The differences between these products for the case and control in a matched pair are included in the model.

Example 4.9 "Hypothetical Data"

In an investigation aimed at assessing the relationship between somatic cell counts (SCC) and the occurrence of mastitis, a 1:1 matched case-control study was postulated. A "case" cow was matched with a control cow from the same farm based on: breed, number of lactations and age as a possible confounder. The data summary is given in Table 4.17.

Table 4.17
"Hypothetical Mastitis Data"

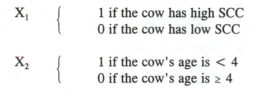

		Case		
		High	Low	
Control	High	5	110	
	Low	216	40	age < 4

		Case		
		High	Low	
Control	High	5	212	
	Low	308	21	age ≥ 4

In this simple example we have one risk factor, namely the SCC which was dichotomised as "high" and "low". Since it is believed that the incidence rate of the disease in younger cows is different from older cows, the matching variable age was divided into two distinct strata, the first for cows whose age is less than 4 years, and the second for those that are at least 4 years old.

As already mentioned, we cannot investigate the association between the disease risk and the age variable since age is a matching variable. However we shall investigate the possible interaction between the SCC and age. Since the risk factor and the confounder are factor variables and each factor has two levels, we define a single dummy variable for each factor. Let X_1 be the indicator variable for SCC and X_2 for age, where

$$X_1 \begin{cases} 1 \text{ if the cow has high SCC} \\ 0 \text{ if the cow has low SCC} \end{cases}$$

$$X_2 \begin{cases} 1 \text{ if the cow's age is } < 4 \\ 0 \text{ if the cow's age is } \geq 4 \end{cases}$$

we also define a third dummy variable X_3, obtained by multiplying X_1 and X_2 for each individual animal in the study. With this coding the data is structured as in Table 4.18.

Table 4.18
Coded Variable for the Data in Table 4.12

Status	X_1(SCC)	X_2(age)	$X_3=X_1X_2$		Number of matched pairs
Case	1	0	0	}	5
Control	1	0	0		
Case	1	0	0	}	216
Control	0	0	0		
Case	0	0	0	}	110
Control	1	0	0		
Case	0	0	0	}	40
Control	0	0	0		
Case	1	1	1	}	5
Control	1	1	1		
Case	1	1	1	}	308
Control	0	1	0		
Case	0	1	0	}	212
Control	1	1	1		
Case	0	1	0	}	21
Control	0	1	0		

The input variables, $Z1$, $Z2$ and $Z3$ are created using variables as set up in Table 4.19, where,

$Z1 = X1(case) - X1 (control)$
$Z2 = X2(case) - X2 (control)$
$Z3 = X3(case) - X3(control)$.

Table 4.19
Variables for the Logistic Regression Model

y	Z_1	Z_2	Z_3
1	0	0	0
1	1	0	0
1	-1	0	0
1	0	0	0
1	0	0	0
1	1	0	1
1	-1	0	-1
1	0	0	0

The following SAS statements describe how the logistic regression model can be fitted.

```
data  match;
input  y  Z₁ Z₂ Z₃  count;
do  i=1  to count;
output; end;
cards;
```

1	0	0	0	5
1	1	0	0	216
1	-1	0	0	110
1	0	0	0	40
1	0	0	0	5
1	1	0	1	308
1	-1	0	-1	212
1	0	0	0	21

```
proc logistic data = match;
model y = Z₁ Z₃ / noint covb;
run;
```

Note that the response variable $y_i = 1$ for all $i = 1, 2, \ldots 8$. In the model statement of the SAS program we did not include the matching variable Z_2 since it is not possible to investigate its association with the disease status. Moreover the option *noint* is specified, so that the logistic regression is fitted without the intercept parameter. The SAS output are:

$$\hat{\beta}_1 = 0.675 \ (.117) \quad \text{P-value} \approx .0001$$
$$\hat{\beta}_2 = -0.301 \ (.147) \quad \text{P-value} \approx .041$$

The bracketed numbers are the estimated standard error of the corresponding regression parameter estimate. The results indicate that there is a significant association between SCC ($\hat{\beta}_1$) and the disease. There is also significant interaction between the SCC ($\hat{\beta}_2$) and age on the risk of mastitis.

Remark

For 1:M matching, where each case is matched with an arbitrary number of controls, the PROC PHREG in SAS can be used to fit the model (SAS Technical Report P229, Release 6.07).

Chapter 5

THE ANALYSIS OF TIME SERIES

I. INTRODUCTION

A time series is an ordered sequence of observations. Although the ordering is usually through time, particularly in terms of some equally spaced intervals, the ordering may also be taken through other dimensions, such as space. Time series occur in a wide variety of fields. In economics, interest may be focused on the weekly fluctuations in the stock prices and their relationships to unemployment figures. Agricultural time series analyses and forecasting could be applied to annual crop yields or the prices of produce with regard to their seasonal variations. Environmental changes over time, such as levels of air and water pollution measured at different places, can be correlated with certain health indicators. The number of influenza outbreaks in successive weeks during the winter season could be approached as a time series by an epidemiologist. In medicine, systolic and diastolic blood pressures followed over time for a group of patients could be useful for assessing the effectiveness of a drug used in treating hypertension. Geophysical time series (Shumway, 1982) are quite important for predicting earthquakes. From these examples one can see the diversity of fields in which time series can be applied. There are, however, some common objectives which must be achieved in collected time series data:

a)   *To describe the behavior of the series in a concise way.* This is done by first plotting the data and obtaining simple descriptive measures of the main properties of the series. This may not be useful for all time series because there are series that require more sophisticated techniques, and thus more complex models need to be constructed.

b)   *To explain the behavior of the series in terms of several variables.* For example, when observations are taken on more than one variable, it may be feasible to use the variation in one time series to explain the variation in another series.

c)   *We may want to predict (forecast) the future values of the series.* This is an important task for the analysis of economic and agricultural time series. It is desirable particularly if there is sufficient evidence in the system to ensure that future behavior will be similar to the past. Therefore our ability to understand the behavior of the series may provide us with more insight into causal factors and help us make projections into the future.

d)   *Controlling the series by generating warning signals of future fluctuations.* For example, if we are measuring the quality of production process, our aim may be to keep the process under control. Statistical quality control provides us with the tools to achieve such an objective by constructing "control charts". More advanced strategies for control are outlined in Box and Jenkins (1970).

The following example (5.1) gives the data (Table 5.1) and time plot (Figure 5.1) for an epidemiological time series showing the average somatic cell count (SCC) by month over a number of years.

Example 5.1

In plotting the SCC data in Table 5.1 we see large fluctuations in both the mean and the variance over time (Figure 5.1).

Table 5.1
Average SCC per Farm in 1000s of Cells per ml of Milk
1984-1990

	1984	1985	1986	1987	1988	1989	1990
Jan	317	345	370	350	400	370	340
Feb	292	310	360	420	385	335	345
Mar	283	307	300	360	350	305	325
Apr	286	310	310	340	325	325	330
May	314	340	389	335	345	310	360
June	301	325	320	350	350	315	330
July	317	340	340	360	375	350	345
Aug	344	370	400	395	410	370	350
Sept	367	400	395	380	360	350	350
Oct	351	380	350	375	375	345	345
Nov	321	345	400	402	370	355	325
Dec	398	330	350	460	395	340	280

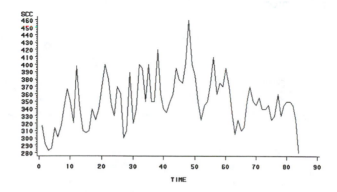

Figure 5.1 A time of average somatic cell counts per month for 84 months.

205

A. SIMPLE DESCRIPTIVE METHODS

In this section we describe some of the simple techniques which will detect the main characteristics of a time series. From a statistical point of view, the description of a time series is accomplished by a time plot; that is, plotting the observations against time and formulating a mathematical model to characterize the behavior of the series. This models the mechanism which governs the variability of the observations over time. Plotting the data could reveal certain features such as trend, seasonality, discontinuities and outliers. The term 'trend' usually means the upward or downward movement over a period of time, thus reflecting the long-run growth or decline in the time series.

In addition to the effect of trend, most economic time series include seasonal variation. To include both the seasonal variation and the trend effect there are two types of models that are frequently used. The first is the Additive Seasonal Variation Model (ASVM) and the second is the Multiplicative Seasonal Variation Model (MSVM). If a time series displays additive seasonal variation, the magnitude of the seasonal swing is independent of the mean. On the other hand, in a multiplicative seasonal variation series we see that the seasonal swing is proportional to the mean. The two models are represented by the following equations

$$y_t = T_t + S_t + \epsilon_y \qquad \text{(ASVM)}$$
$$y_t = T_t S_t + \epsilon_y \qquad \text{(MSVM)}$$

where y_t = the observed value at time t
$\quad\quad T_t$ = the mean trend factor at time t
$\quad\quad S_t$ = seasonal effect at time t
$\quad\quad \epsilon_t$ = the irregular variation of the time series at time t

The two models are illustrated in Figures 5.2a and 5.2b

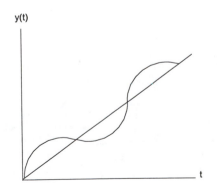
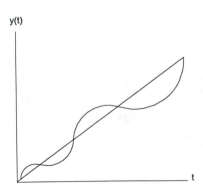

Figure 5.2a Additive seasonal variation model.   **Figure 5.2b** Multiplicative seasonal variation model.

Note that the MSVM has an additive irregular variation term. If such a term is multiplicative, that is if $y_t = T_t S_t \epsilon_t$, then this model can be transformed to ASVM by taking the logarithm of both sides.

1. Multiplicative Seasonal Variation Model

In this section we will be concerned with simple methods of decomposing MSVM into its trend, seasonal and random components. The model is somewhat different from the previous model,

$$y_t = (T_t)(S_t) + \epsilon_t \tag{5.1}$$

and is usually written as

$$y_t = (T_t)(S_t)(C_t)(I_t) \tag{5.2}$$

where y_t, T_t and S_t are as previously defined. Here, C_t represents the cyclical effect on the series at time t, and I_t is the irregular variation.

We will explain how to decompose the multiplicative model in the following example.

Example 5.2

The following data represent the number of cases with bovine respiratory disease in particular feedlots reported in eastern Alberta counties over a period of 4 years in each of the 4 quarters.

Quarter	Year 1	Year 2	Year 3	Year 4
1	21	25	25	30
2	14	16	18	20
3	5	7	9	10
4	8	9	13	15

The first step in the analysis of this time series is the estimation of seasonal factors for each quarter. To do this one has to calculate the moving average (MA), in order to remove the seasonal variation from the series. A moving average is calculated by adding the observations for a number of periods in the series and dividing the sum by the number of periods. In the example above we have a four-period series since we have quarterly data. If the time series consists of data collected every 4 months, we have a three-period time series, and hence a three period moving average should be used. In the above example the average for the four observations in the first year is

$$\frac{21+14+5+8}{4} = 12$$

The second average is obtained by eliminating the first observation in year 1 from the average and including the first observation in year 2 in the new average. Hence

$$\frac{14+5+8+25}{4} = 13$$

The third average is obtained by dropping the second observation in year 1 and adding the second observation in year 2. This gives

$$\frac{5+8+25+16}{4} = 13.5$$

Continuing in this manner, these moving averages are as found in Table 5.2a. Note that, since the first average is the average of the observations in the 4 quarters, it corresponds to a midpoint between the second and third quarter.

To obtain the average corresponding to one of the time periods in the original time series, we calculate a centered moving average. This is obtained by computing a two-period moving average of the moving averages previously calculated (Table 5.2a).

Note that since the moving average is computed using exactly one observation from each season, the seasonal variation has been removed from the data. It is also hoped that this averaging process has removed the irregular variation I_t. This means that the centered moving averages in column 6 in Table 5.1a represent the trend (T_t) and cycle (C_t). Now, since

$$y_t = (T_t)(S_t)(C_t)(I_t)$$

then the entries in column 7 of Table 5.2a are computed as:

$$(S_t)(I_t) = \frac{y_t}{(T_t)(C_t)} = \frac{\text{column (3)}}{\text{column (6)}}$$

The seasonal coefficients ($S_t I_t$) are summarized in Table (5.2b).

Table 5.2a
Moving Average of the Time Series of Example 5.2

1	2	3	4	5	6	7
Year	Quarter	y_t	Moving Total	Moving Average	Centered Moving Average	S_tI_t
1	1	21				
	2	14				
			48	12		
	3	5			12.5	.4
			52	13		
	4	8			13.25	.604
			54	13.5		
2	1	25			13.75	1.818
			56	14		
	2	16			14.125	1.133
			57	14.25		
	3	7			14.25	.491
			57	14.25		
	4	9			14.5	.621
			59	14.75		
3	1	25			15.0	1.667
			61	15.25		
	2	18			15.75	1.143
			65	16.25		
	3	9			16.875	.533
			70	17.5		
	4	13			17.75	.732
			72	18		
4	1	30			18.125	1.655
			73	18.25		
	2	20			18.5	1.081
			75	18.75		
	3	10				
	4	15				

Table 5.2b
Seasonal Coefficients for Each Quarter by Year

Quarter (1)	Quarter (2)	Quarter (3)	Quarter (4)
1.818	1.133	.400	.604
1.667	1.143	.491	.621
1.655	1.081	.533	.732

The seasonal effects for each quarter can be computed by summing and dividing by the number of coefficients. Thus, for quarter 1,

$$\hat{S}_1 = \frac{1.818 + 1.667 + 1.655}{3} = 1.713 \quad .$$

Similarly

$$\hat{S}_2 = 1.119, \ \hat{S}_3 = .475, \ and \ \hat{S}_4 = .652$$

are the estimated seasonal effects for quarters 2, 3 and 4.

Once the estimates of the seasonal factors have been calculated we may obtain an estimate of the trend T_t of the time series. This is done by first estimating the deseasonalized observations.

The deseasonalized observations are obtained by dividing y_t by S_t. That is,

$$d_t \equiv \text{deseasonalized series} = \frac{y_t}{S_t}$$

These values should be close to the trend value T_t. To model the trend effect, as a first step one should plot d_t against the observation number t. If the plot is linear it is reasonable to assume that

$$T_t = \beta_0 + \beta_1 t \ ;$$

on the other hand if the plot shows a quadratic relationship then we may assume that

$$T_t = \beta_0 + \beta_1 t + \beta_2 t^2$$

and so on. Table 5.2c gives the deseasonalized observations; Figure 5.3 is a scatter plot of these observations against time.

Table 5.2c
Deseasonalized Observations

Year	Quarter	t	y_t	S_t	$d_t = y_t/S_t$
1	1	1	21	1.713	12.26
	2	2	14	1.119	12.51
	3	3	5	.475	10.53
	4	4	8	.652	12.27
2	1	5	25	1.713	14.59
	2	6	26	1.119	23.24
	3	7	7	.475	14.74
	4	8	9	.652	13.80
3	1	9	25	1.713	14.59
	2	10	18	1.119	16.09
	3	11	6	.475	12.63
	4	12	13	.652	19.94
4	1	13	30	1.713	17.51
	2	14	20	1.119	17.87
	3	15	10	.475	21.05
	4	16	15	.652	23.01

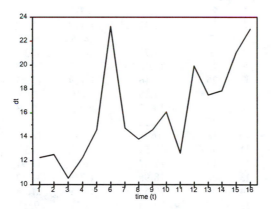

Figure 5.3. Deseasonalized observations over time; BRD data.

The estimated trend is found to be:

$$\hat{d}_t = \hat{T}_t = 10.05 + 0.685t. \qquad , \ t = 1,2,\ldots 16$$

211

To compute the cyclical effect, recall that

$$y_t = (T_t)(S_t)(C_t)(I_t),$$

hence

$$(C_t)(I_t) = \frac{y_t}{(\hat{T}_t)(\hat{S}_t)} \ .$$

We summarize these computations in Table 5.2d

Table 5.2d
Computations of Cyclical Effect and Irregular Variations

(1)	(2)	(3)	(4)	(5)	(6)	(7)	(8)
Year	Quarter	t	y_t	$T_t = 10.05 + .685t$	S_t	$(T_t)(S_t)$	$(C_t)(I_t)$
1	1	1	21	10.74	1.713	18.39	1.14
	2	2	14	11.42	1.119	12.78	1.10
	3	3	5	12.11	.475	5.75	0.87
	4	4	8	12.79	.652	8.34	0.96
2	1	5	25	13.48	1.713	23.08	1.08
	2	6	26	14.16	1.119	15.85	1.64
	3	7	7	14.85	.475	7.05	0.99
	4	8	9	15.53	.652	10.13	0.89
3	1	9	25	16.22	1.713	27.78	0.90
	2	10	18	16.9	1.119	18.91	0.95
	3	11	6	17.59	.475	8.35	0.72
	4	12	13	18.27	.652	11.91	1.09
4	1	13	30	18.96	1.713	32.47	0.92
	2	14	20	19.64	1.119	21.98	0.91
	3	15	10	20.33	.475	9.65	1.04
	4	16	15	21.01	.652	13.70	1.10

$$C_tI_t = \frac{column \ 4}{column \ 7}$$

Once $(C_t)(I_t)$ has been obtained, a three quarter moving average may remove the effect of irregular variation. The results are summarized in Table 5.2e.

Table 5.2e
Estimated Cyclical Effect

Year	Quarter	t	$(C_t)(I_t)$	3-period moving average C_t
1	1	1	1.14	
	2	2	1.10	1.037
	3	3	0.87	0.977
	4	4	0.96	0.970
2	1	5	1.08	1.230
	2	6	1.64	1.24
	3	7	0.99	1.173
	4	8	0.89	0.927
3	1	9	0.90	0.913
	2	10	0.95	0.857
	3	11	0.72	0.920
	4	12	1.09	0.910
4	1	13	0.92	0.973
	2	14	0.91	0.957
	3	15	1.04	1.017
	4	16	1.10	

$$C_t I_t = \frac{column\ 4}{column\ 7}$$

The previous example shows how a time series can be decomposed into its components. Most econometricians use the trend and seasonal effect in their forecast of time series, ignoring the cyclical and irregular variations. Clearly irregular ups and downs cannot be predicted, however, cyclical variation can be forecasted and is treated in the same manner as the seasonal effects shown in Table 5.2b. In our example, the average effect of the cycle at each period is as found in Table 5.2f.

Table 5.2f
Cycle's Effect for Different Periods

quarter (1)	quarter (2)	quarter (3)	quarter (4)
1.23	1.04	0.98	0.97
0.92	1.24	1.17	0.93
0.97	0.86	0.92	0.91
	0.96	1.02	

It should be noted that the estimated cycles are useful if a well-defined repeating cycle of reasonable fixed duration can be recognized. In many 'real life' data this may not be possible. In order to obtain reliable estimates of the cyclical effect, data with several cycles should be available. Since cyclical fluctuations have a duration of 2 to 7 years or more, more than 25 years of data may be needed to estimate the cycle effect and make accurate forecasts. For these reasons, the cyclical variation in time series cannot be accurately predicted. In such situations, forecasts are based on the trend and seasonal factors only. Having obtained $\hat{T}_t$ and $\hat{S}_t$, the forecast of a future observation is given by

$$\hat{y}_t = (\hat{T}_t)(\hat{S}_t) \ .$$

2. Additive Seasonal Variation Model

For this type of model we shall assume, for simplicity, that the series is composed of trend, seasonal effect, and error component, so that

$$y_t = T_t + S_t + I_t$$

As before the trend effect can be modelled either linearly: $T_t = \beta_0 + \beta_1 t$, quadratically: $T_t = \beta_0 + \beta_1 t + \beta_2 t^2$, or exponentially: $T_t = \beta_0 + \beta_1^t$ (which can be linearized through the logarithmic transformation).

The seasonal pattern may be modelled by using dummy variables. Let L denote the number of periods or seasons (quarter, month,...etc.) in the year. S_t can be modelled as follows

$$S_t = \gamma_1 X_{1,t} + \gamma_2 X_{2,t} + \cdots \gamma_{L-1} X_{L-1,t}$$

where $X_{1t} = \begin{cases} 1 & \text{if period t is season 1} \\ 0 & \text{otherwise} \end{cases}$

where $X_{2t} = \begin{cases} 1 & \text{if period t is season 2} \\ 0 & \text{otherwise} \end{cases}$

where $X_{L-1,t} = \begin{cases} 1 & \text{if period t is season L-1} \\ 0 & \text{otherwise} \end{cases}$

214

For example, if L=4 (quarterly data) we have

$$y_t = T_t + S_t + I_t$$

$$= T_t + \gamma_1 X_{1t} + \gamma_2 X_{2t} + \gamma_3 X_{3t} + I_t \qquad (5.3)$$

Similarly, if L=12 (monthly data) we have

$$y_t = T_t + S_t + I_t$$

$$= T_t + \sum_{i=1}^{11} \gamma_i X_{it} + I_t \qquad (5.4)$$

Clearly T_t can be represented by either a linear, quadratic or exponential relationship.

Example 5.3

To decompose an additive time series we use the data of Example 5.2 (BRD occurrence in eastern Alberta) to estimate the trend and seasonal effects under the model,

$$y_t = (\beta_0 + \beta_{1t}) + \gamma_1 X_{1t} + \gamma_2 X_{2t} + \gamma_3 X_{3t} \qquad (5.5)$$

Using SAS, the INPUT statement is given as:

```
INPUT       t      y      X1     X2     X3;
CARDS ;
```

t	y	X1	X2	X3
1	21	1	0	0
2	14	0	1	0
3	5	0	0	1
4	8	0	0	0
5	25	1	0	0
6	16	0	1	0
7	7	0	0	1
8	9	0	0	0
9	25	1	0	0
10	18	0	1	0
11	9	0	0	1
12	13	0	0	0
13	30	1	0	0
14	20	0	1	0
15	10	0	0	1
16	15	0	0	0

The SAS program is:

```
PROC REG;
MODEL y = t    X1_t  X2_t  X3_t;
RUN;
```

The fitted series is given by

$$\hat{y}_t = 5.69+.55t+15.67\ X_{1t}+6.86\ X_{2t}-2.94\ X_{3t}$$

from which

$$\hat{T}_t = 5.69+.55t$$

and

$$\hat{S}_t = 15.67\ X_{1t}+6.86\ X_{2t}-2.94\ X_{3t}$$

the estimated components of the series are summarized in Table 5.3.

Table 5.3
Estimated Trend and Seasonal Effect of the Series (data from example 5.2)

Year	Quarter (Period)	t	y_t	$T_t=5.69 + .55t$	S_t	$\hat{y}_t=T_t+S_t$	$e_t=y_t-\hat{y}_t$
1	1	1	21	6.24	15.67	21.91	-0.91
	2	2	14	6.79	6.86	13.65	0.34
	3	3	5	7.34	-2.94	4.40	0.60
	4	4	8	7.89	0	7.89	0.11
2	1	5	25	8.44	15.69	24.11	0.89
	2	6	26	8.99	6.86	15.85	10.15
	3	7	7	9.54	-2.94	6.60	0.40
	4	8	9	10.09	0	10.09	-1.09
3	1	9	25	10.64	15.67	26.31	-1.31
	2	10	18	11.19	6.86	18.05	-0.05
	3	11	6	11.74	-2.94	8.80	-2.80
	4	12	13	12.29	0	12.29	0.71
4	1	13	30	12.84	15.67	28.51	1.49
	2	14	20	13.39	6.86	20.25	-0.25
	3	15	10	13.94	-2.94	11.00	-1
	4	16	15	14.49	0	14.49	0.51

Figure 5.4 gives the actual series and the fitted series $\hat{y}_t$, for the additive models.

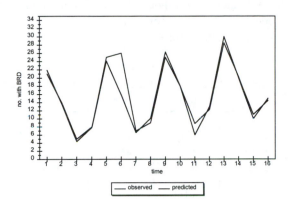

Figure 5.4.   Observed values of y and predicted values of y from the additive model plotted over time.

3. Detection of Seasonality : Nonparametric Test

Seasonality may be tested using a test based on ranks. The test is a simple adaptation of the nonparametric analysis of variance procedure initially proposed by Friedman (1937). After removing a linear trend, if desired, we rank the values within each year from 1 (smallest) to 12 (largest) for monthly data. In general, let the years represent c columns and the months r (=12) rows. Then each column represents a permutation of the integers 1, 2, ...,12. Summing across each row gives the monthly score M_j , j = 1, 2, ..., 12. Under the null hypothesis H_0 : no seasonal pattern, the test statistic

$$T = 12 \sum_{j=1}^{r} \left\{ M_j - \frac{c(r+1)}{2} \right\}^2 / cr(r+1) \qquad (5.6)$$

$$= 12 \left[\sum_{j=1}^{r} M_j^2 / cr(r+1) + \frac{c(r+1)}{4} - \sum_{j=1}^{r} M_j / r \right]$$

is approximately distributed as χ^2 with (r-1) degrees of freedom.

217

Example 5.4 The data from this example were kindly provided by Dr. J. Mallia of the Ontario
Veterinary College.

Cyanosis is one of the leading causes of condemnation of poultry in Canada. To investigate
seasonal patterns in the proportion of turkeys condemned we use the statistic (5.6). The data are
summarized in Table 5.4 and plotted in Figure 5.5.

Table 5.4
Number of Turkeys Condemned (per 100, 1000) Because of Cyanosis

	1987	1988	1989	1990	1991	1992	1993
Month							
Jan	64.3	1168.7	1173.7	1140.4	691.2	1154.4	556.7
Feb	508.6	1422.4	1492.3	1446.4	370.9	683.0	489.3
March	646.2	1748.4	1600.5	1002.7	454.3	535.6	466.2
April	849.1	1226.9	1141.0	999.5	393.9	351.6	448.9
May	710.2	1061.0	861.0	485.1	374.0	430.2	302.1
June	653.0	905.6	706.3	416.9	253.2	371.5	260.3
July	542.2	875.7	537.7	562.6	428.2	317.1	215.6
Aug	502.6	943.0	583.3	483.7	429.5	425.2	272.9
Sept	789.5	1228.2	810.8	490.4	393.7	332.5	286.0
Oct	409.5	1286.0	750.0	670.5	387.9	327.0	270.8
Nov	836.4	1434.8	1137.6	605.6	587.0	427.6	373.3
Dec	792.4	860.3	1178.7	618.5	618.5	381.8	259.6

In Table 5.5 we provide the ranks, M_j and M_j^2 .

Table 5.5
Year

	1987	1988	1989	1990	1991	1992	1993	M_j	M_j^2
Month									
1	1	6	9	11	12	12	12	63	3969
2	4	10	11	12	2	11	11	61	3721
3	6	12	12	10	9	10	10	69	4761
4	12	7	8	9	6	4	9	55	3025
5	8	5	6	3	3	9	7	41	1681
6	7	3	3	1	1	5	3	23	529
7	5	2	1	5	7	1	1	22	484
8	3	4	2	2	8	7	5	31	961
9	9	8	5	4	5	3	6	40	1600
10	2	9	4	8	4	2	4	33	1089
11	11	11	7	6	10	8	8	61	3721
12	10	1	10	7	11	6	2	47	2209

$$\sum_{j=1}^{12} M_j = 546, \quad \sum_{j=1}^{12} M_j^2 = 27{,}750 \text{ and } T = 31.94$$

Since $\chi^2_{.05,11} = 19.67$, the null hypothesis of no seasonal pattern is not supported by the data.

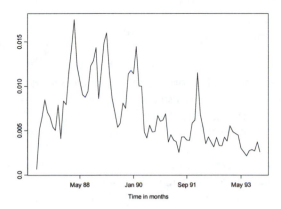

Figure 5.5 Time series plot of proportion of turkeys condemned. The vertical axis is the proportion.

4. Autoregressive Errors: Detection and Estimation

One of the main characteristics of a true series is that adjacent observations are likely to be correlated. One way to detect such correlation is to plot the residuals $e_t = y_t - \hat{y}_t$ against time. This is illustrated using the additive model, where the residuals (from Table 5.3) are plotted in time order as in Figure 5.6.

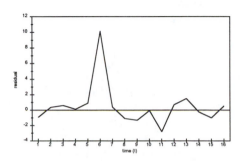

Figure 5.6 Plot of residuals over time

From the plot one can see that the residuals have the signs $-,+,+,+,+,+,+,-,-,-,-,+,+,-,-,+$. This shows a tendency for residuals to be followed by residuals of the same sign, an indication of possible autocorrelation. Also evident from the plot of these residuals is a potential outlier $(e_t = 10.15)$ from the 2nd quarter. Usually, the Durbin-Watson statistic given by

$$d = \frac{\sum_{i=2}^{n} \left(e_t - e_{t-1}\right)^2}{\sum_{i=1}^{n} e_t^2} = 0.5624 \tag{5.7}$$

is used to test for the significance of this correlation. From Durbin and Watson (1951), the upper and lower critical points of the Durbin-Watson statistic d at $\alpha = 0.05$, $n = 16$ and 5 parameters in the model are 0.74 and 1.93 respectively. Since our value of d (0.5624) is less than the lower critical value, the hypothesis of no autocorrelation between the residuals is rejected.

When a significant autocorrelation has been detected, ignoring its effect would produce unrealistically small standard errors for the regression estimates in the fitted model (5.6). Therefore one has to account for the effect of this correlation to produce accurate estimates of the standard error. Our approach to modelling this autocorrelation at present will still be at a descriptive level. More rigorous treatment of the autocorrelation structure will be presented in the next section.

The simplest autocorrelation structure that we shall examine here is called the "first order autoregression process". This model assumes that successive errors are linearly related through the relationship

$$\epsilon_t = \rho \ \epsilon_{t-1} + u_t \tag{5.8}$$

It is assumed that $\{u_t; t = 1, 2, \dots n\}$ are independent and identically distributed $N(0, \sigma^2)$. Under the above specifications we have the model

$$y_t = \beta_0 + \beta_1 X_{1t} + \dots \beta_k X_{kt} + \epsilon_t \tag{5.9}$$

where

$$\epsilon_t = \rho \epsilon_{t-1} + u_t$$

and

$$\rho = Corr(\epsilon_t, \epsilon_{t-1}) \ .$$

Note that

$$\rho y_{t-1} = \rho\beta_0 + \rho\beta_1 X_{1\,t-1} + \ldots + \rho\beta_k X_{k\,t-1} + \varepsilon_{t-1} \qquad (5.10)$$

Subtracting (5.10) from (5.9) we have:

$$y_t - \rho y_{t-1} = \beta_0(1-\rho) + \beta_1(X_{1\,t} - \rho X_{1\,t-1}) + \ldots + \beta_k(X_{k\,t} - \rho X_{k\,t-1}) + U_t \qquad (5.11)$$

The last equation has u_t as an error term that satisfies the standard assumptions of inference in a linear regression model. The problem now is that the left hand side of (5.11) has a transformed response variable that depends on the unknown parameter ρ. A commonly used procedure to estimate the model parameters is to use a procedure known as Cochran and Orcutt procedure which we outline in the following 4 steps

Step (1)   Estimate the parameters of the model (5.9) using least squares or PROC REG from SAS, and compute the residuals $e_1, e_2, \ldots e_n$

Step (2)   From $e_1 \ldots e_n$, evaluate the moment estimator of ρ as

$$\hat{\rho} = \frac{\sum\limits_{t=2}^{n} e_t e_{t-1}}{\sum\limits_{t=2}^{n} e_{t-2}^2} \qquad (5.12)$$

Step (3)   Substitute $\hat{\rho}$ in place of ρ in model (5.11) which has an error term that satisfies the standard assumptions and compute revised least square estimates.

Step (4)   From the least square estimates obtained in step (3), compute the revised residuals and return to step (2); find an updated estimate of ρ using (5.12). We iterate between step (2) and step (4) until the least square estimate has an insignificant change between successive iterations.

5. Modelling Seasonality and Trend Using Polynomial and Trigonometric Functions

It is desirable, in many applications of time series models, to estimate both the trend and seasonal components in the series. This can be done quite effectively by expressing the series y_t as a function of polynomials in t and a combination of sin, cos and trigonometric functions. Therefore

$$y_t = Q(t) + F(t) + e_t$$

where

221

$$Q(t) = \sum_{j=0}^{p} \beta_j t^j$$

models the trend component and

$$F(t) = \sum_{j=1}^{q} \left[a_j \sin(2\pi j t / L) + b_j \cos(2\pi j t / L) \right]$$

models the seasonal components, and L is the number of seasons in a year. Thus $L = 4$ for quarterly data, and $L = 12$ for monthly data. We may fit one of the following models:

(1)   p = q = 1

$$y_t = \hat{\beta}_0 + \hat{\beta}_1 t + a_1 \sin(2\pi t/L) + b_1 \cos(2\pi t/L)$$

or

(2)   p = 1, q = 2

$$y_t = \beta_0 + \beta_1 t + a_1 \sin(2\pi t / L) + b_1 \cos(2\pi t / L) + a_2 \sin(4\pi t / L) + b_2 \cos(4\pi t / L)$$

which may be suitable for modelling additive seasonal variation. Multiplicative time series may be modelled by extending either model (1) or model (2):

For model (1), a time series with multiplicative seasonal variation becomes

$$y_t'' = y_t + c_1 t \sin(2\pi t/L) + c_2 t \cos(2\pi t/L)$$

whereas model (2) becomes

$$y_t''' = y_t'' + d_1 t \sin(4\pi t/L) + d_2 t \cos(4\pi t/L)$$

Example 5.5

The condemnation rate series of example 5.4 showed significant seasonal effect. Here we show how to model both seasonality and trend in the series. Several models have been fitted using SAS Proc Reg. After removing several outliers, the best model is

$$\log(\text{rate}) = y_t$$

$$= \beta_0 + \beta_1 t + \beta_2 t^2 + \beta_3 t^3 + a_1 \sin(2\pi t / 12) + b_1 \cos(2\pi t / 12)$$

The estimated coefficients are

$$\hat{\beta}_0 = 6.197 \ (.096), \qquad \hat{\beta}_1 = .076 \ (.0097), \qquad \hat{\beta}_2 = -.002 \ (.0003)$$

$$\hat{\beta}_3 = 15 \times 10^{-6} \ (2 \times 10^{-6}), \qquad \hat{a}_1 = .195 \ (.0330), \qquad \hat{b}_1 = .206 \ (.0325)$$

$$R^2 = 0.84$$

and the root mean square error $= .202$.

Note the Proc Reg does not account for the correlation in the series. To account for such correlation, we used Proc Genmod. The GEE approach gave similar coefficient estimates with empirical standard errors that are robust against misspecification of the correlation structure (we assumed AR(1)).

Input year month $ t rate;
```
        sin = sin (2*22*t/(7*12));
        cos = cos (2*22*t/(7*12));
        t2 = t**2;
        t3 = t*t2;
        cards;
```

```
proc genmod data = season;
        class year;
                model y = t t2 t3 sin cos/
                        dist = n
                        link = id
                        dscale;
                repeated subject = year / type = AR(1);
run;
```

$$\hat{\beta}_0 = 6.207 \ (.079), \qquad \hat{\beta}_1 = .072 \ (.0082), \qquad \hat{\beta}_2 = -.002 \ (.0003)$$

$$\hat{\beta}_3 = 15 \times 10^{-6} \ (2 \times 10^{-6}), \qquad \hat{a}_1 = .179 \ (.0447), \qquad \hat{b}_1 = .200 \ (.0396)$$

The scale parameter $= 0.2022$.

Figure 5.7 shows the plot of the series (smooth curve) and the fitted series using the above model.

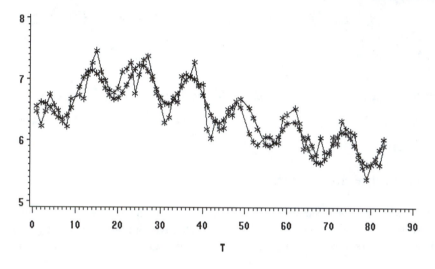

Figure 5.7   Plot of the rate series with predicted series from the regression with polynomial and trigonometric components.

II. FUNDAMENTAL CONCEPTS IN THE ANALYSIS OF TIME SERIES

In order to establish a proper understanding of time series models, we introduce some of the necessary fundamental concepts. Such concepts include a simple introduction to stochastic processes, autocorrelation, and partial autocorrelation functions.

A. STOCHASTIC PROCESSES

As before, y_t denotes an observation made at time t. It is assumed that for each time point t, y_t is a random variable and hence its behavior can be described by some probability distribution. We need to emphasize an important feature of time series models which is that observations made at adjacent time points are statistically correlated. Our main objective is to investigate the nature of this correlation. Therefore for two time points t and S, the joint behavior of (y_t, y_s) is determined from their bivariate distribution. This is generalized to the collection of observations $(y_1, y_2,...y_n)$ where their behavior is described by their multivariate joint distribution.

A stochastic process is a sequence of random variables $\{...y_{-2}, y_{-1}, y_0, y_1, y_2,...\}$. We shall denote this sequence by $\{y_t : t = 0, \pm 1, \pm 2,...\}$. For a given real-valued process we define the mean function of the process:

224

$$\mu_t = E(y_t)$$

the variance function of the process:

$$\sigma_t^2 = E(y_t - \mu_t)^2$$

the covariance function between y_t and y_s:

$$\gamma(t,s) = E[(y_t - \mu_t)(y_s - \mu_s)]$$

and the correlation function between y_t and y_s:

$$\rho(t,s) = \frac{\gamma(t,s)}{\sqrt{\sigma_t^2 \sigma_s^2}} = \frac{\gamma(t,s)}{\sqrt{\gamma(t,t)\gamma(s,s)}}$$

From this definition it is easily verified that

$$\rho(t,t) = 1$$

$$\rho(t,s) = \rho(s,t)$$

$$|\rho(t,s)| \leq 1$$

Values of $\rho(t,s)$ near ± 1 indicate strong dependence, whereas values near zero indicate weak linear dependence.

B. STATIONARY SERIES

The notion of stationarity is quite important in order to make statistical inferences about the structure of the time series. The fundamental idea of stationarity is that the probability distribution of the process does not change with time. Here we introduce two types of stationarity; the first is 'strict' or 'strong' stationarity and the other is 'weak' stationarity.

The stochastic process y_t is said to be strongly stationary if the joint distribution of $y_{t_1} \ldots y_{t_n}$ is the same as the joint distribution of $y_{t_1-k}, \ldots y_{t_n-k}$ for all the choices of the points $t_1 \ldots t_n$ and all the time lags k. To illustrate this concept we examine the two cases n=1 and n=2. For n=1, the stochastic process y_t is strongly stationary if the distribution of y_t is the same as that of y_{t-k}, for any k. This implies

225

$$E(y_t) = E(y_{t-k})$$

and

$$Var(y_t) = Var(y_{t-k})$$

are constant or independent of t. For n=2 the process is strongly stationary if the bivariate distribution of (y_t, y_s) is the same as the bivariate distribution of (y_{t-k}, y_{s-k}), from which we have

$$\gamma(t,s) = Cov(y_t y_s) = Cov(y_{t-k} y_{s-k})$$

Setting k=s, we obtain

$$\gamma(t,s) = Cov(y_{t-s} y_0)$$

$$= \gamma(0, |t-s|) ,$$

hence

$$\gamma\ (t, t-k) = \gamma_k ,$$

and

$$\rho(t, t-k) = \rho_k$$

A process is said to be weakly stationary if

(1) $\mu_t = \mu$ for all t
(2) $\gamma(t, t-k) = \gamma(0, k)$ for all t and k

All the series that will be considered in this chapter are stationary unless otherwise specified.

C. THE AUTOCOVARIANCE AND AUTOCORRELATION FUNCTIONS

For a stationary time series $\{y_t\}$, we have already mentioned that $E(y_t) = \mu$, and Var (y_t) = $E(y_t - \mu)^2$ (which are constant) and Cov (y_t, y_s) is a function of the time difference, $|t-s|$. Hence we can write

$$Cov(y_t y_{t+k}) = E[(y_t - \mu)(y_{t+k} - \mu)] = \gamma_k$$

226

and

$$\rho_k = Corr(y_t, y_{t+k}) = \frac{Cov(y_t, y_{t+k})}{\sqrt{Var(y_t)Var(y_{t+k})}} = \frac{\gamma_k}{\gamma_0}$$

The functions γ_k and ρ_k are called the autocovariance and autocorrelation functions (ACF) respectively. Since the values of μ, γ_k and ρ_k are unknown, the moment estimators of these parameters are as follows:

1. $\bar{y} = \dfrac{1}{n} \sum\limits_{i=1}^{n} y_i$ is the sample mean estimator of μ. It is unbiased and has variance given

by

$$Var(\bar{y}) = \frac{1}{n^2} \sum_{t=1}^{n} \sum_{s=1}^{n} Cov(y_t, y_s)$$

From the strong stationarity assumption

$$Cov(y_t, y_s) = \gamma(t-s).$$

Hence, letting k=t-s

$$Var(\bar{y}) = \frac{1}{n^2} \sum_{t=1}^{n} \sum_{s=1}^{n} \gamma(t-s) = \frac{\gamma_0}{n} \sum_{k=-(n-1)}^{n-1} \left(1 - \frac{|k|}{n}\right) \rho_k$$

$$= \frac{\gamma_0}{n} \left[1 + 2 \sum_{k=1}^{n-1} \left(1 - \frac{k}{n}\right) \rho_k\right].$$

When $\rho_k = 0$ for $k = 2,3,... n-1$ then, for large n

$$Var(\bar{y}) \doteq \frac{\gamma_0}{n}\left[1 + 2\left(\frac{n-1}{n}\right) \rho_1\right] \doteq \frac{\gamma_0}{n} \left[1 + 2\rho_1\right].$$

2. $\hat{\gamma}_k = \dfrac{1}{n} \sum\limits_{t=1}^{n-k} (y_t - \bar{y})(y_{t+k} - \bar{y})$

is the moment estimate of the autocovariance function.

A natural moment estimator for the autocorrelation function is defined as

$$\hat{\rho}_k = \frac{\hat{\gamma}_k}{\hat{\gamma}_0} = \frac{\sum_{t=1}^{n-k} \left(y_t - \bar{y}\right)\left(y_{t+k} - \bar{y}\right)}{\sum_{t=1}^{n} \left(y_t - \bar{y}\right)^2} \qquad k=0,1,2,\ldots \qquad (5.13)$$

A plot of $\hat{\rho}_k$ versus k is sometimes called a sample correlogram. Note that $\hat{\rho}_k = \hat{\rho}_{-k}$, which means that the sample ACF is symmetric around k=0.

For a stationary Gaussian process, Bartlett (1946) showed that for k>0 and k+j>0,

$$Cov\left(\hat{\rho}_k, \hat{\rho}_{k+j}\right) \simeq \frac{1}{n} \sum_{t=-\infty}^{\infty} (\rho_t \rho_{t+j} + \rho_{t+k+j}\rho_{t-k}$$

$$(5.14)$$

$$- 2\rho_k \rho_t \rho_{t-k-j} - 2\rho_{k+j}\rho_t\rho_{t-k} + 2\rho_k\rho_{k+j}\rho_t^2)$$

For large n, $\hat{\rho}_k$ is approximately normally distributed with mean ρ_k and variance

$$Var\left(\hat{\rho}_k\right) \simeq \frac{1}{n} \sum_{t=-\infty}^{\infty} \left(\rho_t^2 + \rho_{t+k}\rho_{t-k} - 4\rho_k\rho_t\rho_{t-k} + 2\rho_k^2\rho_t^2\right)$$

For processes with $\rho_k=0$ for k>ℓ, Bartlett's approximation becomes

$$Var\left(\hat{\rho}_k\right) \simeq \frac{1}{n} \left(1 + 2\rho_1^2 + 2\rho_2^2 + \ldots 2\rho_\ell^2\right) \ . \qquad (5.15)$$

In practice $\rho_i(i=1,2,\ldots\ell)$ are unknown and are replaced by their sample estimates $\hat{\rho}_i$; the large sample variance of $\hat{\rho}_k$ is approximated on replacing ρ_i by $\hat{\rho}_i$ in (5.15).

Example 5.6

Using data on average milk fat yields over a period of 38 months, we first plot the series over time (Figure 5.8), and then, using the time series programs in SAS (ETS) we can calculate the correlations for up to 4 lags. For this particular example PROC AUTOREG was used to generate the values of ρ_k for k=0,1,..4. Note that from the time plot it appears that the series is not stationary.

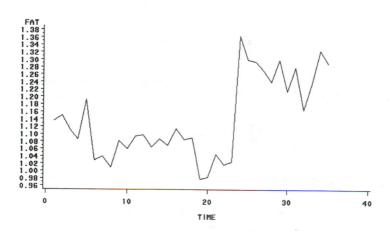

Figure 5.8 Time series plot of average milk fat yield over a period of 38 months.

The estimates of the autocorrelation and the covariance for the 5 lag periods are as follows:

Lag	Covariance	Correlation
0	0.007673	1.0000
1	0.004139	0.5394
2	0.002742	0.3573
3	0.001464	0.1908
4	0.000293	0.0382

III. MODELS FOR STATIONARY TIME SERIES

In this section we consider models based on an observation made by Yule (1921, 1927) that time series in which successive values are autocorrelated can be modelled as a linear combination (or linear filter) of a sequence of uncorrelated random variables. Suppose that $\{a_t a; t=0, \pm1, \pm2,...\}$ are a sequence of identically distributed uncorrelated random variables with $E(a_t)=0$ and $Var(a_t)=\sigma^2$, and $Cov(a_t, a_{t-k})=0$ for all $k \neq 0$. Such a sequence is commonly known as a 'white noise'. With this definition of white noise, we introduce the linear filter representation of the process y_t.

229

A general linear process y_t is one that can be presented as

$$y_t = a_t + \psi_1 a_{t-1} + \psi_2 a_{t-2} + \ldots$$

$$= \sum_{j=0}^{\infty} \psi_j \, a_{t-j} \, , \quad \psi_0 = 1$$

For the infinite series of the right-hand side of the above equation to be meaningful, it is assumed that

$$\sum_{i=1}^{\infty} \psi_j^2 \; < \; \infty \; .$$

A. AUTOREGRESSIVE PROCESSES

As their name implies, autoregressive processes are regressions on themselves. To be more specific, the pth order autoregressive process y_t satisfies,

$$y_t = \phi_1 y_{t-1} + \phi_2 y_{t-2} + \ldots \phi_p y_{t-p} + a_t \qquad (5.16)$$

In this model, the present value y_t is a linear combination of its p most recent values plus an "innovation" term a_t, which includes everything in the series at time t that is not explained by the past values. It is also assumed that a_t is independent of y_{t-1}, y_{t-2},....

Before we examine the general autoregressive process, we first consider the first-order autoregressive model which is denoted by AR(1).

1. The AR(1) Model

Let y_t be a stationary series such that

$$y_t = \phi y_{t-1} + a_t \; . \qquad (5.17)$$

Most text books write the above model as

$$y_t - \mu = \phi(y_{t-1} - \mu) + a_t$$

where μ is the mean of the series. However, we shall use (5.17) assuming that the mean has been subtracted from the series. The requirement $|\phi| < 1$ is a necessary and sufficient condition for stationarity.

From (5.17), $\text{Var}(y_t) = \phi^2\,\text{Var}(y_{t-1}) + \text{Var}(a_t)$ or $\gamma_0 = \phi^2\,\gamma_0 + \sigma_a^2$ from which

$$\gamma_0 = \frac{\sigma_a^2}{1-\phi^2} \quad . \tag{5.18}$$

Multiplying both sides of (5.17) by y_{t-k} and taking the expectation, the result is,

$$E(y_t y_{t-k}) = \phi E(y_{t-1} y_{t-k}) + E(y_{t-k} a_t) \quad .$$

By the stationarity of the series, and the independence of y_{t-1} and a_t,

$$\gamma_k = \phi\gamma_{k-1} \qquad k=1,2,\ldots$$

For k=1,

$$\gamma_1 = \phi\gamma_0 = \phi\,\frac{\sigma_a^2}{1-\phi^2} \quad ;$$

for k=2,

$$\gamma_2 = \phi\gamma_1 = \phi\left(\phi\frac{\sigma_a^2}{1-\phi^2}\right)$$

$$= \phi^2\,\frac{\sigma^2}{1-\phi^2} = \phi^2\gamma_0 \quad .$$

By mathematical induction one can show that

$$\gamma_k = \phi^k\gamma_0$$

or

$$\rho_k = \frac{\gamma_k}{\gamma_0} = \phi^k \quad . \tag{5.19}$$

Note that since $|\phi|<1$, the autocorrelation function is exponentially decreasing in k. For $0<\phi<1$, all ρ_k are positive. For $-1<\phi<0$, $\rho_1<0$ and the sign of successive autocorrelations alternate (positive if k is even and negative if k is odd).

Figures 5.9a and 5.9b are graphs of ρ for $\phi=0.8$, and $-.5$.

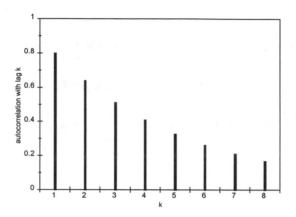

Figure 5.9a Autocorrelation plot for a ϕ of 0.8 and k=1,2,...8.

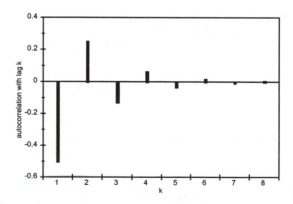

Figure 5.9b Autocorrelation plot for a ϕ of -0.5 and k=1,2,...8.

2.   AR(2) Model (Yule's Process)

The second order autoregressive process AR(2) is a stationary series y_t that is a linear combination of the two preceding observations and can be written as

$$y_t = \phi_1 y_{t-1} + \phi_2 y_{t-2} + a_t \; .$$

(5.20)

To ensure stationarity the coefficients ϕ_1 and ϕ_2 must satisfy

$$\phi_1 + \phi_2 < 1$$

$$\phi_2 - \phi_1 < 1$$

$$\text{and} \quad -1 < \phi_2 < 1$$

The above conditions are called the stationarity conditions for the AR(2) model.

To derive the autocorrelation function for the AR(2) we multiply both sides of (5.20) by y_{t-k}, $(k=1,2,...,)$ and take the expectations. Under the assumptions of independence of y_t and a_t and the stationarity of the series we have

$$E(y_t y_{t-k}) = \phi_1 E(y_{t-1} y_{t-k}) + \phi_2 E(y_{t-2} y_{t-k})$$

$$+ E(a_t y_{t-k})$$

from which

$$\gamma_k = \phi_1 \gamma_{k-1} + \phi_2 \gamma_{k-2}$$

(5.21)

and dividing by γ_0 we get

$$\rho_k = \phi_1 \rho_{k-1} + \phi_2 \rho_{k-2}$$

(5.22)

Equation 5.22 is called Yule-Walker equation. For k=1

$$\rho_1 = \phi_1 \rho_0 + \phi_2 \rho_{-1}$$

Since $\rho_0 = 1$ and $\rho_{-1} = \rho_1$ we have

$$\rho_1 = \frac{\phi_1}{1 - \phi_2} \; .$$

For k=2

$$\rho_2 = \phi_1 \rho_1 + \phi_2$$

or

$$\rho_2 = \phi_2 + \frac{\phi_1^2}{1-\phi_2} \quad .$$

Note also that the variance of the AR(2) process can be written in terms of the model parameters. In fact, from (5.20) we have

$$Var(y_t) = \phi_1^2 Var(y_{t-1}) + \phi_2^2 Var(y_{t-2})$$

$$+ 2\phi_2 \phi_2 Cov(y_{t-1}, y_{t-2}) + \sigma_a^2$$

or

$$\gamma_0 = \phi_1^2 \gamma_0 + \phi_2^2 \gamma_0 + 2\phi_1 \phi_2 \gamma_1 + \sigma_a^2 \quad . \tag{5.23}$$

Setting k=1 in equation (5.21) we have

$$\gamma_1 = \phi_1 \gamma_0 + \phi_2 \gamma_{-1}$$

$$= \phi_1 \gamma_0 + \phi_2 \gamma_1$$

which gives

$$\gamma_1 = \phi_1 \frac{\gamma_0}{1-\phi_2} \tag{5.24}$$

Substituting in (5.23),

$$\gamma_0 = (\phi_1^2 + \phi_2^2)\gamma_0 + 2\phi_1^2 \phi_2 \frac{\gamma_0}{1-\phi_2} + \sigma_a^2$$

and hence

$$\gamma_0 = \frac{\sigma_a^2(1-\phi_2)}{(1-\phi_2)(1-\phi_1^2-\phi_2^2) - 2\phi_1^2 \phi_2} \tag{5.25}$$

It should be noted that, for $-1 < \phi_2 < 0$ the AR(2) process tends to exhibit sinusoidal behaviour, regardless of the value of ϕ_1. When $0 < \phi_2 < 1$, the behavior of the process will depend on the sign of ϕ_1. For $\phi_1 < 0$, the AR(2) process tends to oscillate and the series shows ups and downs.

3. Moving Average Processes

Another type of stochastic model that belongs to the class of linear filter models is called a moving average process. This is given as:

$$y_t = a_t - \theta_1 a_{t-1} - \theta_2 a_{t-2} \ldots - \theta_q a_{t-q}$$

This series is called a "Moving Average" of order q and is denoted by MA(q).

4. First Order Moving Average Process MA(1)

Here we have

$$y_t = a_t - \theta a_{t-1} \qquad (5.26)$$

$$E(y_t) = 0$$

$$\gamma_0 = Var(y_t) = \sigma_a^2 + \theta^2 \sigma_a^2 = \sigma_a^2 (1 + \theta^2) \quad .$$

Moreover,

$$Cov(y_t, y_{t-1}) = E(y_t y_{t-1})$$

$$= E[(a_t - \theta a_{t-1})(a_{t-1} - \theta a_{t-2})]$$

$$= E[a_t a_{t-1}] - \theta \left[E[a_{t-1}^2] + E[a_t a_{t-2}] \right]$$

$$+ \theta^2 E[a_{t-1} a_{t-2}] \quad .$$

Since $a_1, a_2 \ldots$ are independent with $E(a_t) = 0$ for all t, then

$$\gamma_1 = Cov(y_t, y_{t-1}) = -\theta \sigma_a^2$$

$$Cov(y_t, y_{t-k}) = 0 \quad . \qquad k = 2, 3, \ldots$$

Furthermore, the autocorrelation function is

$$\rho_1 = \frac{\gamma_1}{\gamma_0} = -\frac{\theta}{1+\theta^2} \quad . \tag{5.27}$$

$(\rho_k=0$ for k=2,3,...)

Note that if θ is replaced by $1/\theta$ in (5.27), we get exactly the same autocorrelation function. This lack of uniqueness of MA(1) models must be rectified before we estimate the model parameters.

Rewriting equation (5.26) as

$$a_t = y_t + \theta a_{t-1}$$

$$= y_t + \theta(y_{t-1} + \theta a_{t-2})$$

$$= y_t + \theta y_{t-1} + \theta^2 a_{t-2}$$

and continuing this substitution,

$$a_t = y_t + \theta y_{t-1} + \theta^2 y_{t-2} + ... \tag{5.28}$$

or $\qquad y_t = -(\theta y_{t-1} + \theta^2 y_{t-2} + ...) + a_t .$

If $|\theta|<1$, we see that the MA(1) model can be inverted into an infinite-order AR process. It can be shown (see Box & Jenkins 1970) that there is only one invertible MA(1) model with the given autocorrelation function ρ_1.

5. The Second Order Moving Average Process MA(2)

An MA(2) process is defined by:

$$y_t = a_t - \theta_1 a_{t-1} - \theta_2 a_{t-2} \quad . \tag{5.29}$$

The autocovariance functions are given by

$$\gamma_1 = Cov(y_t, y_{t-1}) = E\left[(a_t - \theta_1 a_{t-1} - \theta_2 a_{t-2})(a_{t-1} - \theta_1 a_{t-2} - \theta_2 a_{t-3})\right]$$

$$= -\theta_1 \sigma_a^2 + \theta_1 \theta_2 \sigma_a^2$$

$$= (-\theta_1 + \theta_1 \theta_2)\sigma_a^2$$

236

$$\gamma_2 = Cov(t, y_{t-2}) = E\left[(a_t - \theta_1 a_{t-1} - \theta_2 a_{t-2})(a_{t-2} - \theta_1 a_{t-3} - \theta_2 a_{t-4})\right]$$

$$= -\theta_2 \sigma_a^2$$

and,

$$\gamma_0 = Var(y_t) = \sigma_a^2 + \theta_1^2 \sigma_a^2 + \theta_2^2 \sigma_a^2$$

$$= (1 + \theta_1^2 + \theta_2^2)\sigma_a^2$$

Therefore for an MA(2) process

$$\rho_1 = \frac{\gamma_1}{\gamma_0} = \frac{(-\theta_1 + \theta_1 \theta_2)}{1 + \theta_1^2 + \theta_2^2} \qquad (5.30)$$

$$\rho_2 = \frac{\gamma_2}{\gamma_0} = -\frac{\theta_2}{1 + \theta_1^2 + \theta_2^2} \qquad (5.31)$$

$$\rho_k = 0 \qquad , k=3,4,...$$

6. The Mixed Autoregressive Moving Average Processes

In modelling time series we are interested in constructing a parsimonious model. One type of such a model is obtained from mixing an AR(p) with an MA(q). The general form of this is given by:

$$y_t = (\phi_1 y_{t-1} + \phi_2 y_{t-2} + ... \phi_p y_{t-p}) + (a_t - \theta_1 a_{t-1} - \theta_2 a_{t-2}$$
$$- ... - \theta_q a_{t-q}) \qquad (5.32)$$

The process y_t defined in (5.32) is called the mixed autoregressive moving average process of orders p and q, or ARMA (p,q).

An important special case of the ARMA(p,q) is ARMA(1,1) which can be obtained from (5.32) for p=q=1. Therefore an ARMA(1,1) is

237

$$y_t = \phi y_{t-1} + a_t - \theta a_{t-1} \qquad\qquad (5.33)$$

For stationarity we assume that $|\phi|<1$ and for invertibility, we require that $|\theta|<1$. When $\phi=0$, (5.33) is reduced to an MA(1) process, and when $\theta=0$, it is reduced to an AR(1) process. Thus the AR(1) and MA(1) may be regarded as special processes of the ARMA(1,1).

To obtain the autocovariance for the ARMA(1,1), we multiply both sides of (5.33) by y_{t-k} and take the expectations

$$E(y_t y_{t-k}) = \phi E(y_{t-1} y_{t-k}) + E(a_t y_{t-k})$$

$$-\theta E(a_{t-1} y_{t-k})$$

from which

$$\gamma_k = \phi \gamma_{k-1} + E(a_t y_{t-k}) - \theta E(a_{t-1} y_{t-k}) \quad . \qquad\qquad (5.34)$$

For k=0

$$\gamma_0 = \phi \gamma_1 + E(a_t y_t) - \theta E(a_{t-1} y_t) , \qquad\qquad (5.35)$$

and

$$E(a_t y_t) = \sigma_a^2 .$$

Noting that

$$E(a_{t-1} y_t) = \phi E(a_{t-1} y_{t-1}) + E(a_t a_{t-1}) - \theta E(a_{t-1}^2)$$

$$= \phi \sigma_a^2 - \theta \sigma_a^2 = (\phi - \theta) \sigma_a^2$$

and substituting in (5.35) we see that,

$$\gamma_0 = \phi \gamma_1 + \sigma_a^2 - \theta(\phi - \theta)\sigma_a^2 \qquad\qquad (5.36)$$

and from (5.34) when k=1,

$$\gamma_1 = \phi \gamma_0 + E(a_t y_{t-1}) - \theta E(a_{t-1} y_{t-1})$$

$$= \phi \gamma_0 + E(a_t y_{t-1}) - \theta \sigma_a^2 . \qquad\qquad (5.37)$$

But,

$$E(a_t y_t) = \phi E(a_t y_{t-1}) + E(a_t^2) - \theta E(a_t a_{t-1})$$

or

$$\sigma_a^2 = \phi E(a_t y_{t-1}) + \sigma_a^2$$

so,

$$E(a_t y_{t-1}) = 0 \tag{5.38}$$

Substituting (5.38) in (5.37),

$$\gamma_1 = \phi \gamma_0 - \theta \sigma_a^2 \tag{5.39}$$

and using this in (5.36) we have

$$\gamma_0 = \phi^2 \gamma_0 - \phi \theta \sigma_a^2 + \sigma_a^2 - \theta \phi \sigma_a^2 + \theta^2 \sigma_a^2$$

from which

$$\gamma_0 = \frac{(1 + \theta^2 - 2\phi\theta)}{1 - \phi^2} \sigma_a^2 \quad.$$

Thus

$$\gamma_1 = \phi \left[\frac{1 + \theta^2 - 2\phi\theta}{1 - \phi^2} \right] \sigma_a^2 - \theta \sigma_a^2$$

$$= \frac{(\phi - \theta)(1 - \phi\theta)}{1 - \phi^2} \sigma_a^2 \quad.$$

From (5.34), we have

$$\gamma_k = \phi \gamma_{k-1} \quad.$$

Hence, the ARMA(1,1) has the following autocorrelation function

$$\rho_k = \begin{cases} 1 & k=0 \\ \dfrac{(\phi-\theta)(1-\phi\theta)}{1+\theta^2-2\phi\theta} & k=1 \\ \phi\,\rho_{k-1} & k\geq 2 \end{cases} \qquad (5.40)$$

IV. MODELS FOR NONSTATIONARY TIME SERIES

The time series processes we have introduced (AR, MA, and ARMA) thus far are all stationary processes. This implies that the mean, variance and autocovariances of the process are invariant under time translations. Thus the mean and the variance are constants and the autocovariance depends on the time lag.

Many applied time series, however, are not stationary. For example most economic time series exhibit changing levels and or variances over time. Examples of nonstationary time series plots can be found in Section I (Figure 5.1) and Section II (Figure 5.5) of this chapter.

A. NONSTATIONARITY IN THE MEAN

A time series which is nonstationary in the mean could pose some serious problems for estimation of the time dependent mean function. However, there are two classes of models that have been proven to model nonstationary in the mean time series. These two models are

1. Deterministic Trend Models

For a time series with changing mean, that is, when the mean is time dependent, a standard regression model may be used. For example, one may model the mean as

$$E(y_t) = \alpha_0 + \alpha_1 t$$

and the corresponding linear regression trend model would be given as

$$y_t = \alpha_0 + \alpha_1 t + a_t$$

with a_t being a zero mean white noise series. More generally, we may describe the deterministic trend as

$$y_t = \alpha_0 + \alpha_1 t + \ldots + \alpha_k t^k + a_t . \qquad (5.42)$$

240

Alternatively, we may model the trend by a sine-cosine curve so that

$$y_t = c_0 + c\ Cos(wt+\theta) + \alpha_t$$

$$= c_0 + \alpha\ Cos\ wt + \beta\ \sin\ wt + a_t$$

where

$$\alpha = c\ Cos\ \theta\quad,\quad \beta = -c\ \sin\ \theta$$

$$c = \sqrt{\alpha^2 + \beta^2}$$

and $\qquad \theta = \arctan(-\beta\ /\ \alpha)$

The parameters c, w, and θ are called the amplitude, the frequency, and the phase of the curve respectively.

2. Stochastic Trend Models

Nonstationary time series that are described by a stochastic trend are usually referred to as "homogeneous nonstationarity". Its main characteristic is that, apart from local trend, one part of the series behaves like the other.

Titner (1940), Yaglom (1951), and Box and Jenkins (1976) claim that a homogeneous nonstationarity sequence can be transformed into a stationary sequence by taking the successive differences of the series. To illustrate this idea let us consider the following example.

Example 5.6

We have shown that the AR(1) model

$$y_t = \phi y_{t-1} + a_t$$

is stationary if and only if $|\phi| < 1$. The question is what happens when $|\phi| \geq 1$. For example if we let $\phi = 2$ so that

$$y_t = 2\ y_{t-1} + a_t$$

where a_t is a white noise with $E(a_t) = 0$, $Var(a_t) = \sigma_a^2$ and y_t and a_t are independent, and since

$$y_t = 2\left[2\ y_{t-1} + a_{t-1}\right] + a_t$$

$$= 4\ y_{t-2} + 2\ a_{t-1} + a_t$$

$$= 4\left[2\ y_{t-3} + a_{t-3}\right] + 2\ a_{t-1} + a_t$$

$$= 8\ y_{t-3} + 4\ a_{t-2} + 2\ a_{t-1} + a_t$$

then if we continue in this manner we have

$$y_t = 2^t y_0 + 2^{t-1} a_1 + 2^{t-2} a_2 + \ldots 2\ a_{t-1} + a_t$$

$$= 2^t y_0 + \sum_{j=0}^{t-1} 2^j\ a_{t-j}\ . \tag{5.43}$$

Clearly, since $E(y_t) = 2^t y_0$, which depends on t, then the series is nonstationary. The variance of y_t is,

$$Var(y_t) = \sum_{j=0}^{t-1} 2^{2j}\ Var(a_{t-j})$$

$$= \sigma_a^2 \sum_{j=0}^{t-1} 4^j = \sigma_a^2 \left(\frac{4^t - 1}{3}\right)\ . \tag{5.44}$$

Multiplying both sides of (5.43) by y_{t-k} and taking expectation we get:

$$E(y_t\ y_{t-k}) = 2^{2t-k}\ y_0^2 + 2^k \left(\frac{4^{t-k} - 1}{3}\right)\ \sigma_a^2\ .$$

Hence

$$Cov\left(y_t, y_{t-k}\right) = 2^{2t-k}\ y_0^2 + 2^k \left(\frac{4^{t-k} - 1}{3}\right)\ \sigma_a^2 - (2^t y_0)(2^{t-k} y_0)$$

$$= \frac{2^k}{3}\left(4^{t-k} - 1\right)\ \sigma_a^2\ , \tag{5.45}$$

which depends on t.

B. DIFFERENCING

To illustrate how successive differencing can be used to achieve stationarity of a nonstationary series let us consider the following simple model

$$y_t = y_{t-1} + a_t \qquad (5.46)$$

which is known as the random walk model. Note that the model (5.46) is an AR(1) with $\phi \to 1$. Since the ACF of the AR(1) is $\rho_k = \phi^k$, then as $\phi \to 1$, the random walk model is described by large nondying spikes in the sample ACF.

Next, consider the following modification of the model (5.46) so that

$$y_t = y_{t-1} + \beta + \alpha_t \qquad (5.47)$$

The conditional expectation of this series given y_{t-1}, y_{t-2},... is

$$\mu_t = y_{t-1} + \beta$$

Apart from the constant slope β, the conditional expectation μ_t depends on the previous observation y_{t-1}. Since y_{t-1} is subject to random variation, the trend changes stochastically. It is clear that the first difference

$$w_t = y_t - y_{t-1} = \beta + a_t$$

would produce a stationary series.

Another example is the AR(2) series

$$y_t = \phi_1 y_{t-1} + \phi_2 y_{t-2} + a_t$$

If $\phi_1 = 2$ and $\phi_2 = -1$, henceforth violating the stationarity assumptions of the AR(2) model, we have

$$y_t = 2 y_{t-1} - y_{t-2} + a_t .$$

The first difference is

$$W_t = \nabla y_t = y_t - y_{t-1} = y_{t-1} - y_{t-2} + a_t$$

243

or

$$w_t = w_{t-1} + a_t \; ;$$

this means that the difference $(w_t - w_{t-1})$, which is the second difference of the original series, is stationary.

Referring to the data of example 5.1 (SCC data) we try to stationarize the mean through differencing by 1 lag. This is actually quite effective as can be seen in the resulting plot of the differenced data (Figure 5.10).

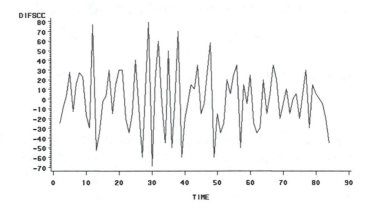

Figure 5.10 Plot of the SCC data - differencing by one lag to stabilize the mean.

C. ARIMA MODELS

A series y_t is said to follow an autoregressive integrated moving average (ARIMA) model of order d if the dth difference denoted by $\nabla^d y_t$ is a stationary ARMA model. The notation used for this model is ARMA (p,d,q), where p is the order of the autoregressive component, q is the order of the moving average component, and d is the number of differences performed to produce a stationary series. Fortunately, for practical reasons we can take d=1 or 2.

An ARIMA (p,1,q) process can be written as

$$w_t = \phi_1 w_{t-1} + \phi_2 w_{t-2} + ... \phi_p w_{t-p} + a_t - \theta_1 a_{t-1} ... - \theta_q a_{t-q} \qquad (5.48)$$

244

where

$$w_t = y_t - y_{t-1}$$

As an example, the ARIMA (0,1,1) or IMA(1,1) is given by

$$w_t = a_t - \theta_1 a_{t-1}$$

(5.49)

or $\quad y_t = y_{t-1} + a_t - \theta_1 a_{t-1}$

which means that the first difference (d=1) would produce a stationary MA(1) series as long as $|\theta_1| < 1$.

D. NONSTATIONARITY IN THE VARIANCE

A process that is stationary in the mean is not necessarily stationary in the variance and the autocovariance. However, a process that is nonstationary in the mean will also be nonstationary in the variance and the autocovariance. Clearly, the ARIMA model is nonstationary in the mean; before we show that it is nonstationary in the covariance, we demonstrate that the complete properties of the process are determined by a few parameters. To show this, suppose that the model IMA(1,1) which satisfies,

$$\nabla y_t = a_t - \theta a_{t-1}$$

or $\quad y_t = y_{t-1} + a_t - \theta a_{t-1}$

is fitted to a series of m observations. By successive substitutions we can write

$$y_t = y_m + a_t + (1-\theta)a_{t-1} + ... + (1-\theta)a_{m+1} - \theta a_m$$

Similarly, for t-k>m

$$y_{t-k} = y_m + a_{t-k} + (1-\theta)a_{t-k-1} + ... + (1-\theta)a_{m+1} - \theta a_m$$

Hence, with respect to the time origin m

$$Var(y_t) = \left[1 + (t-m-1)(1-\theta)^2\right]\sigma_a^2$$

(5.50)

$$Var(y_{t-k}) = \left[1 + (t-k-m-1)(1-\theta)^2\right]\sigma_a^2$$

(5.51)

245

$$Cov(y_t, y_{t-k}) = \left[(1-\theta)+(t-k-m-1)(1-\theta)^2\right]\sigma_a^2 \qquad (5.52)$$

$$Corr(y_t, y_{t-k}) = \frac{(1-\theta)+(t-k-m-1)(1-\theta)^2}{\sqrt{\left[1+(t-m-1)(1-\theta)^2\right]\left[1+(t-k-m-1)(1-\theta)^2\right]}} \qquad (5.53)$$

From equations (5.50) to (5.53) one can see that Var(y_t) increases with t. Moreover, the autocovariance and autocorrelation functions depend on the lag k, the time origin t and the reference point m. Finally, Corr (y_t, y_{t-k}) $\approx$ 1 for large m and moderate k, which implies that the autocorrelation function vanishes slowly as k increases.

Remark: "Variance Stabilizing Transformation"

Not all nonstationary series can be stationarized through differencing. In fact there are many time series that are stationary in the mean but nonstationary in the variance. To overcome this problem, transformations such as the logarithm or square root, perhaps followed by differencing are useful methods in certain situations. It is very common to encounter series where increased variance seems to be associated with increased levels (means) of the series. This relationship between the variance and the mean can be modelled as

$$Var(y_t) = C\ V(\mu_t)$$

for some constant C and function V($\cdot$). To find the approximate transformation F that stabilizes the variance of y_t, we approximate the function F by the first term of a Taylor's series expansion about μ_t. That is

$$F(y_t) \approx F(\mu_t)+(y_t-\mu_t)F'(\mu_t)$$

where F'(μ_t) is the first partial derivative of T with respect to y_t, evaluated at $y_t=\mu_t$. Now,

$$Var(F(y_t)) = Var(y_t)\ (F'(\mu_t))^2$$

$$= C\ V(\mu_t)(F'(\mu_t))^2 . \qquad (5.54)$$

Thus in order that Var (F(y_t)) be constant we must have

$$\left[F'(\mu_t)\right]^2 = \frac{1}{V(y_t)}$$

or

$$F'(\mu_t) = \frac{1}{\sqrt{V(y_t)}}$$

or

$$F(\mu_t) = \int \frac{1}{\sqrt{V(\mu_t)}} \, d\mu_t \, .$$ (5.55)

As an example, if $V(\mu_t) = \mu_t^2$, then

$$F(\mu_t) = \ell n(\mu_t)$$

$$F(\mu_t) = 2\sqrt{\mu_t} \, .$$

Hence the logarithmic transformation $\ell n(y_t)$ will give a constant variance. If $V(\mu_t) = \mu_t$, then $\sqrt{y_t}$ will give a constant variance. If $V(\mu_t) = \mu_t^4$, then $T(\mu_t) = -\dfrac{1}{\mu_t}$ which means that the reciprocal transformation $1/y_t$ will have a constant variance.

A general family of transformations, known as the "Power Transformation" was introduced by Box and Cox (1964). For a given value of the parameter λ, the transformation is defined by

$$T(y_k) = \left\{ \begin{array}{ll} \dfrac{y_t^{\lambda} - 1}{\lambda} & \lambda \neq 0 \\ \ln(y_t) & \lambda = 0 \end{array} \right.$$ (5.56)

Note that $\lambda = 1/2$ produces a square root transformation useful with Poisson like data, and $\lambda = -1$ corresponds to a reciprocal transformation. For all practical purposes, a precise estimate of λ is not needed and a grid of values in the interval [-2,2] may suffice.

V. MODEL SPECIFICATION AND PARAMETER ESTIMATION

A. Specification

We have discussed some of the parametric models for stationary and nonstationary time series. To model time series data it is quite important to identify and build an appropriate model. The objectives of this section are: (i) how to select appropriate values of p,d, and q of an ARIMA model, for a given series, (ii) how to estimate the parameters of the specified model.

As we have already indicated, many of the characteristics of the time series can be studied in terms of the ACF ρ_k. Another function which is quite useful in identifying a time series model, is the partial autocorrelation function (PAC). This function will be discussed in this section. First we start by defining what is meant by model specification.

Model specification or model identification is the methodology followed by researchers in identifying the required transformations such as variance stabilizing transformations and/or differencing transformations, and the proper values of p and q.

The following steps are helpful in identifying a time series model:

(i)   Plot the data. Careful examination of the plot may reveal seasonality, trend, nonconstancy of variance, nonstationarity and other features of the series.

(ii)   For the series $y_1, y_2, \ldots y_n$, construct the sample ACF $\hat{\rho}_k$ (see 5.13). The large sample variance of the $\hat{\rho}_k$ is given by (5.15). For some of the models considered in this chapter the following approximate results were given by Cryer (1986).

1)

$$y_t = a_t \qquad \text{white noise model}$$

$$Var(\hat{\rho}_k) \approx \frac{1}{n}$$

2)

$$y_t = \phi \, y_{t-1} + a_t \qquad \text{(AR(1) model)}$$

$$\rho_k = \phi^k$$

and

$$Var(\hat{\rho}_k) \approx \frac{1}{n} \left[\frac{(1+\phi^2)(1-\phi^{2k})}{1-\phi^2} - 2k \, \phi^{2k} \right] \tag{5.57}$$

For k=1

$$Var(\hat{\rho}_1) \approx \frac{1-\phi^2}{n} \tag{5.58}$$

3) For the MA(q) process we have

$$Var(\hat{\rho}_k) \simeq \frac{1+2\sum_{i=1}^{q} \rho_i^2}{n} \qquad k>q$$

To test the hypothesis that the series is an MA(q) we compare

$$Z = \frac{\sqrt{n}\ \hat{\rho}_{q+1}}{\sqrt{1+2\sum_{i=1}^{q} \hat{\rho}_i^2}} \tag{5.59}$$

to the critical point of a standard normal distribution. The model MA(q) is rejected for large values of Z.

(iii)   Since for the MA(q) process the correlation function is zero for lags larger than q, the sample autocorrelation function may be a good indicator of the order of the process. However, the autocorrelation of an AR(p) process does not stay at zero after a certain number of lags. The partial autocorrelation function can be useful for determining the order p. This partial autocorrelation function (PAC) describes the correlation between y_t and y_{t+k} after removing the effect of the intervening variables $y_{t+1}, y_{t+2}, \ldots y_{t+k-1}$. It is defined as

$$\phi_{kk} = Corr\ (y_t y_{t+k} \mid y_{t+1}, \ldots y_{t+k-1}) \tag{5.60}$$

Consider a stationary process y_t and assume that its values have already been subtracted from the mean so that $E(y_t) = 0$. Suppose that y_{t+k} is regressed on k lagged variables $y_t, y_{t+1}, \ldots y_{t+k-1}$. That is

$$y_{t+k} = \phi_{k1}\ y_{t+k-1} + \phi_{k2}\ y_{t+k-2} + \ldots \phi_{kk}\ y_t + e_{t+k} \tag{5.61}$$

where ϕ_{kj} is the jth regression parameter and e_{t+k} is a normal error term not correlated with y_{t+k-i} for $i \geq 1$. Multiplying both sides of equation (5.61) by y_{t+k-i} and taking the expectation we obtain

$$\gamma_i = \phi_{k1}\ \gamma_{i-1} + \phi_{k2}\ \gamma_{i-2} + \ldots \phi_{kk}\ \gamma_{i-k}$$

and when both sides are divided by γ_0 we get

$$\rho_i = \phi_{k1}\ \rho_{i-1} + \phi_{k2}\ \rho_{i-2} + \ldots \phi_{kk}\ \rho_{i-k}\ .$$

249

For i=1,2,...k we have the following system of equations

$$\rho_1 = \phi_{k1}\rho_0 + \phi_{k2}\rho_1 + ... + \phi_{kk}\rho_{k-1}$$

$$\rho_2 = \phi_{k1}\rho_1 + \phi_{k2}\rho_2 + ... + \phi_{kk}\rho_{k-2}$$

$$.$$

$$(5.62)$$

$$.$$

$$\rho_k = \phi_{k1}\rho_{k-1} + \phi_{k2}\rho_{k-2} + ... + \phi_{kk}\rho_0$$

Levinson (1947) gave an efficient method for obtaining the solutions of equations (5.62). He showed that

$$\phi_{kk} = \frac{\rho_k - \sum_{j=1}^{k-1} \phi_{k-1,j}\, \rho_{k-j}}{1 - \sum_{j=1}^{k-1} \phi_{k-1,j}\, \rho_j} \qquad (5.63)$$

$$\phi_{kj} = \phi_{k-1,j} - \phi_{kk}\phi_{k-1,k-j}$$

$$j = 1,2,...k-1$$

In particular

$$\phi_{11} = \rho_1$$

$$\phi_{22} = \frac{\rho_2 - \rho_1^2}{1 - \rho_1^2}$$

$$\phi_{21} = \phi_{11} - \phi_{22}\phi_{11}$$

and $\qquad \phi_{33} = \dfrac{\rho_3 - \phi_{21}\rho_2 - \phi_{22}\rho_1}{1 - \phi_{21}\rho_1 - \phi_{22}\rho_2}$.

Quenouille (1949) has shown that, under the hypothesis that the correct model is AR(p), the estimated partial autocorrelation at lags larger than p are approximately normally distributed with mean zero and variance $1/n$. Thus $\pm 1.96/\sqrt{n}$ can be used as critical limits on ϕ_{kk} for $k > p$ to test the hypothesis of an AR(p) model.

B. Estimation

After identifying a tentative model, the next step is to estimate the parameters of this model based on the observed time series $y_1, y_2,..., y_n$. With respect to nonstationarity, since the dth difference of the observed series is assumed to be the general stationary ARMA(p,q) model, we need only concern ourselves with estimating the parameters of the model,

$$y_t = \phi_1 y_{t-1} + \phi_2 y_{t-2} + ... + \phi_p y_{t-p} + a_t - \theta a_{t-1} - ... - \theta_q a_{t-q}$$

where we assume that the mean μ is subtracted from the observations and stationarity has been achieved. Moreover $\{a_t\}$ are i.i.d. $N(0, \sigma_a^2)$ white noise.

We now discuss two of the widely used estimation procedures.

a. The Method of Moments

This method is the simplest technique used in estimating the parameters. It consists of equating sample moments to theoretical moments and solving the resulting equations to obtain estimates of the unknown parameters.

For example, in the AR(p) model (5.16) the mean $\mu = E(y_t)$ is estimated by,

$$\bar{y} = \frac{1}{n} \sum_{t=1}^{n} y_t .$$

To estimate $\phi_1, \phi_2, ... \phi_p$, we use the equation

$$\rho_k = \phi_1 \rho_{k-1} + \phi_2 \rho_{k-2} + ... \phi_p \rho_{k-p} \quad \text{for} \quad k > 1$$

to obtain the following system of equations

$$\rho_1 = \phi_1 + \phi_2 \rho_1 + \phi_3 \rho_2 + ... + \phi_p \rho_{p-1}$$

$$\rho_2 = \phi_1 \rho_1 + \phi_2 + \phi_3 \rho_1 + ... + \phi_p \rho_{p-2}$$

.

.

$$\rho_p = \phi_1 \rho_{p-1} + \phi_2 \rho_{p-2} + ... + \phi_p$$

Then replacing ρ_k by $\hat{\rho}_k$ we obtain the moment estimators $\phi_1, \phi_2, \ldots \phi_p$ by solving the above system of linear equations. That is

$$
\begin{vmatrix} \hat{\phi}_1 \\ \hat{\phi}_2 \\ \cdot \\ \cdot \\ \cdot \\ \hat{\phi}_p \end{vmatrix} = \begin{bmatrix} 1 & \hat{\rho}_1 & & \hat{\rho}_{p-1} \\ \hat{\rho}_1 & 1 & & \hat{\rho}_{p-2} \\ \cdot & \cdot & & \cdot \\ \cdot & \cdot & & \cdot \\ \cdot & \cdot & & \cdot \\ \hat{\rho}_{p-1} & \hat{\rho}_{p-2} & & 1 \end{bmatrix}^{-1} \begin{bmatrix} \hat{\rho}_1 \\ \hat{\rho}_2 \\ \cdot \\ \cdot \\ \cdot \\ \hat{\rho}_p \end{bmatrix}
\qquad (5.64)
$$

Having obtained $\phi_1, \phi_2, \ldots \phi_p$ we use the result

$$
\gamma_0 = \phi_1 \gamma_1 + \phi_2 \gamma_2 + \ldots \phi_p \gamma_p + \sigma_a^2
\qquad (5.65)
$$

to obtain the moment estimator of σ_a^2 as

$$
\hat{\sigma}_a^2 = \hat{\gamma}_0 (1 - \hat{\phi}_1 \hat{\rho}_1 - \hat{\phi}_2 \hat{\rho}_2 - \ldots \hat{\phi}_p \hat{\rho}_p)
\qquad (5.66)
$$

We now illustrate the method of moments on the AR, MA, and ARMA models:

    (1) For the AR(1) model

$$
y_t = \phi y_{t-1} + a_t
$$

and from (5.64)

$$
\hat{\phi}_1 = \hat{\rho}_1
$$

$$
\hat{\mu} = \bar{y}
$$

$$
\hat{\sigma}_a^2 = \hat{\gamma}_0 (1 - \hat{\phi}_1 \hat{\rho}_1)
$$

where

$$
\hat{\gamma}_0 = \frac{1}{n-1} \sum_{t=1}^{n} (y_t - \bar{y})^2
$$

is the sample variance of the series

    (2) Consider the MA(1) model

$$
y_t = a_t - \theta \, a_{t-1}
$$

Since

$$\rho_1 = -\frac{\theta}{1+\theta^2}$$

then replacing ρ_1 by $\hat{\rho}_1$ and solving the quadratic for θ_1 we have

$$\hat{\theta} = \frac{-1 \pm \sqrt{1-4\hat{\rho}_1^2}}{2\hat{\rho}_1}$$

If $\hat{\rho} = \pm .5$ we have a unique solution $\hat{\theta} = \pm 1$ but neither is invertible. If $|\hat{\rho}_1| > .5$, no real solution exists. For $|\hat{\rho}_1| < .5$, there exist two distinct real valued solutions and we have to choose the one that satisfies the invertibility condition. Having obtained an estimate $\hat{\theta}$, the white noise variance is estimated as

$$\hat{\sigma}_a^2 = \frac{\hat{\gamma}_0}{1+\hat{\theta}^2}$$

(3) Consider the ARMA (1,1) model

$$y_t = \phi y_{t-1} + a_t - \theta a_{t-1}$$

From (5.40) we have

$$\hat{\rho}_k = \phi_k = \hat{\phi}\hat{\rho}_{k-1} \qquad k \geq 2$$

and

$$\hat{\rho}_1 = \frac{(\hat{\phi}-\hat{\theta})(1-\hat{\phi}\hat{\theta})}{1+\hat{\theta}^2-2\hat{\phi}\hat{\theta}}$$

First we estimate ϕ as $\hat{\phi} = \hat{\rho}_2/\hat{\rho}_1$ and then solve the above quadratic equation for $\hat{\theta}$, and only the solution that satisfies the invertibility condition is kept.

In (2) and (3) it can be seen that the moment estimators for MA and ARMA models are complicated, and are very sensitive to rounding errors. They are usually used to provide initial estimates needed for other estimation procedures such as maximum likelihood estimation, which we discuss in the remainder of this section.

b. Maximum Likelihood Method

To illustrate how the maximum likelihood is used we consider the simplified model AR(1):

253

$$\dot{y}_t = \phi \dot{y}_{t-1} + a_t \qquad |\phi| < 1$$

where $\dot{y}_t = y_t - \mu$ and a_t are iid $N(0, \sigma_a^2)$. Write

$$e_1 = y_1 - \mu$$

$$a_2 = y_2 - \mu - \phi(y_1 - \mu)$$

$$a_3 = y_3 - \mu - \phi(y_2 - \mu)$$

.

.

$$a_n = y_n - \mu - \phi(y_{n-1} - \mu).$$

Note that e_1 follows a normal distribution with mean 0 and variance $\dfrac{\sigma_a^2}{(1-\phi^2)}$,

and $a_t \sim N(0, \sigma_a^2)$, $t = 2, 3, \dots n$.

Since $e_1, a_2, a_3, \dots a_n$ are independent of each other their joint distribution is given by:

$$p(e_1, a_2, a_3, \dots a_n) = f(e_1) \prod_{t=2}^{n} f(a_t)$$

$$= \left(\frac{1-\phi^2}{2\pi\sigma_a^2} \right)^{1/2} \exp\left[\frac{-e_1^2(1-\phi^2)}{2\sigma_a^2} \right] \left(\frac{1}{2\pi\sigma_a^2} \right)^{\frac{n-1}{2}} \exp\left[-\frac{1}{2\sigma_a^2} \sum_{t=2}^{n} a_t^2 \right]$$

Consider the inverse transformations:

$$y_1 - \mu = e_1$$

$$y_2 - \mu = a_2 + \phi(y_1 - \mu)$$

$$y_n - \mu = a_n + \phi(y_{n-1} - \mu) \, ,$$

whose Jacobian is 1. Therefore, the joint distribution of $y_1 - \mu$, $y_2 - \mu$, $\dots y_n - \mu$ is

$$L = \left(\frac{1-\phi^2}{2\pi\sigma_a^2}\right)^{\frac{1}{2}} \exp\left[-\frac{(y_1-\mu)^2(1-\phi^2)}{2\phi_a^2}\right]$$

$$\cdot \left(\frac{1}{2\pi\sigma_a^2}\right)^{\frac{n-1}{2}} \exp\left[-\frac{1}{2\sigma_a^2}\sum_{t=2}^{n}(y_t-\mu-\phi\,(y_{t-1}-\mu))^2\right]$$

The log-likelihood is given by:

$$\ell = -\frac{n}{2}\ln\sigma_a^2 + \frac{1}{2}\ln(1-\phi^2) - \frac{s(\phi,\mu)}{2\sigma_a^2}$$

where

$$s(\phi,\mu) = (y_1-\mu)^2(1-\phi^2) + \sum_{t-2}^{n}\left[y_t-\mu-\phi(y_{t-1}-\mu)\right]^2$$

$$= (y_1-\mu)^2(1-\phi^2) + s^*(\phi,\mu) \qquad (5.67)$$

and $\qquad s^*(\phi,\mu) = \sum_{t=2}^{n}\left[y_t-\mu-\phi(y_{t-1}-\mu)\right]^2$

For given ϕ and μ, ℓ can be maximized with respect to σ_a^2. Setting $\dfrac{\partial\ell}{\partial\sigma_a^2} = 0$ we get

$$\hat{\sigma}_a^2 = \frac{s(\phi,\mu)}{n} \qquad (5.68)$$

Since $s^*(\phi,\mu)$ involves a sum of n-1 similar terms and the term $(y_1-\mu)^2 (1-\phi^2)$ does not depend on n, we have $s(\phi,\mu) \simeq s^*(\phi,\mu)$, or

$$s(\phi,\mu) \simeq \sum_{t=2}^{n}\left[y_t-\mu-\phi(y_{t-1}-\mu)\right]^2$$

Differentiating $s^*(\phi,\mu)$ with respect to μ, equating to zero, and solving for μ we get

$$\hat{\mu} = \frac{\sum\limits_{t=2}^{n} y_t - \phi \sum\limits_{t=2}^{n} y_{t-1}}{(n-1)(1-\phi)} \tag{5.69}$$

Since for large n

$$\sum\limits_{t=2}^{n} \frac{y_t}{n-1} \simeq \sum\limits_{t=2}^{n} \frac{y_{t-1}}{n-1} \simeq \bar{y}$$

then regardless of the value of ϕ, equation (5.69) gives

$$\hat{\mu} \simeq \bar{y} \tag{5.70}$$

Finally differentiating $s^*(\phi,\hat{\mu})$ with respect to ϕ we have

$$\frac{\partial s^*(\phi,\hat{\mu})}{\partial \phi} = -2 \sum\limits_{t=2}^{n} \left[y_t - \bar{y} - \phi(y_{t-1} - \bar{y}) \right](y_{t-1} - \bar{y})$$

which when set to zero and solving for ϕ yields

$$\hat{\phi} = \frac{\sum\limits_{t=2}^{n} (y_t - \bar{y})(y_{t-1} - \bar{y})}{\sum\limits_{t=2}^{n} (y_{t-1} - \bar{y})^2} . \tag{5.71}$$

Remark

The maximum likelihood estimators (5.70) and (5.71) are also referred to as the least squares (LS) estimators as they can be obtained by minimizing the sum of squares of the differences

$$(y_t - \mu) - \phi(y_t - 1) = a_t$$

from the AR(1) model. Clearly this sum of squares is identically $s^*(\phi,\mu)$. This indicates that the MLE and the LS estimates of μ and ϕ are approximately equal.

The large sample properties of the maximum likelihood and least squares estimators are identical and can be obtained by modifying standard maximum likelihood theory for large samples. The details are omitted and can be found in Box and Jenkins (1976).

For large n, the estimators are approximately unbiased, normally distributed and have approximate variance given by:

$$AR(1): \quad Var(\hat{\phi}) \doteq \frac{1}{n}(1-\phi^2)$$

$$MA(1): \quad Var(\hat{\theta}) \doteq \frac{1}{n}(1-\theta^2)$$

$$ARMA(1,1): \quad Var(\hat{\phi}) \doteq \frac{1-\phi^2}{n}\left(\frac{1-\phi\theta}{\phi-\theta}\right)^2$$

$$Var(\hat{\theta}) \doteq \frac{1-\theta^2}{n}\left(\frac{1-\phi\theta}{\phi-\theta}\right)^2$$

$$Cov(\hat{\phi},\hat{\theta}) \doteq \frac{\sqrt{(1-\phi^2)(1-\theta^2)}}{1-\phi\theta} \tag{5.72}$$

Example 5.7 Using the somatic cell count data, we will fit two models: the AR(1) and the MA(1). This will be accomplished by using the ETS program in SAS; more specifically by utilizing the PROC ARIMA procedure. The data were differenced once so as to stabilize the mean.

SAS Plots of the Autocorrelation Function and the Partial Correlation Function

Autocorrelations

```
Lag Covariance Correlation -1 9 8 7 6 5 4 3 2 1 0 1 2 3 4 5 6 7 8 9 1      Std
 0  1007.500   1.00000 |           |********************|      0
 1  -207.536  -0.20599 |       ****|    .            |    0.109764
 2  -293.539  -0.29135 |      *****|    .            |    0.114327
 3   177.445   0.17612 |           .   |****.        |    0.122948
 4   16.087389  0.01597 |          .   |    .        |    0.125951
 5  -196.800  -0.19533 |       ****|    .            |    0.125975
 6   43.477143  0.04315 |          .   |*   .        |    0.129573
 7  -110.087  -0.10927 |         . **|    .          |    0.129746
 8  -90.458840 -0.08979 |         . **|    .          |    0.130850
 9   129.236   0.12827 |          .   |*** .        |    0.131590
10   15.333874  0.01522 |          .   |    .        |    0.133088
11  -103.832  -0.10306 |         . **|    .          |    0.133109
12   244.896   0.24307 |          .   |*****        |    0.134067
13   186.837   0.18545 |          .   |****.        |    0.139276
14  -175.962  -0.17465 |        . ***|    .          |    0.142219
15  -86.242765 -0.08560 |         . **|    .          |    0.144781
16   44.315590  0.04399 |          .   |*   .        |    0.145389
17   12.768269  0.01267 |          .   |    .        |    0.145549
18  -114.418  -0.11357 |         . **|    .          |    0.145563
19  -65.211117 -0.06473 |          . *|    .          |    0.146626
20  -3.510942  -0.00348 |          .   |    .        |    0.146970
                        "." marks two standard errors
```

Partial Autocorrelations

```
Lag Correlation -1 9 8 7 6 5 4 3 2 1 0 1 2 3 4 5 6 7 8 9 1
  1  -0.20599 |           ****|   .         |
  2  -0.34858 |        *******|   .         |
  3   0.02780 |               |*  .         |
  4  -0.03616 |             . *|   .         |
  5  -0.15610 |            .***|   .         |
  6  -0.06452 |             . *|   .         |
  7  -0.26000 |           *****|   .         |
  8  -0.21459 |            ****|   .         |
  9  -0.09149 |             .**|   .         |
 10  -0.09294 |             .**|   .         |
 11  -0.14724 |            .***|   .         |
 12   0.11120 |               |** .         |
 13   0.24885 |               .  |*****       |
 14   0.11414 |               .  |** .        |
 15  -0.00798 |               .  |   .         |
 16  -0.06103 |             . *|   .         |
 17   0.07968 |               .  |** .        |
 18  -0.00731 |               .  |   .         |
 19  -0.02574 |             . *|   .         |
 20  -0.01489 |               |   .         |
```

AR(1)

Estimates of the parameters were computed using two methods: MLE and Least Squares. The results are found below.

	Parameter	Estimate	Std. Error
MLE	μ	-0.30911	2.85377
	AR1,1	-0.21112	0.1100
Least Squares	μ	-0.30061	2.85539
	AR1,1	-0.21011	0.10993

MA(1)

Estimates of the Moving Average model were obtained using two methods; MLE and Least Squares.

	Parameter	Estimate	Std. Error
MLE	μ	-0.01804	1.71859
	MA1,1	0.49064	0.09976
Least Squares	μ	-0.02744	1.73327
	MA1,1	0.48592	0.10055

Remarks :

The autocorrelation function (ACF) lists the estimated autocorrelation coefficients at each lag. The value of the ACF at lag 0 is always 1. The dotted lines provide an approximate 95%

258

confidence limit for the autocorrelation estimate at each lag. If none of the autocorrelation estimates falls outside the strip defined by the two dotted lines (and no outliers in the data) one may assume the absence of serial correlation. In effect the ACF is a measure of how important the sequence of distant observations $Y_{t-1}, Y_{t-2}, \ldots$ are to the current time series value Y_t.

The partial autocorrelation function (PACF) is the ACF at lag p accounting for the effects of all intervening observations. Thus the PACF at lag 1 is identical to the ACF at lag 1, but they are different at higher lags.

VI. FORECASTING

One of the most important objectives of time series analysis is to forecast the future values of the series. The term forecasting is used more frequently in recent time series literature than the term prediction. However, most forecasting results are derived from a general theory of linear prediction developed by Kolmogorov (1939, 1941), Kalman (1960) and Whittle (1983) and many others.

Once a good time series model has become available, it can be used to make inferences about future observations. What we mean by a good model is that identification estimation and diagnostics have been completed. Even with good time series models the reliability of the forecast is based on the assumption that the future behaves like the past. However, the nature of the stochastic process may change in time and the current time series model may no longer be appropriate. If this happens the resulting forecast may be misleading. This is particularly true for forecasts with a long lead time.

Let y_n denote the last value of the time series and suppose we are interested in forecasting the value that will be observed ℓ time periods ($\ell > 0$) in the future; that we are interested in forecasting the future value $y_{n+\ell}$. We denote the forecast of $y_{n+\ell}$ by $\hat{y}_n(\ell)$, where the subscript denotes the forecast origin and the number in parentheses denotes the lead time. Box and Jenkins (1970) showed that the "best" forecast of $y_{n+\ell}$ is given by the expected value of $y_{n+\ell}$ at time n, where best is defined as that forecast which minimizes the mean square error

$$E\left[y_{n+\ell} - \hat{y}_n(\ell)\right]^2$$

It should be noted that the above expectation is in fact a conditional expectation since, in general, it will depend on $y_1, y_2, \ldots y_n$. This expectation is minimized when

$$\hat{y}_n(\ell) = E\left(y_{n+\ell}\right) .$$

We now show how to obtain the forecast for the time series models AR(1), AR(2) and MA(1).

259

(1) AR(1) Model:

$$y_t - \mu = \phi \, (y_{t-1} - \mu) + a_t \tag{5.73}$$

Consider the problem of forecasting 1 time unit into the future. Replacing t by t+1 in equation (5.73) we have

$$y_{t+1} - \mu = \phi(y_t - \mu) + a_{t+1} \, . \tag{5.74}$$

Conditional on $y_1, y_2, \ldots y_{t-1}, y_t$, the expectation of both sides of (5.74) is

$$E \, (y_{t+1}) - \mu = \phi \, \left\{ E \left[y_t | y_t, y_{t-1}, \ldots y_1 \right] - \mu \right\}$$
$$+ E \left[a_{t+1} \mid y_t, y_{t-1}, \ldots y_1 \right] \, .$$

Since

$$E(y_t | y_t, y_{t-1}, \ldots y_1) = y_t$$
$$E(a_{t+1} | y_t, y_{t-1}, \ldots y_1) = E(a_{t+1}) = 0$$

and

$$E(y_{t+1}) = \hat{y}_t(1)$$

then

$$\hat{y}_t(1) = \mu + \phi(y_t - \mu) \tag{5.75}$$

For a general lead time ℓ, we replace t by t+ℓ in equation (5.73) and taking the conditional expectation we get

$$\hat{y}_t(\ell) = \mu + \phi(y_t(\ell-1) - \mu) \qquad \ell \geq 1 \tag{5.76}$$

It is clear now that equation (5.76) is recursive in ℓ. It can also be shown that

$$\hat{y}_t(\ell) = \mu + \phi^\ell(y_t - \mu) \qquad \ell \geq 1 \tag{5.77}$$

Since $|\phi|<1$, we may simply have

$$\hat{y}_t(\ell) \approx \mu \qquad \qquad \text{for large } \ell$$

Now let us consider the one-step ahead forecast error, $e_t(1)$. From equations (5.74) and (5.75)

$$
\begin{aligned}
e_t(1) &= y_{t+1} - \hat{y}_t(1) \\
&= \mu + \phi(y_t - \mu) + a_{t+1} - \left[\mu + \phi(y_t - \mu)\right] \qquad (5.78)\\
&= a_{t+1}
\end{aligned}
$$

This means that the white noise a_{t+1}, can now be explained as a sequence of one-step-ahead forecast errors. From (5.78)

$$Var\left[e_t(1)\right] = Var(a_{t+1}) = \sigma_a^2 .$$

It can be shown (see Abraham and Ledolter page 241; 1983) that for the AR(1)

$$Var\left[e_t(\ell)\right] = \frac{1-\phi^{2\ell}}{1-\phi^2} \sigma_a^2 \qquad (5.79)$$

and for large ℓ

$$Var\left[e_t(\ell)\right] \approx \frac{\sigma_a^2}{1-\phi^2} . \qquad (5.80)$$

(2) AR(2) Model:

Consider the AR(2) model

$$y_t - \mu = \phi_1(y_{t-1} - \mu) + \phi_2(y_{t-2} - \mu) + a_t$$

Setting t=t+ℓ the above equation is written as

$$y_{t+\ell} = \mu + \phi_1(y_{t+\ell-1} - \mu) + \phi_2(y_{t+\ell-2} - \mu) + a_{t+\ell}$$

For the one-step ahead forecast (i.e. $\ell=1$)

$$y_{t+1} = \mu + \phi_1(y_t - \mu) + \phi_2(y_{t-1} - \mu) + a_{t+1} \qquad (5.81)$$

From the observed series, y_t and y_{t-1} are the last two observations in the series. Therefore, for given values of the model parameters, the only unknown quantity on the right-hand side of (5.81) is a_{t+1}. Therefore conditional on $y_t, y_{t-1}, ... y_1$ we have

$$E(y_{t+1}) = \mu + \phi_1(y_t - \mu) + \phi_2(y_{t-1} - \mu) + E(a_{t+1})$$

By assumption, $E(a_{t+1}) = 0$, and hence the forecast of y_{t+1} is

$$\hat{y}_t(1) = E(y_{t+1})$$
$$= \mu + \phi_1(y_t - \mu) + \phi_2(y_{t-1} - \mu)$$

where μ, ϕ_1, and ϕ_2 are replaced by their estimates. In general we have

$$\hat{y}_t(\ell) = E[y_{t-\ell}]$$
$$= \mu + \phi_1(\hat{y}_t(\ell-1) - \mu) + \phi_2(y_t(\ell-2) - \mu) \qquad \ell \geq 3 \qquad (5.82)$$

The forecast error is given by

$$e_t(\ell) = y_t(\ell) - \hat{y}_t(\ell)$$

For $\ell=1$

$$Var\left[e_t(\ell)\right] = \sigma_a^2\left[1 + \psi_1^2 + \psi_2^2 + ... \psi_{\ell-1}^2\right] \qquad (5.83)$$

where

$$\psi_1 = \phi_1, \quad \psi_2 = \phi_1^2 + \phi_2$$

and

$$\psi_j = \phi_1\psi_{j-1} + \phi_2\psi_{j-2} \quad j \geq 2$$

For $\ell = 1$

$$Var\left[e_t(1)\right] = \sigma_a^2\left[1+\phi_1^2+(\phi_1^2+\phi_2)^2\right] \tag{5.84}$$

(see Abraham and Ledolter, page 243)

(3) MA(1) Model:

In a similar manner we show how to forecast an MA(1) time series model

$$y_t = \mu+a_t-\theta a_{t-1}$$

First we replace t by t+ℓ so that

$$y_{t+\ell} = \mu+a_{t+\ell}-\theta a_{t+\ell-1} \ . \tag{5.85}$$

Conditional on the observed series we have

$$\hat{y}_t(\ell) = \mu+a_{t+\ell}-\theta a_{t+\ell-1} \tag{5.86}$$

because for $\ell>1$ both $a_{t+\ell}$ and $a_{t+\ell-1}$ are independent of $y_t, y_{t-1},..., y_1$. Hence

$$\hat{y}_t(\ell) = \begin{cases} \mu & \ell>1 \\ \mu-\theta a_t & \ell=1 \end{cases}$$

$$Var(e_t(\ell)) = \begin{cases} \sigma_a^2(1+\theta^2) & \ell>1 \\ \sigma_a^2 & \ell=1 \end{cases} \tag{5.87}$$

The above results allow constructing $(1-\alpha)$ 100% confidence limits on the future observations $y_{t+\ell}$ as

$$\hat{y}_t(\ell) \pm Z_{1-\alpha/2} \sqrt{Var(e_t(\ell))} \qquad\qquad (5.88)$$

Example 5.8

Again, using the SCC data of example 5.1, we can compute forecasts for the variable by employing the time series programs in SAS. The estimates and 95% upper and lower confidence values are given below.

observation	forecast	std. err.	lower 95%	upper 95%
85	289.126	31.42	227.54	350.72
86	286.83	40.03	208.38	365.27
87	286.94	47.83	193.19	380.69
88	286.54	54.39	179.93	393.14
89	286.25	60.27	168.12	404.37

VII. MODELING SEASONALITY WITH ARIMA:
The condemnation rates series revisited.

The ARIMA models presented in this chapter assume that seasonal effects are removed, or that the series is non-seasonal. However, in many practical situations it is important to model and quantify the seasonal effects. For example, it might be of interest to poultry producers to know which months of the year the condemnation rates are higher. This is important in order to avoid potential losses if the supply falls short of the demand due to excess condemnation.

Box and Jenkins (1970) extend their ARIMA models to describe and forecast time series with seasonal variation. Modeling of time series with seasonal variation using ARIMA is quite complicated and detailed treatment of this topic is beyond the scope of this chapter. However, we shall outline the steps of modelling seasonality in the condemnation rates data of example (5.4).

First we examine the monthly means of the series (Table 5.6):

Table 5.6
Monthly Means of Condemnation Rates per 100,000 Birds

	Jan	Feb	Mar	Apr	May	Jun	Jul	Aug	Sep	Oct	Nov	Dec
mean rate	932.58	916.13	921.98	772.98	603.37	509.54	497.01	520.03	618.73	585.96	771.76	692.69

The data suggest that condemnation rates tend to be relatively high in the winter and low in the summer. Figure 5.11 shows the sample ACF of the series for lags 1 through 21.

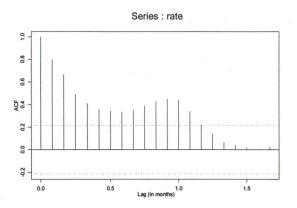

Figure 5.11 The autocorrelation function of the rate.

The sample autocorrelations beyond lag 13 are insignificant. Moreover, we can see that (i) the autocorrelations are not tailing off, which indicates that the series is nonstationary (ii) there is a sinosoidal pattern confirming a seasonal pattern in the series. Note that an AR(1) series $y_t = \mu + \phi y_{t-1} + \rho_t$ is nonstationary if $\phi = 1$. To confirm the nonstationarity of our series, we fitted an AR(1) model. The MLE of ϕ was $\hat{\phi} = 0.961$ with SE = .03, and hence a value of 1.0 for ϕ seems acceptable. To remove nonstationarity, we formed the series $w_t = y_t - y_{t-1}$. Figure 5.12 is the time series plot of w_t.

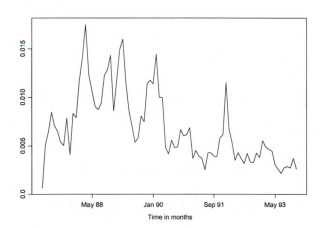

Figure 5.12 Time-series plot of the differenced series of rates.

265

The plot shows that the series is oscillating around a constant mean.

In modeling seasonality we decided to choose between two models; the first is an MA with one non-seasonal component and several seasonal components, the second being an AR with similar structure.

Since the series is monthly, it has 12 periods. Accordingly we constructed the series with 12 periods of differencing. The ACF plot of this series is given in Figure 5.13.

```
                         Autocorrelations

Lag Covariance Correlation -1 9 8 7 6 5 4 3 2 1 0 1 2 3 4 5 6 7 8 9 1
  0  65929.434    1.00000  |                    |********************|
  1 -19692.471   -0.29869  |              ******|                    |
  2   6127.668    0.09294  |                 .  |**      .           |
  3  -8942.191   -0.13563  |                 . ***|       .           |
  4  -5593.375   -0.08484  |                 .  **|       .           |
  5   -817.122   -0.01239  |                 .    |       .           |
  6  -1354.484   -0.02054  |                 .    |       .           |
  7    260.957    0.00396  |                 .    |       .           |
  8  13738.889    0.20839  |                 .    |****.              |
  9  -6842.945   -0.10379  |                 . **|       .           |
 10   6180.334    0.09374  |                 .    |**     .           |
 11  -1751.808   -0.02657  |                 .   *|       .           |
 12 -20075.614   -0.30450  |              ******|       .           |
                          "." marks two standard errors
```

Figure 5.13   Autocorrelation plot of condemnation data.

This plot shows significant autocorrelations at lags 1,3,8, and 12.

The candidate models for this situation is an additive AR model:

$$y_t = \mu + \phi_1 y_{t-1} + \phi_2 y_{t-3} + \phi_3 y_{t-8} + \phi_4 y_{t-12} + a_t$$

or an additive MA model:

$$y_t = \mu + a_t - \theta_1 a_{t-1} - \theta_2 a_{t-3} - \theta_3 a_{t-8} - \theta_4 a_{t-12}.$$

Our choice of an additive structure is for simplicity. The fitted seasonal AR model is

$$y_t = -16.26 - .278\, y_{t-1} - .109\, y_{t-3} - .170\, y_{t-8} - .303\, y_{t-12}$$
$$\quad\ (18.18)\quad (.117)\qquad (.117)\qquad (.116)\qquad (.127)$$

266

and the fitted seasonal MA model is

$$y_t = -15.38 - .260\,\varepsilon_{t-1} - .175\,\varepsilon_{t-3} - .165\,\varepsilon_{t-8} - .467\,\varepsilon_{t-12}$$
$$(12.21) \quad (.115) \qquad (.116) \qquad (.116) \qquad (.112)$$

The bracketed numbers are the standard errors of the estimates. Parameters of both models were estimated via conditional least squares.

As a diagnostic tool that may be used to check the model adequacy we use the Q statistic of Box and Pierce (1970), later modified by Ljung and Box (1978). They suggested a test to contrast the null hypothesis

$$H_0 : \rho_1 = \rho_2 = \ldots = \rho_k = 0$$

against the general alternative

$$H_1: \text{ not all } \rho_j = 0.$$

Based on the autocorrelation between residuals, they suggested the statistic

$$Q = Q(k) = n(n+2)\sum_{j=1}^{k} r_j^2 / (n-j)$$

where n is the length of the series after any differencing and r_j is the residual autocorrelation at lag j. Box and Pierce (1970) showed that under H_0, Q is asymptotically distributed as chi-squared with (k-p-q) degrees of freedom. Typically, the statistic is evaluated for several choices of k. Under H_0, for large n

$$E\big[Q(k_2) - Q(k_1)\big] = k_2 - k_1$$

so that different sections of the correlogram can be checked for departures from H_0.

Tables 5.7 and 5.8 give the Box-Pierce-Ljung (BPL) statistics for the condemnation rates using the AR and MA models.

Table 5.7
BPL Statistic Using AR Seasonal Model

Lag(k)	$Q(k)$	df	Prob	Autocorrelations r_j					
6	1.16	2	0.559	-0.018	-0.014	-0.049	-0.093	-0.035	-0.046
12	1.82	8	0.986	0.006	0.023	-0.003	0.070	0.006	-0.046
18	5.26	14	0.982	0.178	-0.034	0.034	-0.061	0.021	0.009
24	15.80	20	0.729	-0.072	0.003	0.158	-0.014	0.198	-0.171

p = 4, q = 0, df = k - 4

Table 5.8
BPL Statistic Using MA Seasonal Model

Lag(k)	$Q(k)$	df	Prob	Autocorrelations r_j					
6	0.68	2	0.712	0.008	0.042	0.029	-0.063	-0.016	-0.043
12	1.94	8	0.983	0.016	-0.002	-0.049	0.098	0.001	0.050
18	5.61	14	0.975	0.178	-0.046	0.052	-0.055	0.037	0.003
24	13.17	20	0.870	-0.042	-0.011	0.166	0.004	0.189	-0.076

p = 0, q = 4, df = k - 4

Note that all the Q statistics of the two models are non-significant. This means that the models captured the autocorrelation in the data, and that the residual autocorrelations are non-significant. Therefore, both models provide good fit to the condemnation rates times series process. However, we might argue that the MA seasonal model gives a better fit due to the fact that it has smaller Akaike information criterion (AIC = 978.07 for the MA, and AIC = 982.25 for the AR).

The following SAS program produces the parameter estimates, the Q statistics and the AIC's.

```
PROC ARIMA;   /*MA seasonal model*/
     i var = rate (1,12) nlag = 12;
     e q(1,3,8)(12);
     run;
```

Note that q(1,3,8) is replaced by p(1,3,8) to fit an AR seasonal model.

VIII. THE ONE-WAY AND TWO-WAY ANALYSIS OF VARIANCE WITH TIME-SERIES DATA

A. INTRODUCTION

In the previous sections we introduced and discussed some of the statistical properties of the most widely known time series models. We have also shown, through examples, their use in the analysis of time series arising from observational studies. This section discusses the problems of estimation and inference in experimental time series. These series appear in situations where an investigation is repeated over time. For example, in field trials that are aimed at assessing the weight gains of animals randomized into groups, each group represents a specific diet; repeated measures of weight constitute an experimental series. The observations (weights) forming a time series are characterized by a high degree of correlation among contiguous observations, therefore, comparing the group means cannot be routinely analyzed by ANOVA methods.

We shall focus our attention on the analysis of experimental time series (ETS) data under the one-way ANOVA with error variables forming an AR(1) and with the autoregression parameter ϕ assumed common among all the groups. Furthermore, when the time is considered a specific factor, a two-way ANOVA methodology is discussed and an example will be given.

B. ONE-WAY ANOVA WITH CORRELATED ERRORS

Consider the fixed effects models

$$y_{it} = \mu + \alpha_i + u_{it} \tag{5.89}$$

where y_{it} is the observation at time t due to the ith treatment, μ is an overall mean, α_i is the effect of the ith treatment, and u_{it} is a component of an AR(1) process with

$$u_{it} = \phi u_{it-1} + a_{it} \qquad i = 1,2,...k$$

$$t = 1,2,...n$$

where a_{it} are iid $N(0,\sigma_a^2)$, and $|\phi| < 1$. For the usual ANOVA model $y_{it} = \mu + \alpha_i + a_{it}$, we test the hypothesis $H_0 : \alpha_1 = \alpha_2 = ... = \alpha_k = 0$ using the well-known F-statistic,

$$F = \frac{kn(n-1) \sum_{i=1}^{k} (\bar{y}_i - \bar{y}_{..})^2}{(k-1) \sum_{i=1}^{k} \sum_{t=1}^{n} (y_{it} - \bar{y}_{i.})^2} \tag{5.90}$$

where

$$\bar{y}_{i.} = \frac{1}{n} \sum_{t=1}^{n} y_{it} \quad \text{and} \quad \bar{y}_{..} = \frac{1}{nk} \sum_{i=1}^{k} \sum_{t=1}^{n} y_{it} .$$

For the model (5.89) the statistic given in (5.90) is inappropriate because of the dependency among the observations within each treatment.

Recently Sutradhar et al. (1987) proposed the corrected F-statistic

$$F^* = \frac{c_2(\hat{\phi})}{c_1(\hat{\phi})} F \qquad\qquad (5.91)$$

$$c_1(\phi) = \frac{1}{(1-\phi)^2} \left[1 - \frac{2\phi(1-\phi^n)}{n(1-\phi^2)} \right]$$

for testing H_0: $\alpha_1 = \alpha_2 = \ldots = \alpha_k = 0$ where
and

$$c_2(\phi) = \frac{1}{n-1} \left[\frac{n}{(1-\phi)^2} - c_1(\phi) \right]$$

The hypothesis is rejected for large values of F^*. For fixed ϕ, they expressed the probability distribution of F^* as a linear combination of independent central chi-square distributions. In theory, one can find the exact percentage points of this statistic, however, one should be able to provide the value of $\hat{\phi}$ and hence that of the correction factor,

$$\frac{c_1(\hat{\phi})}{c_2(\hat{\phi})} .$$

For practical purposes, we may reject the hypothesis when the value of F given in (5.90) exceeds that of

$$\frac{c_1(\hat{\phi})}{c_2(\hat{\phi})} F_{\alpha, k-1, k(n-1)},$$

where $F_{\alpha, k-1, k(n-1)}$ is obtained from the F-tables at α level of significance and $(k-1, k(n-1))$ degrees of freedom. The estimate $\hat{\phi}$ may be any consistent estimate of ϕ such as the MLE given in (5.71) pooled over all treatments.

C. TWO-WAY ANOVA WITH CORRELATED ERRORS

In the one-way ANOVA with correlated errors we ignored time as a factor. While the primary interest is the comparison between treatments in this section we shall account for time, so that the postulated model is given by

$$y_{it} = \alpha_i + \beta_t + u_{it} , \qquad i = 1,2,...k$$

$$t = 1,2,...n$$

(5.92)

where α_i and u_{it} are as defined previously for the one-way ANOVA, and β_t is the time effect.

Here we assume that series of observations y_{it} on a variable, made at n distinct time points are available from k treatments. It is also assumed that the rows of the matrix u_{it} are iid, and each row is assumed to be distributed as n random variables from a stationary Gaussian process with mean 0

Time (columns)

		1	2	....	n	Total
	1	y_{11}	y_{12}		y_{1n}	$y_{1.}$
	2	y_{21}	y_{22}		y_{2n}	$y_{2.}$
Treatment	.	y_{i1}	y_{i2}		y_{in}	.
(rows)	.					.
	.					.
	k	y_{k1}	y_{k2}		y_{kn}	$y_{k.}$
		$y_{.1}$	$y_{.2}$		$y_{.n}$	

This model has been investigated in an important paper by Box (1954), in which exact formulae for the tail probabilities of the usual F-test were given. An approximation to the distribution of the F-test statistic for the hypothesis of no time effect (β_t is constant) is

$$F_c = \frac{SS\ (Columns)/(n-1)}{SS\ (Residuals)/(k-1)(n-1)}$$

where

$$SS\ (Columns) = k \sum_{t=1}^{n} (\bar{y}_{.t} - \bar{y}_{..})^2$$

and

$$SS \text{ (Residuals)} = \sum_{i=1}^{k} \sum_{t=1}^{n} (y_{it} - \bar{y}_{.t} - \bar{y}_{i.} + \bar{y}_{..})^2$$

Box concluded that the test of no column effects was not seriously affected by the presence of serial correlations. Moreover, the test of no row effects was affected essentially as a consequence of the serious effect on the variance of a mean by serial correlation. However, Box did not go into the details of investigating the validity of the approximation. Furthermore, he did not consider the analogous approximation to the distribution of the F-test statistic of the hypothesis $H_0: \alpha_1 = \alpha_2 = \ldots = \alpha_k$, presumably because the two sums of squares involved are, in general, dependent.

More recently Andersen et al. (1981) studied approximations on the hypothesis $H_0: \beta_1 = \beta_2 = \ldots = \beta_n$ and $H_0: \alpha_1 = \alpha_2 = \ldots = \alpha_k$ for the model (5.92) under two types of stationary Gaussian processes, AR(1) and MA(1). The approximate tests developed by Andersen et.al. are outlined below.

In addition to the sums of squares given in (5.93) we define the sum of squares of row effects as

$$SS \text{ (Rows)} = n \sum_{i=1}^{k} (\bar{y}_{i.} - \bar{y}_{..})^2$$

(a) To test for no column effect $H_0: \beta_1 = \beta_2 = \ldots = \beta_n$ under the model $E(y_{it}) = \alpha_i + \beta_t$ (i.e. in the presence of row effect) use the approximation

$$\frac{SS \text{ (Columns)}/(n-1)}{SS \text{ (Residuals)}/(k-1)(n-1)} \approx F\left\{\tilde{f}, (k-1)\tilde{f}\right\}$$

and in testing the hypothesis $H_0: \beta_1 = \beta_2 = \ldots = \beta_k$ under the model $E(y_{it}) = \alpha + \beta_t$ (i.e. in the absence of row effect) use the approximation

$$\frac{SS \text{ (Columns)}/(n-1)}{(SS \text{ (Residuals)} + SS \text{ (Rows)})/(k-1)n} \approx d_c F\left\{\tilde{f}, (k-1)f\right\}$$

(b) To test for no row effect $H_0: \alpha_1 = \alpha_2 = \ldots = \alpha_k$ under the model $E(y_{it}) = \alpha_i + \beta_t$ (i.e. in the presence of column effect) use the approximation

272

$$\frac{SS \ (\text{Rows})/(k-1)}{SS \ (\text{Residuals})/(k-1)(n-1)} \approx d_r \ F \left\{k-1,(k-1)\tilde{f}\right\}$$

and in testing the hypothesis $H_0: \alpha_1 = \alpha_2 = \ldots = \alpha_k$ under the model $E(y_{it}) = \alpha_i + \beta$ (i.e. in the absence of column effect) use the approximation

$$\frac{SS \ (\text{Rows})/(k-1)}{(SS \ (\text{Residuals}) + SS(\text{Columns}))/k(n-1)} \approx d_r \ F \left\{k-1,k\tilde{f}\right\} \ ,$$

where $F\{a,b\}$ is the F-statistic with a and b degrees of freedom. Note that d_c, d_r, f, and $\hat{f}$ depend on the parameters of the underlying stationary Gaussian processes. Anderson et al (1981) gave the values of these quantities for the AR(1) and MA(1) models. For AR(1) $N(0,1)$, we have

$$f = \frac{n^2(1-\tilde{\phi}^2)^2}{\left\{n-2\tilde{\phi}^2-n\tilde{\phi}^4+2(\tilde{\phi})^{2n+2}\right\}}$$

$$\tilde{f} = \frac{(1+\tilde{\phi})^2\left\{n(n-1)-2(n^2-1)\tilde{\phi}+n(n+1)\tilde{\phi}^2-2(\tilde{\phi})^{n+1}\right\}^2}{A_1+B_1+C_1}$$

where

$$A_1 = \sum_{i=0}^{6} a_i \ \tilde{\phi}^i \ , \quad B_1 = \sum_{i=1}^{5} b_i(\tilde{\phi})^{n+i} \ , \quad C_1 = \sum_{i=2}^{4} c_i(\tilde{\phi})^{2n+i}$$

$$d_r = \frac{1-2n^{-1}\tilde{\phi}-\tilde{\phi}^2+2n^{-1}(\tilde{\phi})^{n-1}}{1-2(n+1)n^{-1}\tilde{\phi}+(n+1)(n-1)^{-1}\tilde{\phi}^2-2n^{-1}(n-1)^{-1}\left(\tilde{\phi}\right)^{n+1}}$$

and

$$d_c = \left(1-\tilde{\phi}\right)^{-2} \left\{1-\frac{2(n+1)}{n} \ \tilde{\phi} + \frac{n+1}{n-1} \ \tilde{\phi}^2 - 2\frac{(\tilde{\phi})^{n+1}}{n(n-1)}\right\}$$

273

The constant a_i, b_i, and c_i are obtained from the following table:

i	a_i	b_i	c_i
0	$n^2(n-1)$	----	----
1	$-2n(n^2+n-2)$	$-4n(n+1)$	----
2	n^3-n^2+4n+4	$-8(n+1)$	$2(n^2+2n+2)$
3	$8(n^2+1)$	$8(n^2-2)$	$-4(n^2-2)$
4	$-n^3-n^2-4n+4$	$8(n-1)$	$2(n^2-2n+2)$
5	$2n(n^2-n-2)$	$-4n(n-1)$	----
6	$-n^2(n+1)$	----	----

To find the moment estimates of the parameters of the AR(1) process, Anderson at el. defined, for the model $y_{it}=\alpha_i+\beta_t+u_{it}$ with $u_{it}=\phi\, u_{it-1}+a_{it}$, where a_{it} are iid $N(0,\sigma^2)$, the residuals $R_{i,t}=y_{it}-\bar{y}_{i.}-\bar{y}_{.t}+\bar{y}_{..}$.

Using the statistics

$$Cov(R_{i,t};\ R_{i,t+1}) = \frac{\sum_{i=1}^{k}\sum_{t=1}^{n}(R_{i,t})(R_{i,t+1})}{n(k-1)}$$

$$Var(R_{it}) = \frac{1}{n(k-1)}\sum_{i=1}^{k}\sum_{t=1}^{n}R_{it}^2$$

and

$$\hat{\rho}_1 = \frac{Cov(R_{i,t};\ R_{i,t+1})}{Var(R_{i,t})}$$

the moment estimators are

$$\tilde{\phi} = \frac{1}{n} + \left(1+\frac{2}{n}\right)\hat{\rho}_1$$

$$\tilde{\phi}^2 = Var(R_{i,t})\left\{1-(\tilde{\phi})^2 + \frac{(1+\tilde{\phi})^2}{n}\right\}$$

Example 5.9

The arrthymogenic dose of epinephrine (ADE) is determined by infusing epinephrine into a dog until arrhythmia criteria are reached. The arrhythmia criteria were the occurrence of 4 intermittent or continuous premature ventricular contractions. Once the criterion has been reached, the infusion required to produce the criteria is recorded. The ADE is then calculated on multiplying the duration of the infusion by the infusion rate. The following table (Table 5.9) gives the ADE for 6 dogs, measured at the 0, ½, 1½, 2, 2½, 3, 3½, 4 and 4½ hours.

Table 5.9
ADE in Dogs

Dog	Time 0	½	1	1½	2	2½	3	3½	4	4½
1	5.7	4.7	4.8	4.9	3.88	3.51	2.8	2.6	2.5	2.5
2	5.3	3.7	5.2	4.9	5.04	3.5	2.9	2.6	2.4	2.5
3	4	4.6	4.1	4.58	3.58	4.0	3.6	2.5	3.5	2.1
4	13.7	8.9	9.6	8.6	7.5	4.0	3.1	4.1	4.08	3.0
5	5	4	4.1	3.9	3.4	3.39	2.95	3.0	3.1	2.0
6	7.1	3	2.4	3.3	3.9	4.0	3	2.4	2.1	2.0

Here we analyze the data under a two-way ANOVA, assuming that the error term follows an AR(1) process.

i) First of all a two-way ANOVA is run in testing
      H_0: no dog effect
      H_0: no time effect

The results obtained from SAS Proc ANOVA show that both these hypotheses are rejected.

Source	df	F value	Pr $>$ F
dog	5	10.10	0.0001
time	9	7.18	0.0001
error	45		

ii) Then we proceed with the computation of $\hat{\phi}$ and $\hat{\sigma}_a^2$ by first calculating $R_{i,t}$, $\gamma(t,t+1)$, $\gamma(0)$ and $\hat{\rho}_1$, followed by the calculation of d_r, d_c, f, $\tilde{f}$. These are

275

$$d_r \ = \ 4.7541$$

$$d_c \ = \ 0.7280$$

$$f \ = \ 4.8332$$

$$\tilde{f} \ = \ 5.1303$$

iii) Now we must adjust the degrees of freedom of the F statistics for testing the null hypotheses,

    i) H_0: no time effect
    ii) H_0: no dog effect

So that for,

    i) the degrees of freedom for the $F_{0.05}$ corrected are 5 and 5*5.1303. This F value (with fractional degrees of freedom) is 2.59.
    Our calculated F value is 7.18, thus we have reason to reject the null hypothesis implying that there is a time effect in the presence of group effect.

and then,

    ii) the corrected F is

$$F_{.05, \ 5, \ 5*5.1303)}*4.7541 = 2.59*4.7541 \ = \ 12.313$$

against which we compare our calculated F value of 10.1. Here there is no reason to reject the null hypothesis which implies that there is no dog effect in the presence of the time effect.

The overall conclusion is that all dogs react consistently through time, but that there is a significant time effect. This suggests that the ADE varies over time.

Chapter 6

REPEATED MEASURES ANALYSIS

I. INTRODUCTION

Experimental designs with repeated measures over time are very common in biological and medical research. This type of design is characterized by repeated measurements on a relatively large number of individuals for a relatively small number of time points, where each individual is measured at each time. The repeated measures design has a clear advantage over completely randomized designs. This is because by measuring individuals repeatedly over time, one can control the biological heterogeneity between individuals. This reduction in heterogeneity makes the repeated measures design more efficient than completely randomized designs.

Repeated measures are very frequent in almost all scientific fields, including agriculture, biology, medicine, epidemiology, geography, demography and many other disciplines. It is not the aim of this chapter to provide a comprehensive coverage of the subject. We refer the reader to the many excellent texts such as Crowder and Hand's (1990) and Lindsay (1993) and many articles that have appeared in scientific journals (*Biometrics and Statistics in Medicine*). In the following section we provide examples of repeated measures experiments that are frequently encountered. In section III we describe the statistical methodologies that are used to answer the scientific questions posed. In section IV we illustrate how the "generalized estimating equations" technique is used to analyse repeated measures or longitudinal studies that include covariate effects.

II. EXAMPLES

A. "EFFECT OF MYCOBACTERIUM INOCULATION ON WEIGHT"

The following experiment was conducted in a veterinary microbiology lab. Its main objective was to determine the effect of mycobacterium inoculation on the weight of immunodeficient mice. Severely immunodeficient beige mice (6-8 weeks old) were randomly allocated to the following groups:

Groups 1:   Control group where animals did not receive any inoculation.

Groups 2:   Animals were inoculated intraperitoneally with live mycobacterium paratuberculosis (MPTB) and transplanted with peripheral blood leucocytes (PBL) from humans with Crohn's disease.

Groups 3:   Animals were inoculated with live MPTB and transplanted with PBL from bovine.

In each group the mice were weighed at baseline (week 0), at week 2 and week 4. The data are given in Table 6.1.

The question of interest concerns differences between the mean weights among the three groups.

Table 6.1
Weights in Grams of Inoculated Immunodeficient Mice

Group	Mice	Time in weeks		
		0	2	4
1	1	28	25	45
	2	40	31	70
	3	31	40	44
	4	27	21	26
	5	27	25	40
2	6	34	25	38
	7	36	31	49
	8	41	21	25
	9	28	22	10
	10	29	24	22
	11	31	18	36
	12	31	15	5
3	13	28	28	61
	14	27	23	63
	15	31	30	42
	16	19	16	28
	17	20	18	39
	18	22	24	52
	19	22	22	25
	20	28	26	53

B. VARIATION IN TEENAGE PREGNANCY RATES (TAPR) IN CANADA

Table 6.2 shows teenage pregnancy rates (per 1000 females aged 15-17 and 18-19). These rates include live births, therapeutic abortions, hospitalized cases of spontaneous and other unspecified abortions, and registered stillbirths with at least 20 weeks gestation at pregnancy termination. (Source: Health Report; Statistics Canada 1991, Vol. 3, No. 4). The Newfoundland data are not included. It is important to detect variations regionally and over time in TAPR.

Table 6.2
Teenage Pregnancy Rates per 1000 Females Aged 15-17 and 18-19

Year	Age group	East					West				Territories	
		PEI	NS	NB	QU	ONT	MAN	SASK	ALTA	BC	Y	NWT
1975	15-17	41.6	43.4	40.5	12.8	39	41.7	48.5	50.5	48.6	70	77.5
	18-19	96.6	98.6	103.7	43.4	91.5	98.8	116.1	120.9	102	110	192.5
1980	15-17	28	34.9	28.9	12.9	32.8	34.2	46.3	43.2	41.0	65	104
	18-19	59.6	81.4	73.6	39.8	74.4	85.4	106.6	111.4	89.4	105	165
1985	15-17	20.3	28.6	21.6	11.9	26	35.4	37.9	34.5	29.9	50	109.3
	18-19	64.5	65.8	59.5	37.7	63	88.9	88.4	85.8	72	77.5	186
1989	15-17	21.1	30.8	22.6	16.6	27.2	40.5	37.7	34.4	31.6	38.3	97.8
	18-19	60.5	69.4	56.4	48.0	64.8	96.9	95.0	87.6	77.2	120	220

PEI = Prince Edward Island, NS = Nova Scotia, NB = New Brunswick, QU = Quebec,
ONT = Ontario, MAN = Manitoba, SASK = Saskatchewan, ALTA = Alberta, BC = British Columbia, Y = Yukon,
NWT = North West Territories

III. METHODS FOR THE ANALYSIS OF REPEATED MEASURES EXPERIMENTS

UNIVARIATE ANALYSIS OF VARIANCE OF REPEATED MEASURES EXPERIMENTS

There are two approaches for the analysis of repeated measures data, the first derived from the general setting of the linear mixed model which is known as Univariate Analysis of Variance, and will henceforth be referred to as ANOVA. The second approach attempts to use the multivariate normal model as the joint distribution of the measurements over time, and is known as Multivariate Analysis of Variance, or MANOVA.

The main focus of this section is to provide the researcher with the most practical procedure for the analysis of repeated measures data. It will be shown that in some situations the scientific question can be answered by employing just the ANOVA, while in other situations one may use both ANOVA and MANOVA procedures.

A. BASIC MODELS

Suppose that observations are obtained on n time points for each of k subjects that are divided into g groups, with k_i subjects in the ith group.

Let y_{ij} denote the jth measurement made on the ith subject. The approach to analysing data from the type of repeated measures designs that will be considered in this chapter is to proceed as in a linear mixed model analysis of variance. Therefore, the components of variation in y_{ij} are such that

$$y_{ij} = \mu_{ij} + \alpha_{ij} + e_{ij} \tag{6.1}$$

where

(i) μ_{ij} is the overall mean response of the ith individual measured at the jth time point. The μ_{ij} is called the fixed effect because it takes one unique value irrespective of the subject being observed and irrespective of the time point.

(ii)   the α_{ij} represent the departure of y_{ij} from μ_{ij} from a particular subject; this is called a random effect.

(iii) e_{ij} is the error term representing the discrepancy between y_{ij} and $\mu_{ij} + \alpha_{ij}$.

Further to the above set-up, some assumptions on the distributions of the random components of the mixed model (6.1) are needed. These assumptions are

A. $E(\alpha_{ij}) = E(e_{ij}) = 0$,
      $Var(\alpha_{ij}) = \sigma_\alpha^2$, and $Var(e_{ij}) = \sigma_e^2$.

      It is also assumed that $Cov(\alpha_{ij}, \alpha_{i'j'}) = 0$ and $Cov(\alpha_{ij}, \alpha_{ij'}) = \sigma_\alpha^2$.

B. $Cov(e_{ij}, e_{i'j'}) = 0$ $i \neq i'$ or $j \neq j'$.

C.   The random effect α_{ij} is uncorrelated with e_{ij}.

D. α_{ij} and e_{ij} are normally distributed.

Consequently, it can be shown that the covariance between any pair of measurements is

$$Cov(y_{ij}, y_{i'j'}) = \begin{cases} 0 & i \neq i' \\ \sigma_\alpha^2 & i = i' \ j \neq j' \\ \sigma_\alpha^2 + \sigma_e^2 & i = i' \text{ and } j = j' . \end{cases} \tag{6.2}$$

Therefore, any pair of measurements on the same individual are correlated, and the correlation is given by

$$Corr(y_{ij}, y_{ij'}) = \rho = \frac{\sigma_\alpha^2}{\sigma_\alpha^2 + \sigma_e^2} \tag{6.3}$$

which we have already discussed in Chapter 2 (known as the intraclass correlation). The assumptions of common variance of the observations and common correlation among all pairs of observations on the same subject can be expressed as an $n_i \times n_i$ matrix Σ, where n_i is the number of observations taken repeatedly on the ith subject and

$$Cov\ (y_{ij}, y_{ij'}) = \Sigma = \sigma^2 \begin{bmatrix} 1 & \rho & \cdots & \rho \\ \rho & 1 & \cdots & \rho \\ \cdot & \cdot & \cdots & \cdot \\ \cdot & \cdot & \cdots & \cdot \\ \cdot & \cdot & \cdots & \cdot \\ \rho & \rho & \cdots & 1 \end{bmatrix}. \tag{6.4}$$

where $\sigma^2 = \sigma_\alpha^2 + \sigma_e^2$.

A matrix of the form (6.4) is said to possess the property of compound symmetry or uniformity (Geisser, 1963). It is also assumed that the variance-covariance matrices associated with each level of group m, Σ_m, being common to all g groups (i.e., $\Sigma_m = \Sigma$).

Unfortunately, the data obtained in many experimental settings (e.g., longitudinal) rarely satisfy the assumption of compound symmetry. In such designs, it is common to find that the adjacent observations or successive measurements on adjacent time points are more highly correlated than non-adjacent time points, with the correlation between these measurements decreasing the further apart the measurement.

It should be noted that the assumptions of compound symmetry of Σ and homogeneity of Σ_m's are sufficient, but not necessary, conditions for the F-statistics to be exact in the repeated measures ANOVA.

The effects of heterogeneity of Σ_i's was first realized by Kogan (1948) who suggested that positive intraclass correlations between measurements on time points would result in a liberal F-test of the time effect.

In the next section, the recommended approach to hypothesis testing in repeated measures, including situations when the compound symmetry and homogeneity conditions are not satisfied, will be discussed following recommendations made by Looney and Stanley (1989).

B. HYPOTHESIS TESTING

Looney and Stanley (1989) recommended that hypothesis testing in the repeated measures experiment proceeds as follows:

(H-1) Test for no group by time interaction or H_{gxT}.
(H-2) Test for no group effect, or H_g.
(H-3) Test for no time effect, or H_T.

Looney and Stanley (LS) emphasized the importance of the hierarchical nature of the tests H_{gxT}, H_g, and H_T; namely, H_{gxT} must always be tested first. If H_{gxT} is supported based on the data (i.e., there is no group by time interaction) then one proceeds to test H_g (no group effect) and H_T (no time effect). On the other hand, if H_{gxT} is not supported by data, then the presence of significant interaction would require change in the strategy of hypothesis testing (Harris, 1975, p. 81). This does mean that tests for group differences and time effect cannot be performed, but alternative techniques for testing H_g and H_T should be followed. Before we outline LS recommendations, we give a brief literature review of the methods proposed to adjust the F-statistic to account for the departure of the variance-covariance matrix from compound symmetry.

1. Adjustment in the Two-Way ANOVA

Perhaps the most effective way of explaining the technical procedures suggested by statisticians, to adjust the standard F-test, is through an example. For this we look at the ADE data given in Chapter 5. The general data layout would be as in Table 6.3.

Table 6.3
General Data Layout

Time

Subject	1	2	3	n
1	y_{11}	y_{12}		y_{1n}
2	y_{21}	y_{22}		y_{2n}
$\vdots$	$\vdots$	$\vdots$		$\vdots$
k	y_{k1}	y_{k2}		y_{kn}

The two-way ANOVA for this data is summarized in Table 6.4.

Table 6.4
ANOVA of Two-Way Design

S.O.V.	d.f	S.O.S
Subjects	k-1	$n \sum_{i=1}^{k} \left(\overline{y}_{i.} - \overline{y}_{..}\right)^2 = SSBS$
Time	n-1	$k \sum_{j=1}^{k} \left(\overline{y}_{j.} - \overline{y}_{..}\right)^2 = SSBT$
Error	(k-1)(n-1)	$\sum_{i}^{k} \sum_{j}^{n} \left(y_{ij} - \overline{y}_{.j} - \overline{y}_{i.} + \overline{y}_{..}\right)^2 = SSE$

282

From the standard ANOVA, the "no time effect" is rejected if $\dfrac{SSBT/n-1}{SSE/(k-1)(n-1)}$ exceeds

$F_{\alpha,(n-1),\,(k-1)(n-1)}$.

The above test statistic is only valid if the covariance homogeneity is satisfied. That is, if the covariance matrix of the within-subjects measurements is the same for all subjects.

Box (1954b) and Imhof (1962) found that as the degree of covariance homogeneity decreased, the within-subject F-test became increasingly liberal.

In order to circumvent the problem associated with violating the compound symmetry assumption, various adjustments have been made.

Box (1954b) demonstrated that under the hypothesis H_T (no time effect), the true distribution of the F-statistic is no longer $F_{\alpha,\,n-1(k-1)(n-1)}$, but rather is approximately given by:

$$F_T = F_{\alpha,\,(n-1)\epsilon,\,(k-1)(n-1)\epsilon}$$

where the correction factor, ϵ (see equation [6.5]), depends on the elements of the variance-covariance matrix Σ. When the n variances and $\dbinom{n}{2}$ covariances are constant, that is, when Σ is of compound symmetry, ϵ equals its upper bound of unity and Box's (1954b) approximation yields exact results. However, as Σ departs from uniformity (compound symmetry), the value of ϵ decreases from one, and F_T is distributed with a decreased number of degrees of freedom. The correction factor ϵ is given by

$$\epsilon = \frac{n^2(\bar{\sigma}_{jj}-\bar{\sigma})^2}{(n-1)\left[\displaystyle\sum_{\ell}\sum_{j}\sigma_{\ell j}^2 - 2n\sum_{j=1}^{n}\bar{\sigma}_j^2 + n^2\,\bar{\sigma}^2\right]} \ . \qquad (6.5)$$

To show how ϵ is calculated, for example, let

$$\Sigma = \begin{bmatrix} 4 & 1.5 & 0.5 \\ 1.5 & 4 & 3 \\ 0.5 & 3 & 2.5 \end{bmatrix}.$$

Here,

$$n = 3 \quad , \quad \overline{\sigma}_{jj} = \frac{4+4+2.5}{3} = 3.5$$

$$\overline{\sigma} = \frac{1}{9} \, [4+1.5+0.5+1.5+4+3+.5+3+2.5] = 2.28$$

$$\overline{\sigma}_1 = \frac{4+1.5+.5}{3} = 2 \ , \overline{\sigma}_2 = \frac{1.5+4+3}{3} = 2.83 \ , \overline{\sigma}_3 = \frac{.5+3+2.5}{3} = 2$$

and

$$\epsilon = \frac{9(3.5-2.28)^2}{2\left[77.25-6(2^2+2.83^2+2^2)+9(2.28)^2\right]}$$

$$= 0.48 \ .$$

In designs with more than two factors, Geisser and Greenhouse (1958) and Greenhouse and Geisser (1959) extended the work of Box (1954b) to the analysis of groups by time, showing that the F ratio for the tests on H_T and H_{gxT} could be adjusted by the same approximate method. Moreover, these authors showed that for designs with data layout as in Table 6.3, the lower bound of ϵ was equal to $(n-1)^{-1}$. Thus,

$$\frac{1}{n-1} \leq \epsilon \leq 1 \ . \tag{6.6}$$

Geisser and Greenhouse's (1958) results can be extended to mixed models, where one of the factors is "subjects" subdivided into groups and the remaining factors are fixed. We shall investigate the circumstances under which the F ratio in question needs to be adjusted as we move along in the chapter examining each of the examples discussed in section II.

C. RECOMMENDED ANALYSIS FOR EXAMPLE A

In this data set we have three groups; subjects (mice) are nested within groups, and each mouse has been measured on three occasions.

The linear model upon which the analysis will be based has the following structure:

$$y_{ijk} = \mu + G_i + S_{j(i)} + T_k + (GT)_{ik} + TS_{kj(i)} + e_{m(ijk)} \ . \tag{6.7}$$

The notation $S_{j(i)}$ indicates that the effect of subject j is nested within the ith level of group G_i, T_k is the time effect, GT is the group by time interaction, and TS is the time by subject interaction; e is the error term (Table 6.5).

Table 6.5 ANOVA of Table 6.1

S.O.V.	df		S.O.S.	EMS
G_i	2	g-1	Between groups	$\sigma_e^2 + 3\sigma_s^2 + 18\,\phi_g^2$
$S_{j(i)}$	15	g(k-1)	Subjects within groups	$\sigma_e^2 + 3\sigma_s^2$
T_k	2	(n-1)	Time effect	$\sigma_e^2 + \phi^2_{TS} + 18\,\phi_t^2$
$(GT)_{ik}$	4	(g-1)(n-1)	Group by time interaction	$\sigma_e^2 + \sigma^2_{TS} + 6\phi^2_{gt}$
$TS_{kj(i)}$	30	g(k-1)(n-1)	Time x subject (group)	$\sigma_e^2 + \sigma^2_{TS}$
$\epsilon_{m(ijk)}$	0		Error	σ_e^2

The EMS column is easily obtainable by following Cornfield-Tukey algorithms (see Winer, 1972, p. 371). This column aids in determining the error term needed to compute the F-test on the hypothesis in question.

Step 1:

Following Looney and Stanley (1989), the first step in the exploratory analysis is to test H_{gxT} or no group by time interaction. From the EMS table the test statistic is

$$F_{gxT} = \frac{(Group \ * \ Time \ Sum \ of \ Squares)/(g-1)(n-1)}{Time \ * \ Subject \ within \ Groups \ Sum \ of \ Squares)/g(k-1)} \tag{6.8}$$

where g ≡ number of groups
n ≡ number of occasions
k ≡ number of subjects within each group

which is compared with $F_{\alpha;(g-1)(n-1),g(k-1)(n-1)}$. If the F_{gxT} given in (6.8) is not significant using the unadjusted degrees of freedom, then the result also would not be significant using any adjusted degrees of freedom and the test stops here. If the calculated F_{gxT} is significant using the unadjusted degrees of freedom, then H_{gxT} is tested using the adjusted F-test given as

$$F_{\alpha;(g-1)(n-1)\epsilon,g(k-1)(n-1)\epsilon} \ . \tag{6.9}$$

Remark

Greenhouse and Geisser (1959) suggested setting ϵ at its lower bound, $\frac{1}{n-1}$. This produces a very conservative test and the resulting $F_{\alpha;(g-1),g(k-1)}$, instead of the degrees of freedom indicated in (6.9). Setting ϵ equal to its lower limit $\left(\dfrac{1}{n-1}\right)$ tends to make the test procedure extremely conservative relative to using the F-ratio suggested in the ANOVA table. Monte Carlo studies carried out by Collier et al. (1967) indicated that the usual tests suggested by the ANOVA tend to give results closer to the nominal level of significance than do results under the Greenhouse-Geisser (1959) conservative approach, provided that the degree of heterogeneity of the covariances is relatively moderate.

Step 2: (A)

What do we do when H_{gxT} (no significant group by Time interaction) is not rejected before or after adjusting the degrees of freedom of the F-statistic?

If H_{gxT} is _not_ rejected, then H_g and H_T can be accomplished by performing tests on the marginal means. For testing no time effect (H_T) and if the matrix Σ satisfies the required pattern, then we compare the standard F-ratio (based on the EMS rule)

$$F_T = \frac{\text{(Time Sum of Squares)}/(n-1)}{\text{(Time} * \text{Subject (Group) Sum of Squares)}/g(k-1)(n-1)}$$

with

$$F_{\alpha;(n-1),g(k-1)(n-1)} \qquad\qquad (6.10)$$

However, if the covariance matrix departs from the given requirement, the Greenhouse-Geisser procedure calls for using the critical values

$$F_{\alpha;(n-1)\epsilon,g(k-1)(n-1)\epsilon} \qquad\qquad (6.11)$$

instead of (6.10).

Remarks

1. Setting $\epsilon = 1/(n-1)$ gives a very conservative test and the resulting critical value is $F_{\alpha;1,g(k-1)}$.

2. We may now wish to make comparisons between time means, noting that there is no significant group by time interaction.

To investigate a contrast between time means

$$\hat{\eta}_t = c_1 \overline{Y}_{..1} + c_2 \overline{Y}_{..2} + \ldots + c_n \overline{Y}_{..n}$$

where $\hat{y}_{..i}$ is the estimated i^{th} time mean with estimated standard error given by

$$\hat{SE}(\hat{\eta}_t) = \sqrt{\frac{\text{time*subject(group) sum of squares}/g(k-1)(n-1)}{gk} \sum_{i=1}^{n} c_i^2} \qquad (6.12)$$

Two important inference procedures based on the sampling distribution of $\hat{\eta}_T$ can be constructed:

(i)   To test the hypothesis

$$H_0 : \eta_T = \sum_{i=1}^{n} c_i \mu_{..i} = 0 , \left(\sum_{i=1}^{n} c_i = 0 \right)$$

versus $H_1 : \eta_t \neq 0$, one may use the F-statistic

$$F = \frac{(\hat{\eta}_t)^2}{\left[\hat{SE}(\hat{\eta}_t) \right]^2}$$

which is compared to $F_{\alpha;1,v_n}$, where $v_n = g(k-1)(n-1)$; the degrees of freedom of the error term "Time * Subject (Groups)" sum of squares.

(ii)   One may also construct $(1-\alpha)$ 100% confidence interval about η_t which is

$$\hat{\eta}_t - t_{\alpha/2,v_n} \cdot \hat{SE}(\hat{\eta}_t) < \eta_t < \hat{\eta}_t + t_{\alpha/2,v_n} \cdot \hat{SE}(\hat{\eta}_t). \qquad (6.13)$$

(iii)   It should be noted that if there is interest to construct simultaneous tests of the contrasts among the marginal time means, one can apply Bonferroni's approach to control the overall level of significance. For instance, since in the example being discussed we have 3 time points: 0, 2, and 4 weeks, if we are to compare the mean response at time 0 to time 2 and time 4; and time 2 to time 4 then the number of pairwise comparisons is $\binom{3}{2} = 3$. In this case, one has to divide the level of significance associated with the F-test (6.13) by $\binom{3}{2}$. That is, one should test at $\alpha^* = \alpha/3$.

3. We may wish to make comparisons between group means, noting that there is no significant group by time interaction.

A contrast between group means can be written as

$$\hat{\xi}_g = c_1 \, \bar{y}_{1..} + c_2 \, \bar{y}_{2..} + ... c_g \, \bar{y}_{g..}$$

where $\bar{y}_{i..}$ is the estimated i^{th} group mean. The estimated standard error of ξ_g is given by

$$\hat{SE} (\hat{\xi}_g) = \sqrt{\frac{\text{Subjects (Group) Sum of Squares} / g(k-1)}{kn} \sum_{i=1}^{g} c_i^2} \; .$$

Similarly, such a contrast can be used to:

(i)   Test the hypothesis

$$H_0: \xi_g = \sum_{i=1}^{g} c_i \, \mu_{i..} = 0 \quad \left(\sum_{i=1}^{g} c_i = 0 \right)$$

versus $H_1: g \neq 0$ using the F-statistic

$$F = \frac{\left(\hat{\xi}_g \right)^2}{\left[\hat{SE} \left(\hat{\xi}_g \right) \right]^2} \tag{6.14}$$

which is compared with $F_{\alpha, 1, v_g}$, where $v_g = g(k-1)$ are the degrees of freedom associated with the error term "Subject (Groups)" sum of squares.

(ii)   One may also construct a $(1-\alpha)$ 100% confidence interval about ξ_g which is

$$\hat{\xi}_g - t_{\frac{\alpha}{2}, v_g} \cdot \hat{SE} \left(\hat{\xi}_g \right) < \xi_g < \hat{\xi}_g + t_{\frac{\alpha}{2}, v_g} \cdot \hat{SE} \left(\hat{\xi}_g \right) \tag{6.15}$$

(iii)   Construction of simultaneous tests of the contrasts of interest can be applied following Bonferroni's approach.

Step 2: (B)

What do we do when H_{gxT} (no significant group by time interaction) is rejected? That is, we have significant group by time interaction.

If H_{gxT} is rejected, then it is not appropriate to test either H_g or H_T as we suggested. One alternative is to attempt to find a transformation of the multivariate data so that the test of H_{gxT} using the transformed data is no longer significant, as was suggested by Andrews et al. (1971). If such a transformation can be found, we may then proceed as in Step 2: (A). We do not recommend transforming the data, but rather recommend the procedure suggested by Morrison (1976, p. 208). Morrison suggested that one tests the hypothesis of equal group means separately for each time point using the usual ANOVA test. Similarly, test the hypothesis of equal time means for each group.

(a)   *Testing equality of time means at each group.* Let θ_T denote the contrast between time means at the same group. Then

$$\hat{\theta}_T = c_1 \bar{y}_{.1} + c_2 \bar{y}_{.2} + \ldots + c_n \bar{y}_{.n}$$

where

$$\sum_{i=1}^{n} c_i = 0$$

the estimated standard error of θ_T is given by

$$\hat{SE}\left(\hat{\theta}_T\right) = \sqrt{\frac{\text{Time} * \text{Subject (Group) Sum of Squares}/g(k-1)(n-1)}{k} \sum_{i=1}^{n} c_i^2} \, .$$

Before we construct test statistic and confidence intervals on the appropriate contrasts one should realize that for a test of overall size α, it is suggested that $\alpha^* = \alpha/g$ be used for tests of the time effects at each group.

We can now test the hypothesis

$$H_0: \theta_T = 0 \quad \text{versus} \quad H_1: \theta_T \neq 0$$

by comparing the statistic

$$F = \frac{\left[\hat{\theta}_T\right]^2}{\left[\hat{SE}\left(\hat{\theta}_T\right)\right]^2}$$

to $F_{\alpha^*, 1, v_n}$ where $v_n = g(k-1)(n-1)$.

(b)   *To compare two group means at the same time.* For test of size α, it is suggested to use $\alpha^* = \alpha/n$ to control the overall level of significance. The estimated contrast is given by

289

$$\hat{\theta}_g = c_1 \bar{y}_{1..} + ... + c_g \bar{y}_{g..}$$

with estimated standard error

$$S\hat{E}(\hat{\theta}_g) = \sqrt{M \sum_{i=1}^{g} c_i^2}$$

where

$$M = \frac{E_1 + (n-1) E_2}{n},$$

$E_1 =$ Subject (Group) sum of squares / g(k-1), and

$E_2 =$ Time * Subject (Group) sum of squares / g(k-1) (n-1) (for details, see Milliken & Johnson, 1984).

To test the hypothesis H_0: $\theta_g = 0$ versus H_1: $\theta_g \neq 0$, we use the approximate F-statistic

$$F = \frac{[\hat{\theta}_g]^2}{[S\hat{E}(\hat{\theta}_g)]^2}$$

with $F_{\alpha^*, 1, r}$, where r is determined via the Satterwaite approximation as

$$r = [E_1 + (n-1)E_2]^2 / w \tag{6.16}$$

where

$$w = \frac{E_1^2}{g(k-1)} + \frac{[(n-1)E_2]^2}{g(k-1)(n-1)}.$$

D.   RECOMMENDED ANALYSIS FOR EXAMPLE B

This example has three factors; region (3 levels), age (2 levels) and year (4 levels). Provinces or subjects are nested within region; the observations are repeated on levels of age and years. Our first attempt to analyze this data is under the assumption that the underlying model may be written as

$$y_{ijkl} = \mu + R_i + A_j + (RA)_{ij} + P_{k(i)} + T_l + TR_{il} + (RT)_{il} + (AT)_{jl} + (RAT)_{ijk} + e_{ijklm}$$

where μ is the overall mean, R_i is the effect of the ith region (i=1,2,3), A_j is the effect of the jth age (j=1,2), $P_{k(i)}$ are the provinces within region and T_l are the years or time effect (4 levels). The other terms represent interactions and e_{ijkl} is the error term. Clearly, under this model, one should not consider "province" (subjects) as a random effect since all available provinces within a particular region have been included, hence should be considered "fixed". However, in many investigations, sampling from a large number of subjects within a particular region is more efficient, in which case "subjects" within region must be considered "random".

Under the above model, the G-G estimate of ϵ was $\hat{\epsilon}=0.52$. Following the hypothesis testing strategy that we have outlined, the following interactions were tested:

Source of variation	df	G-G F statistic	G-G p-value
Year*region	6	2.54	0.078
Year*age	3	3.98	0.041
Year*age*region	6	3.52	0.0287

Because of the significant interaction of age*region and year*region*age, we decided to reanalyze the data after stratifying on the levels of age; *age1* is the 15-17 age group and *age2* is the 18-19 age group. This results in "two" data sets, one for each age level, each of which can be analyzed under a model similar in structure to the model of Example A. That is,

$$y = \mu + Region + Province(Region) + Year + Year*Region + error \ .$$

We now follow the hypothesis testing hierarchy:

i) H_0: no region by year interaction
ii) H_0: no region effect
iii) H_0: no year effect

for each age level.

1. Age 15-17

The G-G estimate ϵ is 0.36 and the p-value on the hypothesis of no time*region interaction is 0.3009. Thus, there is no need to look for regional differences at each year. In general, there was a significant year effect (p-value of 0.0001) and a significant region effect for which the F statistic is $F = 18128.02/2440.055 = 7.429$ this value being larger than $F_{.05,2,8} = 4.46$. By region, the mean pregnancy rates are 42.82, 76.49 and 39.74 for the East, Territories and West, respectively. To look for regional differences, we use a pairwise comparison. Since the error term is Province (Region), with a mean square value of 563.27, the standard error of the contrast,

$$\hat{\xi}_g = (mean\ region)_i - (mean\ region)_j$$

is

$$S\hat{E}(\hat{\xi}_g) = \sqrt{2[\frac{563.27}{44}]}$$

$$= 5.06$$

Therefore, the Territories differ significantly from both the East and the West, and there is no significant difference between the East and the West.

2. Age 18-19

The G-G estimate of ϵ is 0.54 and the p value on the hypothesis of no time*region interaction is 0.0305. Because there is a significant interaction, we compare between the regions at each time point (year). Under the hypothesis H_0: no regional effect, the resulting p-values by year are:

Year	p-value
1975	0.0664
1980	0.0102
1985	0.0444
1989	0.0034

The pairwise comparisons showed significant differences between the Territories and both the East and the West, but no significant difference between the East and the West. Without going into the details, the standard error of a contrast between two regions is

$$S\hat{E} = \sqrt{2M}$$

where

$$M = \frac{E_1 + (n-1)E_2}{n}$$

$$E_1 = 2440.055$$

$$E_2 = 89.957 \ .$$

We now compare between Years (time) for each region. The following tables gives the p values for the hypothesis H_0: no time effect :

Region	p-value
East	0.0512
Territory	0.9225
West	0.0201

Now, since the Western provinces show strong variation over time, it may be worthwhile to show the differences in pregnancy rates between time points.

Contrast	Mean Differences	F
1975 - 1980	109.45 - 98.2	2.81
1975 - 1985	109.45 - 83.78	14.64
1975 - 1989	109.45 - 89.175	9.13
1980 - 1985	98.2 - 83.78	4.62
1980 - 1989	98.2 - 89.175	1.8
1985 - 1989	83.78 - 89.175	0.646

The test-wise Type I error rate is $0.10/(6)(3) = 0.005$ for an overall level of 10%. Note that the standard error for any contrast is

$$\hat{SE}(\hat{\theta}_T) = \sqrt{2[\frac{89.957}{4}]} = 6.71$$

with 24 degrees of freedom. Since $F_{0.005,1,24} = 9.55$, the only significant difference is between 1975 and 1985.

IV. MISSING OBSERVATIONS

The methods for the analysis of repeated measures experiments that we have discussed depend on a balanced design where all subjects are observed at the same times and there are no missing observations. In this section, we examine the problems caused by missing observations and unbalanced designs.

It is important to consider the reasons behind missing observations. If the reason that an observation is missing is related to the response variable that is missed, the analysis will be biased. Rubin (1976, 1994) discusses the concept of data that are missing at random in a general statistical setting. Little and Rubin (1987) distinguish between observations that are missing completely at random, where the probability of missing an observation is independent of both the observed responses and the missing responses, and missing at random, where the probability of missing an observation is independent of the responses. Laird (1988) used the term "ignorable missing data" to describe observations that are missing at random.

Diggle (1989) discussed the problem of testing whether dropouts in repeated measures investigations occur at random. Dropouts are a common source of missing observations, and it is important for the analysis to determine if the dropouts occur at random. In the subsequent section, we will assume that the mechanism for missing observations is ignorable. Moreover, the analyses recommended are only appropriate for repeated measures experiments that satisfy the sphericity conditions.

It is perhaps most appropriate to explain how an unbalanced repeated measures experiment can be analyzed by introducing an example.

Suppose that a feed trial aims at comparing between two diets (1 and 2). Three animals were assigned to diet 1 and four animals were assigned to diet 2. Also suppose that three animals died after the first 2 weeks. Their weights in kilograms are shown in the following table (Table 6.6).

Example 6.1

Table 6.6
Weights (in kg) of Animals on Two Diets Over 4 Weeks

Week	Diet (1) Animal			Diet (2) Animal			
	1	2	3	4	5	6	7
1	2.5	2	1.5	2	1	3	4
2	3	2.5	2	4	2	3	6
3	4		2			4	8

In general, let y_{hij} denote the score at the jth week for the ith individual assigned to the hth diet group.

$$y_{hij} = \mu + G_h + \delta_{i(h)} + w_j + \epsilon_{ij(h)} \qquad (6.17)$$

The "mixed effects" model under which this experiment is analyzed can be written as $h=1,2,...M$; $i=1,2,...k_h$; $j=1,2,...n_{hi}$ where M is the number of diet groups, k_h is the number of subjects receiving the hth diet, n_{hi} is the number of time points at which the ith subject is measured under the hth group (diet). Here μ is an overall mean, G_h is a fixed effect representing the effect of the hth group. It is assumed that within groups subjects $\delta_{i(h)} \sim$ i.i.d. $N(0,\sigma_\delta^2)$ which in fact represents the error distribution of the ith subject in the hth group. Also, as before, it is assumed that $\epsilon_{ij(h)} \sim$ i.i.d. $N(0,\sigma_\epsilon^2)$ which represents the error distribution of the jth week to the ith patient within the hth group. It is also assumed that these two error components are distributed independently of each other. Finally, w_j is the time effect.

Under the model (6.17),

$$E(y_{hij}) = \mu + G_h + w_j \ .$$

and

$$Var(y_{hij}) = \sigma_\delta^2 + \sigma_\epsilon^2 \ .$$

Furthermore, the correlation between any two measurements on the same subject is

$$\rho = Corr(y_{hij}, y_{him}) = \frac{\sigma_\delta^2}{\sigma_\delta^2 + \sigma_e^2} \qquad j \neq m$$

$$i = 1,2,...k_h$$
$$j,m = 1,2,...n_{hi} \ .$$

We shall use the following notation in the remainder of this section:

$$K = \sum_{h=1}^{M} k_h \qquad = \qquad \text{Total number of subjects}$$

$$N_h = \sum_{i=1}^{k_h} n_{hi} \qquad = \qquad \text{Number of repeated measures in the } h^{th} \text{ group}$$

$$N = \sum_{h=1}^{M} N_h \qquad = \qquad \text{Total number of observations in the entire data set}$$

SSG = The group sum of squares

$$= \sum_{h=1}^{M} \sum_{i=1}^{k_h} n_{hi}(\bar{y}_{hi.} - \bar{y}_{h..})^2$$

which carries (M-1) degrees of freedom.

$$E\left(\frac{SSG}{M-1}\right) = \sigma_e^2 + \lambda_1 \sigma_\delta^2 + \frac{1}{M-1} Q$$

where

$$\lambda_1 = \frac{1}{M-1}\left[\sum_{h=1}^{M} \sum_{i=1}^{k_h} \frac{n_{hi}^2}{N_h} - \sum_{h=1}^{M} \sum_{i=1}^{k_h} \frac{n_{h_i}^2}{N}\right]$$

and

$$Q = \sum_{h=1}^{M} N_h^2 G_h^2 .$$

The subjects within groups sum of squares is given as

$$Subject(Group) = \sum_{h-1}^{M} \sum_{i=1}^{k_h} N_h \left(\bar{y}_{hi.} - \bar{y}_{h..}\right)^2$$

which carries K-M degrees of freedom.

$$E\left[\frac{Subject(Group)}{K-M}\right] = \sigma_e^2 + \lambda_2 \sigma_\delta^2$$

where

$$\lambda_2 = \frac{1}{K-M}\left[N - \sum_{h=1}^{M} \sum_{i=1}^{k_h} \left(\frac{n_{hi}^2}{N_h}\right)\right] .$$

The error sum of squares is

$$SSE = \sum_{h=1}^{M} \sum_{i=1}^{k_h} \sum_{j=1}^{n_{hi}} \left(y_{hij} - \bar{y}_{hi.} \right)^2$$

which carries N-K degrees of freedom. Moreover,

$$E \left[\frac{SSE}{N-K} \right] = \sigma_e^2 \; .$$

The above results can be summarized in the following ANOVA table (Table 6.7).

Table 6.7
ANOVA for Unbalanced Repeated Measures Experiment

S.O.V.	d.f	S.O.S.	M.S.	EMS
Group	M-1	SSG	$SG = \dfrac{SSG}{M-}$	$\sigma_e^2 + \lambda_1 \sigma_\delta^2 + \dfrac{Q}{M-1}$
Subjects (Group)	K-M	Subj(Group)	$MSSubj(Group)$ $= \dfrac{Subj(Group)}{K-M}$	$\sigma_e^2 + \lambda_2 \sigma_\delta^2$
Residuals	N-K	SSE	$SE = \dfrac{SSE}{N-K}$	σ_e^2
Total	N-1			

As can be seen from the EMS column of Table 6.7, the only meaningful exact F-test that can be constructed is on the hypothesis $H_0 : \sigma_\delta^2 = 0$ versus $H_1 : \sigma_\delta^2 > 0$ which tests the variability among subjects. This is given as

$$F = MSSubj(Group)/MSE \sim F_{K-M, N-K} \; .$$

We now show how a repeated measures experiment with missing data can be analysed in SAS, using PROC GLM including an interaction term:

```
PROC GLM;
CLASS DIET ANIMAL WEEK;
Model Y = DIET | WEEK ANIMAL (DIET) / E SS3;
RANDOM ANIMAL (DIET);
```

The EMS table produced by SAS is given below (Table 6.8).

Table 6.8
The EMS Table for the Data of
Example 6.1

Source	Type III Expected Mean Square
DIET	$\sigma_e^2 + 2.2105\,\sigma_\delta^2 + Q\,(\text{DIET, DIET*WEEK})$
WEEK	$\sigma_e^2 + Q\,(\text{WEEK, DIET*WEEK})$
DIET*WEEK	$\sigma_e^2 + Q\,(\text{DIET*WEEK})$
ANIMAL(DIET)	$\sigma_e^2 + 2.4\,\sigma_\delta^2$
ERROR	σ_e^2

The ANOVA results are given in Table 6.9.

Table 6.9
ANOVA of the Mixed Repeated Measures Experiment

Source	df	Type III SS	Mean Square
DIET	1	6.32	6.32
WEEK	2	7.67	3.83
DIET*WEEK	2	1.52	0.76
ANIMAL(DIET)	5	21.71	4.34
ERROR	7	3.04	0.44

From Table 6.9, the F-statistic on $H_0 : \sigma_\delta^2 = 0$ is

$$F = \frac{4.34}{0.44} = 9.98$$

since $F_{.05,5,7} = 3.97$, the hypothesis is rejected.

To test no time effect (H_T), the ratio

$$F = \frac{MS\ Week}{MSE} = \frac{3.83}{0.44} = 8.81$$

is larger than $F_{.05,2,7} = 4.74$, and we conclude that there is a significant time effect. To test the hypothesis of no diet (group) by week interaction (i.e., H_{gxT}) we compare the ratio

$$F = \frac{MS\ Diet*Week}{MSE} = \frac{.76}{.44} = 1.75$$

to $F_{.05,2,7} = 4.74$, and we conclude that there is no diet by week interaction.

Recall that for the balanced data, we used the ratio

$$MS(DIET) / MSSubject(DIET)$$

to test for significant diet effect (i.e., H_0:No diet effect). This is not appropriate for unbalanced data since the coefficient of σ_δ^2 in the expected values of the numerator and denominator mean square are different. In fact, in Table 6.7 these coefficients are given respectively by λ_1 and λ_2, and in the given example $\lambda_1 = 2.21$, and $\lambda_2 = 2.4$. A statistic to test H_0: (no diet effect) should be

$$\tilde{F} = \frac{6.32}{\hat{\sigma}_e^2 + 2.21\ \hat{\sigma}_\delta^2} \ . \qquad\qquad (6.18)$$

Therefore, one has to find estimates for the variance components σ_e^2 and σ_δ^2. This is usually obtained by equating the mean squares from the ANOVA to their expected values. Since

$$E\left[MSE\right] = \sigma_e^2$$

and

$$E\left[MSSubj(Diet)\right] = \sigma_e^2 + 2.4\ \sigma_\delta^2$$

we need to solve the equations

$$0.44 = \hat{\sigma}_e^2$$
$$4.34 = \hat{\sigma}_e^2 + 2.4\ \hat{\sigma}_\delta^2$$

therefore,

$$\hat{\sigma}_e^2 = 0.44$$
$$\hat{\sigma}_\delta^2 = 1.63 \ .$$

Substituting in (6.18) we get

$$\tilde{F} = \frac{6.32}{0.44 + 2.21(1.63)} = 1.57 \ .$$

The statistic $\tilde{F}$ does not have an exact F-distribution for two reasons: (i) The statistic in the denominator, $\hat{\sigma}_e^2 + \lambda_2\ \hat{\sigma}_\delta^2$, does not have a distribution proportional to an exact chi-square distribution, and (ii) the numerator and denominator of $\tilde{F}$ may not be statistically independent.

However, we may use the Satterwaite approximation to the distribution of $\widetilde{F}$, in order to obtain approximate critical points. The approximation is quite simple and proceeds as follows:

Let

$$\frac{v_i \, \hat{\sigma}_i^2}{\sigma_i^2} \sim \chi_{v_i}^2 \qquad (i=1,2) \ .$$

If $\hat{\sigma}_1^2$ and $\hat{\sigma}_2^2$ are independent, we need to find an approximation to the distribution of $D=c_1 \, \hat{\sigma}_1^2 + c_2 \, \hat{\sigma}_2^2$, where c_1 and c_2 are positive constants. Let us assume that the distribution of D can be approximated by a multiple of χ^2 with v degrees of freedom. That is,

$$c_1\hat{\sigma}_1^2 + c_2\hat{\sigma}_2^2 \ \dot{\sim} \ a \, \chi_v^2 \ . \qquad (6.19)$$

The idea is to obtain a and v by equating the mean and variance of both sides of (6.19).

Since

$$E(D) \ = \ c_1 \ E(\hat{\sigma}_1^2) + c_2 \ E(\hat{\sigma}_2^2)$$

$$= \ c_1 \ \sigma_1^2 + c_2 \ \sigma_2^2$$

and

$$Var(D) \ = \ c_1^2 \ Var(\hat{\sigma}_1^2) + c_2^2 \ Var(\hat{\sigma}_2^2)$$

$$= \ \frac{2 \ c_1^2 \ \sigma_1^4}{v_1} + \frac{2 \ c_2^2 \ \sigma_2^4}{v_2} \ .$$

Therefore,

$$c_1 \ \sigma_1^2 + c_2 \ \sigma_2^2 \ = \ av$$

and

$$2(\frac{c_1^2 \ \sigma_1^4}{v_1} + \frac{c_2^2 \ \sigma_2^4}{v_2}) \ = \ 2a^2v$$

from which

$$a = \frac{\dfrac{c_1^2 \, \sigma_1^4}{v_1} + \dfrac{c_2^2 \, \sigma_2^4}{v_2}}{c_1\sigma_1^2 + c_2\sigma_2^2}$$

and

$$v = \frac{(c_1 \, \sigma_1^2 + c_2\sigma_2^2)^2}{\dfrac{c_1^2 \, \sigma_1^4}{v_1} + \dfrac{c_2^2 \, \sigma_2^4}{v_2}} \; . \tag{6.20}$$

Therefore,

$$\frac{v(c_1 \, \hat{\sigma}_1^2 + c_2\hat{\sigma}_2^2)}{c_1 \, \sigma_1^2 + c_2 \, \sigma_2^2} \sim X_v^2 \tag{6.21}$$

approximately.

We shall now use the above approximation to derive an approximate test on group (diet) effect. Under H_0: No diet effect,

$$(1) \; \frac{\left[\textit{Type III Diet Mean Squares}\right]}{\sigma_e^2 + 2.21 \, \sigma_\delta^2} \sim X_{(1)}^2$$

and

$$(5) \; \frac{\left[\textit{Animal (Diet) Mean Squares}\right]}{\sigma_e^2 + 2.4 \, \sigma_\delta^2} \sim X_{(5)}^2$$

and

$$(7) \; \frac{\left[\textit{Error Mean Squares}\right]}{\sigma_e^2} \sim X_{(7)}^2 \; .$$

Let $\sigma_1^2 = \sigma_e^2 + 2.4 \, \sigma_\delta^2$, and $\sigma_2^2 = \sigma_e^2$. Then from (6.20) and (6.21)

$$\hat{\sigma}_1^2 = \hat{\sigma}_e^2 + 2.4 \, \hat{\sigma}_\delta^2, \quad \hat{\sigma}_2^2 = \hat{\sigma}_e^2,$$
$$v_1 = 5, \text{ and } v_2 = 1 \; .$$

Noting that

$$\sigma_e^2 = \sigma_2^2$$

and

$$\sigma_\delta^2 = \frac{\sigma_1^2 - \sigma_2^2}{2.4}$$

then

$$\sigma_e^2 + 2.21\ \sigma_\delta^2 = \sigma_2^2 + 2.21\left(\frac{\sigma_1^2 - \sigma_2^2}{2.4}\right)$$

$$= \frac{2.21}{2.4}\ \sigma_1^2 + \left(1 - \frac{2.21}{2.4}\right)\ \sigma_2^2$$

$$= .92\ \sigma_1^2 + 0.08\ \sigma_2^2$$

which means that $c_1 = 0.92$ and $c_2 = 0.08$. Using Satterwaite's approximation we have

$$\frac{v(0.92\ \hat{\sigma}_1^2 + 0.08\ \hat{\sigma}_2^2)}{\sigma_e^2 + 2.21\ \sigma_\delta^2} \sim X_v^2$$

where

$$v = \frac{[(0.92)(4.34) + (0.08)(0.44)]^2}{(0.92)\dfrac{(4.34)^2}{5} + (0.08)\dfrac{(0.44)^2}{7}}$$

$$= \frac{16.27}{3.47 + .002} = 4.68\ .$$

Therefore, $\widetilde{F}$ is approximately distributed as $F_{\alpha,1,4.68}$, under H_0. The p-value is given by $\Pr[F > F_{\alpha,1,4.68}] = 0.269$ and we conclude that there is no significant difference between diets in the way they affect growth.

The analysis of repeated measures with missing observations is one of the most ubiquitous models in biometrical and medical research. However, significance testing under such models is not straightforward, even under the assumption that the sphericity condition is satisfied. However, using PROC GLM in SAS we can obtain some useful information from the Type III sum of squares that helps with the testing for group effect. The procedure which was discussed in the above example was recommended by Milliken and Johnson (1984) and can be summarized as follows:

(1)   In the model statement of PROC GLM in SAS, we should list the fixed effects first, followed by listing the random effects. One should also use the RANDOM option in order to obtain expected mean squares, which are required to construct tests of hypotheses and confidence intervals.

(2)   From the table of expected mean squares (Table 6.8), we decide on the one corresponding to the effect being considered. This will be in the form $\sigma_e^2 + c\sigma_\delta^2 + Q$. Therefore, an approximate divisor for obtaining an approximate F-statistic will be $\hat{\sigma}_e^2 + c\hat{\sigma}_\delta^2$. An approximate F-statistic is

$$\tilde{F} = \frac{\textit{Mean Square of effect}}{\hat{\sigma}_e^2 + c\,\hat{\sigma}_\delta^2}$$

which should be compared with F_{α, d_1, d_2} where d_1 are the degrees of freedom of the numerator of $\tilde{F}$ and

$$d_2 = \frac{\left| c\,\hat{\sigma}_1^2 + (q-c)\,\hat{\sigma}_2^2 \right|^2}{\dfrac{c^2\,\hat{\sigma}_1^4}{\nu_1} + \dfrac{(q-c)^2\,\hat{\sigma}_2^2}{\nu_2}} \quad .$$

$\hat{\sigma}_1^2 = \hat{\sigma}_e^2 + q\,\hat{\sigma}_\delta^2$, $\hat{\sigma}_2^2 = \hat{\sigma}_e^2$. Note that $\hat{\sigma}_1^2$ is the Subject (Group) mean square, which carries ν_1 degrees of freedom, and $\hat{\sigma}_2^2$ is the error mean squares, which carries ν_2 degrees of freedom.

V. MIXED LINEAR REGRESSION MODELS

A. FORMULATION OF THE MODELS

A general linear mixed model for longitudinal and repeated over time data has been proposed by Laird and Ware (1982);

$$y_i = x_i\beta + z_i\alpha_i + \varepsilon_i \tag{6.22}$$

where y_i is an $n_i \times 1$ column vector of measurements for subject (cluster) i, X_i is an $n_i \times p$ design matrix, β is a $p \times 1$ vector of regression coefficients assumed to be fixed, z_i is an $n_i \times q$ design matrix for the random effects, α_i, which are assumed to be independently distributed across subjects with distribution $\alpha_i \sim N(0, \sigma^2 B)$ where B is an arbitrary covariance matrix. The within subjects errors, ε_i are assumed to be distributed $\varepsilon_i \sim N(0, \sigma^2 W_i)$. In many applications, it is assumed that $W_i = I$, the identity matrix. It is also assumed that ε_i and α_i are independent of each other. The fact that Y_i, X_i, Z_i and W_i are indexed by i means that these matrices are subject specific. Note that, the model (6.17) is a special case of (6.22). To illustrate this point, we consider example (6.1) with the missing values being imputed, so that we have balanced data. Table (6.6) becomes

Week	Diet (1) Animal			Diet (2) Animal			
	1	2	3	4	5	6	7
1	2.5	2	1.5	2	1	3	4
2	3	2.5	2	4	2	3	6
3	4	3	2	3	3	4	8

To show how model (6.17) is a special case of the general linear mixed model (6.22), we define an indicator variable for each animal to indicate group membership.

$$\text{Let} \quad x_{(i)jk} = \begin{cases} 0 & \text{if group is 1} \\ 1 & \text{if group is 2} \end{cases}$$

The $(i)jk$ subscript indicates that animal j is nested within the diet group i, and k indicates repeated measures on the jth animal. Hence equation (6.17) can now be written as

$$Y_{(i)jk} = \beta_1 + \beta_2 x_{(i)jk} + \alpha_{(j)k} + \varepsilon_{(i)jk} \qquad (6.23)$$

The model for animal j in diet group 1 is

$$\begin{bmatrix} y_{(1)j1} \\ y_{(1)j2} \\ y_{(1)j3} \end{bmatrix} = \begin{bmatrix} 1 & 0 \\ 1 & 0 \\ 1 & 0 \end{bmatrix} \begin{bmatrix} \beta_1 \\ \beta_2 \end{bmatrix} + \begin{bmatrix} 1 \\ 1 \\ 1 \end{bmatrix} \alpha_{(1)j} + \begin{bmatrix} \varepsilon_{(1)j1} \\ \varepsilon_{(1)j2} \\ \varepsilon_{(1)j3} \end{bmatrix}$$

and for animals in the second diet group

$$\begin{bmatrix} y_{(2)j1} \\ y_{(2)j2} \\ y_{(2)j3} \end{bmatrix} = \begin{bmatrix} 1 & 1 \\ 1 & 1 \\ 1 & 1 \end{bmatrix} \begin{bmatrix} \beta_1 \\ \beta_2 \end{bmatrix} + \begin{bmatrix} 1 \\ 1 \\ 1 \end{bmatrix} \alpha_{(2)j} + \begin{bmatrix} \varepsilon_{(2)j1} \\ \varepsilon_{(2)j2} \\ \varepsilon_{(2)j3} \end{bmatrix}.$$

The β_1 is the mean of diet group 1, and $\beta_1 + \beta_2$ is the mean of diet group 2.

There are several ways of modelling the means or fixed effects. We show two of them:

(i) Model (1) for fixed effects:

In this model, the fixed part $x_i\beta$, for the first diet group can be written as

$$x_j\beta = \begin{bmatrix} 1 & 0 & 0 & 0 & 0 & 0 \\ 0 & 1 & 0 & 0 & 0 & 0 \\ 0 & 0 & 1 & 0 & 0 & 0 \end{bmatrix} \begin{bmatrix} \beta_1 \\ \beta_2 \\ \beta_3 \\ \beta_4 \\ \beta_5 \\ \beta_6 \end{bmatrix} \qquad (a)$$

$$= \begin{bmatrix} \beta_1 & \beta_2 & \beta_3 \end{bmatrix}'$$

and for the second diet group is

$$x_j\beta = \begin{bmatrix} 0 & 0 & 0 & 1 & 0 & 0 \\ 0 & 0 & 0 & 0 & 1 & 0 \\ 0 & 0 & 0 & 0 & 0 & 1 \end{bmatrix} \begin{bmatrix} \beta_1 \\ \beta_2 \\ \beta_3 \\ \beta_4 \\ \beta_5 \\ \beta_6 \end{bmatrix} \qquad (b)$$

$$= \begin{bmatrix} \beta_4 & \beta_5 & \beta_6 \end{bmatrix}'$$

The above two equations represent a model in which each diet group has a different growth curve. The number of parameters of the fixed effect part of the model equals the number of diet groups multiplied by the number of time points.

Another possible reparameterization is, for the diet group 1 we keep the representation (a) but for diet group 2 we have

$$x_j \beta = \begin{bmatrix} 1 & 0 & 0 & 1 & 0 & 0 \\ 0 & 1 & 0 & 0 & 1 & 0 \\ 0 & 0 & 1 & 0 & 0 & 1 \end{bmatrix} \begin{bmatrix} \beta_1 \\ \beta_2 \\ \beta_3 \\ \beta_4 \\ \beta_5 \\ \beta_6 \end{bmatrix}$$

$$= \begin{bmatrix} \beta_1 + \beta_4 & \beta_2 + \beta_5 & \beta_3 + \beta_6 \end{bmatrix}$$

Under this parameterization, the coefficients β_1, β_2 and β_3 still generate growth curve for diet group 1, while $(\beta_4, \beta_5, \beta_6)$ are the differences between the two growth curves.

(ii) Model (2) for fixed effects :

The second model for the fixed effects is a straight line for each group. One possibility for group 1 is

$$x_j \beta = \begin{bmatrix} 1 & 1 & 0 \\ 1 & 2 & 0 \\ 1 & 3 & 0 \end{bmatrix} \begin{bmatrix} \beta_1 \\ \beta_2 \\ \beta_3 \end{bmatrix} = \begin{bmatrix} \beta_1 + \beta_2 \\ \beta_1 + 2\beta_2 \\ \beta_1 + 3\beta_2 \end{bmatrix} \qquad (c)$$

and for group 2

$$x_j \beta = \begin{bmatrix} 1 & 1 & 1 \\ 1 & 2 & 1 \\ 1 & 3 & 1 \end{bmatrix} \begin{bmatrix} \beta_1 \\ \beta_2 \\ \beta_3 \end{bmatrix} = \begin{bmatrix} \beta_1 + \beta_2 + \beta_3 \\ \beta_1 + 2\beta_2 + \beta_3 \\ \beta_1 + 3\beta_2 + \beta_3 \end{bmatrix} \qquad (d)$$

Note that for group 1, β_1 is the intercept and β_2 is the slope for the line. The hypothesis that the two lines are parallel is

$$H_0: \beta_3 = 0$$

Note also that the second column in the X_i matrix represents the time points at which the measurements were taken. The third column is for the dummy variable denoting group membership.

(iii) Modelling the random effects :

There are many ways to model the random components of the between subjects random effects and the within subjects error structure. Here we present 4 models as suggested by Jennrich and Schluchter (JS) (1986).

1. Unstructured Covariance (arbitrary covariance structure):

Here, the covariance matrix of the observations on each animal is an arbitrary (3×3) covariance matrix. This arbitrary error structure can be specified under the model (6.22) following one of the two approaches :

- eliminate ε_i term, and set z_i equal to 3×3 identity matrix. The resulting error structure has arbitrary covariance matrix:

$$E\left[\left(Y - E(Y) \right)\left(Y - E(Y) \right)' \right] = E[\alpha\alpha'] = \sigma_\alpha^2 B$$

- eliminate α_i which is equivalent to setting $B = 0$. The resulting covariance structure is

$$E\left[\left(Y - E(Y) \right)\left(Y - E(Y) \right)' \right] = E[ee'] = \sigma^2 W_i$$

These models are equivalent to models 1, 2 and 3 of JS (1986).

2. Compound Symmetry (CS model):

This covariance structure is obtained by setting $W_i = \sigma_e^2 I$, z_i as a 3×3 matrix of 1's and $B = \sigma_a^2 I$. Hence

$$E\left[(Y - E(Y))(Y - E(Y))'\right] = E\left[(1a + e)(1a + e)'\right]$$
$$= \sigma_a^2 11' + \sigma_e^2 I .$$

Model 7 of JS is the CS model..

3. Random Coefficients Model (RCM):

For this model, the fixed effects are a separate line for each group as in equations (c) and (d). The between animals (clusters) component is

$$z_j \alpha_j = \begin{bmatrix} 1 & 1 \\ 1 & 2 \\ 1 & 3 \end{bmatrix} \begin{bmatrix} \alpha_{1j} \\ \alpha_{2j} \end{bmatrix} = \begin{bmatrix} \alpha_{1j} + \alpha_{2j} \\ \alpha_{1j} + 2\alpha_{2j} \\ \alpha_{1j} + 3\alpha_{2j} \end{bmatrix}$$

The random components α_{1j} and α_{2j} represent the random deviation of the intercept and slope for the jth animal from the animal's group mean intercept and slope. Moreover, it is assumed that $W_i = I$.

4. Autoregressive Covariance Structure (AR)

JS (1986) model 5 is a first order autoregression where the within animal error structure is a first order autoregressive process, and the between subjects components is set equal to zero $(B = 0)$. For the 3 time points example:

$$W_i = \begin{bmatrix} 1 & \rho & \rho^2 \\ \rho & 1 & \rho \\ \rho^2 & \rho & 1 \end{bmatrix}$$

with $-1 \le \rho \le 1$.

B. MAXIMUM LIKELIHOOD (ML) AND RESTRICTED MAXIMUM LIKELIHOOD (REML) ESTIMATION

The estimation of the parameters of the general mixed model (6.22) is obtained using the method of maximum likelihood, which is described in full detail in Jennrich and Schluchter (1986) and Jones (1993). The solution of the likelihood equations are:

$$\hat{\beta} = \sum_j \left(x_j' V_j^{-1} x_j \right)^{-1} \left(\sum_j x_j' V_j^{-1} x_j \right) \qquad (6.24)$$

and

$$Cov\left(\hat{\beta}\right) = \hat{\sigma}^2 \left(\sum x_j' V_j^{-1} x_j \right)^{-1}$$

for given B and W_i, where

$$V_j = Z_j B Z_j' + W_j$$

is the covariance matrix for the jth subject. The maximum likelihood estimator (MLE) of the scale parameter σ^2 is

$$\hat{\sigma}^2 = \frac{1}{N} \sum_j \left(Y_j - X_j \hat{\beta} \right)' V_j^{-1} \left(Y_j - X_j \hat{\beta} \right) \qquad (6.25)$$

where $N = \sum_j n_j$ is the total number of observations on all subjects. When the number of parameters is not small relative to the total number of observations, the estimated variances of the MLEs become seriously biased. To reduce the bias the restricted maximum likelihood (REML) is used. For repeated measures experiments, Diggle (1988) showed that the REML estimate of β is the same as in (6.24), however, the unbiased estimate of σ^2 is

$$\hat{\sigma}^2 = \frac{1}{n-p} \sum_j \left(Y_j - X_j \hat{\beta} \right)' V_j^{-1} \left(Y_j - X_j \hat{\beta} \right) \qquad (6.26)$$

C.   MODEL SELECTION

As we have already discussed in section V of Chapter 4, the likelihood ratio test (LRT) is the appropriate tool to perform model comparisons. When many models are fitted to the same data, an LRT can be calculated between any two "nested models", where nested means one of the models is a reduced version of the other (see Chapter 4, Section B). If we are to compare two models that are not nested the LRT cannot be used to test their difference. An alternative model selection procedure is Akaike's Information Criterion (AIC) where

$$AIC = -2 \ln L + 2(\text{number of estimated parameters})$$

The model that has the lowest value of AIC is selected as the best model.

Example 6.2 : Potthoff and Roy (1964) present a set of growth data for 11 girls and 16 boys. For each subject, the distance (mm) from the center of the pituitary to the pterygomaxillary fissure was recorded at the ages of 8, 10, 12, and 14. None of the data are missing. The questions posed by the authors were

(i) Should the growth curves be presented by second degree equation in age, or are linear equations adequate?

(ii) Should two separate curves be used for boys and girls, or do both have the same growth curve?

(iii) Can we obtain confidence bands(s) for the expected growth curves?

Before we address these questions, we should emphasize that little is known about the nature of the correlation between the $n_i = 4$ observations on any subject, except perhaps that they are serially correlated. The simplest correlation model is the one in which the correlation coefficient between any two observations t periods of time apart is equal to ρ^t, and in which the variance is constant with respect to time, under this model, the covariance matrix is:

$$w_i \sigma^2 = \begin{bmatrix} 1 & \rho & \rho^2 & \rho^3 \\ \rho & 1 & \rho & \rho^2 \\ \rho^2 & \rho & 1 & \rho \\ \rho^3 & \rho^2 & \rho & 1 \end{bmatrix} \sigma^2$$

This is the AR(1) correlation structure.

For the ith subject we fitted the following model

310

$$Y_i = \beta_0 + \beta_1 x_{i1} + \beta_2 x_{i2} + z_i \alpha_i + \varepsilon_i$$

where

$$x_{i1} = \begin{cases} 0 & \text{if gender is boy} \\ 1 & \text{if gender is girl} \end{cases}$$

and

$$x_{i2} = \begin{pmatrix} 8 & 10 & 12 & 14 \end{pmatrix}'$$

is treated as a continuous covariate. When a quadratic component of age was added to the model, its β-coefficient was not significant and thus a linear effect was deemed sufficient.

We used PROC MIXED provided in the SAS package to fit the above model. The SAS program is :

```
PROC MIXED;
CLASS GENDER SUBJECT;
MODEL DISTANCE = GENDER AGE|S;
REPEATED/TYPE = AR(1)      SUB = SUBJECT(GENDER) R CORR;

RUN;
```

The SAS output summary is

$$\hat{\rho} = 0.6259 \qquad \hat{\sigma}^2 = 5.2969$$

$$\hat{\beta}_0 = 17.8787 \quad (1.0909) \quad p \cong 0.0001$$

$$\hat{\beta}_1 = -2.4187 \quad (0.6933) \quad p \cong 0.0018$$

$$\hat{\beta}_2 = 0.6529 \quad (1.0909) \quad p \cong 0.0001$$

Based on the fitted model, it seems that a linear function in age is adequate, and that separate curves be used for boys and girls (significant coefficient of x_{i1}).

To address question (iii) posed by Potthoff and Roy, we follow the approach suggested by Jones (1993). Since for each subject in the study, there is an X_i matrix, let x denote a possible row of X_i for any subject. For example, for a subject who is a girl at age 14, then $x = \begin{pmatrix} 1 & 1 & 14 \end{pmatrix}'$ or for a subject who is a boy at age 10, then $x = \begin{pmatrix} 1 & 0 & 10 \end{pmatrix}'$. The estimated population mean for a given x vector is

$$\hat{Y} = X\hat{\beta} \qquad\qquad (6.27)$$

and has estimated variance

$$Var(\hat{y}) = xCov(\hat{\beta})x'$$

$$= \hat{\sigma}^2 x \left(\sum_i x_i' V_i^{-1} x_i \right)^{-1} x' \qquad\qquad (6.28)$$

By varying the elements of the x vector, estimated population curves can be generated for different values of the covariates. The confidence limits are thus $\hat{Y} \pm z_{\frac{\alpha}{2}} \sqrt{Var(\hat{Y})}$.

As we have indicated, there are several approaches to model the within subject correlation structure. In addition to the AR(1), we fitted two linear mixed models, the first with unstructured correlation and the other has the form of compound symmetry. Table 6.10 gives a summary of fitting the three models for the sake of comparison.

Table 6.10
Summary of Fitting Linear Mixed Models to the Repeated Measures Data of Example 6.2
UN=unstructured, AR(1)=first order autoregressive process, CS=compound symmetry

Parameter Estimate	Model		
	UN	AR(1)	CS
$\hat{\beta}_0$	17.417 (0.866)	17.879 (1.091)	17.707 (0.834)
$\hat{\beta}_1$	-2.045 (0.736)	-2.418 (0.693)	-2.321 (0.761)
$\hat{\beta}_2$	0.674 (0.070)	0.653 (0.091)	0.660 (0.062)
AIC	224.347	224.724	220.756
Non-linear parameters		$\hat{\rho} = 0.626$	$\sigma_b^2 = 3.267$
		$\hat{\sigma}^2 = 5.297$	$\sigma_e^2 = 2.049$

Clearly, there are little or no differences among the three models. Since none of the above models is considered nested within either of the others, model comparisons should be confined to the AIC. Clearly, UN and AR(1) are indistinguishable, and both are worse than the CS model (since the CS has smaller AIC value).

The PROC MIXED programs to fit the UN and CS models are similar to the above programs, with AR(1) being replaced by UN for unstructured covariance and CS for compound symmetry.

Example 6.3 : **"Multiple levels of nesting"**

Situations in which more than two factors would be nested within each other are of frequent occurrence in repeated measures experiments. Example 6.3 illustrates this situation. Pens of animals are randomized into two diet groups. Animals in each pen are approximately of the same age and initial weight. Their weights were measured at weeks 1,2 and 3.

The data are presented in Table 6.11.

Table 6.11
Data for the Feed Trial with Pens Nested in the
Diet Groups and Animals Nested in Pens.

	Diet (1) Pen Animal											Diet (2) Pen Animal										
Pen	1	1	1	1	2	2	2	2	3	3	3	4	4	4	4	5	5	5	6	6	6	6
Week / Animal	1	2	3	4	5	6	7	8	9	10	11	12	13	14	15	16	17	18	19	20	21	22
1	2.5	2	1.5	1.5	2	2.5	1.5	1	1.5	2	1.5	3	3	3	4	3	4	3	3	4	4	4
2	3	2.5	2	2.5	4	3.5	2	1.5	1.5	2	2	3	6	4	5	3	4.5	3	7	6	6	6
3	4	2.5	2	3	4	2.5	2.5	1	2	2.5	2	4	8	7	6	5	5.5	4	9	7	8	7

The data in Table 6.11 show that pens 1, 2 and 3 are nested within the diet group 1 while pens 4, 5 and 6 are nested in diet group 2 (i.e. different pens within each group) and animals are nested within pens. The main objectives of the trial were to see if the weights of animals differed between groups, and if such differences were present over time.

To answer the above questions, we follow a strategy similar to that of Looney and Stanley (1989). The general mixed models were fitted using PROC MIXED in SAS. The first model included diet by week interaction term:

Model with interaction:

```
PROC MIXED;
        CLASS DIET PEN ANIMAL WEEK;
                MODEL WEIGHT = DIET|WEEK;
                RANDOM PEN(DIET) ANIMAL(PEN DIET);
    RUN;
```

There was significant interaction. We therefore sorted the data by week:

Model to compare between diets at each week:

```
PROC SORT;
        BY WEEK;
PROC MIXED;
        BY WEEK;
CLASS DIET PEN;
        MODEL WEIGHT = DIET;
        RANDOM PEN(DIET);
    RUN;
```

The results are summarized in table 6.12.

Table 6.12
Data Analysis of Example 6.3
Comparing Between Diets at Each Week

		Week 1	Week 2	Week 3
Covariance	σ_p^2	0.000	0.858	0.755
Parameters	σ_e^2	0.245	0.761	1.224
Test for fixed effects				
F statistic		63.88	8.100	19.78
p value		0.0001	0.0120	0.0004

There are significant differences between the two diets at each week. Bonferroni corrected error rate is $\cong 0.02$.

VI. THE GENERALIZED ESTIMATING EQUATIONS APPROACH

It is apparent that the ANOVA approach to the analysis of repeated measures data is quite simple, particularly when the data are balanced; that is, when each subject is measured the same number of times. The use of PROC ANOVA with the "repeated" statement in SAS is quite effective in deriving the necessary information to carry on with the hypothesis testing. PROC GLM with the Type III sum of squares using the option E, provide sufficient information to identify the error structure corresponding to the hypothesis in question. Efficient fitting procedures such as PROC MIXED in SAS are available for more general structure of the linear mixed model.

However, there are limitations to the use of ANOVA which prevent its recommendation as a general tool for the analysis of repeated measures or longitudinal data (see Diggle et al., 1994). The first limitation is that it fails to exploit the efficiency that may be gained if the covariance among repeated observations is modeled. The second limitation is that it assumes that the mechanism of missing data is ignorable, and does not produce reliable results if the missing data are excessive. Thirdly, most biomedical data are not normally distributed, and may not satisfy the sphericity assumptions, and therefore the ANOVA and PROC MIXED procedures may no longer be valid. Liang and Zeger (1986) and Zeger and Liang (1986) presented a unified approach to analyzing longitudinal data, which models the covariance among the repeated observations. Their approach models both discrete and continuous outcomes based on the application of generalized estimating equations (see Chapter 4). To explain how their approach works for repeated measures observations, let us consider the following example.

Example 6.4:

Suppose that $I=2k$ subjects have been randomized into two treatment groups, each has k individuals. Suppose that each subject was measured twice, at baseline (0) and two hours later. Let y_{ij} represent the measurement on the ith subject at the jth time point ($j=0,1$). Note that if the number of repeated measures per subject is 2, then no assumptions are needed on the nature of the correlation between the observations over time. Now, we distinguish between two situations:

(i)   Modeling the group (treatment effect): Let $y_{ij}=\beta_0+\beta_1 X_i+e_{ij}$, where X_i is a dummy variable coded as 0 if the ith subject belongs to the first group, and 1 if it belongs to the second group; e_{ij} is a random error, so that $E(e_{ij})=0$ and

$$Cov(e_{ij}, e_{ij'}) = \begin{bmatrix} \sigma^2 & j=j' \\ \rho\sigma^2 & j \neq j' \end{bmatrix}.$$

Under the above representation, β_0 is the average response of a subject in group 1, and $\beta_0 + \beta_1$ is the average response for a subject in group 2. Therefore, β_1 is the group effect. It is known that the least squares estimate of β_1 is

$$\hat{\beta}_1 = \frac{\sum_{i=1}^{I} \sum_{j=0}^{1} (y_{ij} - \bar{y})(x_i - \bar{x})}{\sum_{i=1}^{n} \sum_{j=0}^{1} (x_i - \bar{x})^2}$$

$$= \frac{1}{I} \left[\sum_{group\ 1} (y_{i0} + y_{i1}) - \sum_{group\ 2} (y_{i0} + y_{i1}) \right]$$

Hence

$$Var(\hat{\beta}_1) = \frac{2\sigma^2(1+\rho)}{I} .$$

(ii)   Modeling the time effect: Again let $y_{ij} = \gamma_0 + \gamma_1 j + e_{ij}$, where γ_1 represents the time effect. The least squares estimate of γ_1 is

$$\hat{\gamma}_1 = \frac{\sum_{i=1}^{I} \sum_{j=0}^{1} (y_{ij} - \bar{y})(j - \bar{j})}{\sum_{i=1}^{n} \sum_{j=0}^{1} (j_i - \bar{j})^2}$$

$$= \frac{1}{I} \sum_{i=1}^{I} (y_{i1} - y_{i0})$$

and

$$Var(\hat{\gamma}_1) = \frac{2\sigma^2}{I} (1-\rho) .$$

Note that, if the data were analyzed under independence ($\rho=0$), a test on $\beta_1=0$ will reject the null hypothesis of no group effect too often, and a test on $\gamma_1=0$ will result in accepting the hypothesis of no time effect too often.

In addition to the assumption of independence required by the classical regression models considered above, one also assumes that the group and time effects, when modeled simultaneously, should be additive, and this could be quite restrictive. Moreover, one also assumes that the observations are normally distributed, an assumption that is violated by many types of data. The introduction of the Generalized Linear Models (GLM) alleviates these restrictions. A key component of such models is a link function $g(E(y_{ij})) = X_{ij}^T \beta$ that relates a monotone differentiable function of $E(y_{ij})$ to $X_{ij}^T \beta$. In Chapter 4 we showed that the conventional link for binary data was the logit transformation.

316

Another feature of GLM is the relaxation of the assumption that y_{ij} have a constant variance. Instead, it is assumed that the variance of y_{ij} is a known function of $\mu_{ij}=E(y_{ij})$. The variance of y_{ij} is written as $\phi Var(\mu_{ij})$ where ϕ is a scalar parameter.

The extension of GLM by Liang and Zeger (1986) was to account for the correlation among repeated observations. Let subject i (i=1,2,...k) be observed at times j=1,2,...n_i, which results in a total of $N = \sum_{i=1}^{k} n_i$ observations. The response vector for the ith individual $y_j = (y_{i1}, y_{i2},...y_{in_i})^T$ and the associated covariate information $x_j = (x_{i1}, x_{i2},...x_{in_i})^T$. It is also assumed that y_{ij} and y_{ij}' (j$\neq$j') are correlated, while y_{ij} and $y_{i'j}''$ are (i$\neq$i') uncorrelated. The covariance matrix has the form ϕV_i, where

$$V_i = A_i^{1/2} R_i(\alpha) A_i^{1/2} \qquad (6.29)$$

$R_i(\alpha)$ is a working correlation matrix where the parameter α fully specifies the form of $R_i(\alpha)$, and $A_i=diag[V(\mu_{i1}),...V(\mu_{in_i})]$. When $R_i(\alpha)$ is in fact the true correlation matrix, then $V_i=Cov(y_i)$. The covariance matrix of all N observations $y = [y_1^T,...y_J^T]^T$ is block diagonal; $\phi V = \phi$ diag $[V_1,...V_k]$. The regression parameters β are estimated by solving:

$$\sum_{i=1}^{k} D_i^T V_i^{-1} (y_i - \mu_i) = 0 \qquad (6.30)$$

where

$$D_i = \frac{\partial \mu_i}{\partial \beta}.$$

These are generalized estimating equations (GEE) and are identical in form to the weighted least squares estimating equations (see McCullagh and Nelder, 1989, p. 339). The solution to the GEE (6.24) gives a consistent estimate of β that is asymptotically multivariate normal with covariance matrix

$$\phi \left\{ \sum_{i=1}^{k} D_i^T V^{-1} D_i \right\}^{-1}. \qquad (6.31)$$

Let us consider situation (ii) of the previous example, where

$$\mu_{ij} = \gamma_0 + \gamma_{1j}$$

$$\mu_{i0} = \gamma_0 , \quad \mu_{i1} = \gamma_0 + \gamma_1$$

$$\frac{\partial \mu_{i0}}{\partial \gamma_0} = 1 \quad \frac{\partial \mu_{i1}}{\partial \gamma_0} = 1$$

$$\frac{\partial \mu_{i0}}{\partial \gamma_1} = 0 \quad \frac{\partial \mu_{i1}}{\partial \gamma_1} = 1$$

which means that

$$D_i^T = \begin{bmatrix} 1 & 1 \\ 0 & 1 \end{bmatrix} .$$

Moreover,

$$V_i = \begin{bmatrix} 1 & \rho \\ \rho & 1 \end{bmatrix}$$

$$V_i^{-1} = \frac{1}{(1-\rho^2)} \begin{bmatrix} 1 & -\rho \\ -\rho & 1 \end{bmatrix} .$$

The regression coefficients can be estimated by solving

$$\begin{pmatrix} 0 \\ 0 \end{pmatrix} = \frac{1}{(1-\rho^2)} \sum_{i=1}^{k} \begin{pmatrix} 1 & 1 \\ 0 & 1 \end{pmatrix} \begin{bmatrix} 1 & -\rho \\ -\rho & 1 \end{bmatrix} \begin{bmatrix} y_{i0} - \mu_{i0} \\ y_{i1} - \mu_{i1} \end{bmatrix}$$

or

$$\begin{pmatrix} 0 \\ 0 \end{pmatrix} = \begin{bmatrix} \sum_{i=1}^{k} (y_{i0} + y_{i1}) - 2 k \gamma_0 - k \gamma_1 \\ \sum_{i=1}^{k} (-\rho y_{i0} + y_{i1}) - (1-\rho) k \gamma_0 - k \gamma_1 \end{bmatrix}$$

from which

318

$$\hat{\gamma}_0 = \frac{1}{k} \sum_{i=1}^{k} y_{i0}$$

$$\hat{\gamma}_1 = \frac{1}{k} \sum (y_{i1} - y_{i0}) .$$

The covariance matrix is

$$\phi \left[\sum_{i=1}^{k} \begin{pmatrix} 1 & 1 \\ 0 & 1 \end{pmatrix} \frac{1}{(1-\rho^2)} \begin{pmatrix} 1 & -\rho \\ -\rho & 1 \end{pmatrix} \begin{pmatrix} 1 & 0 \\ 1 & 1 \end{pmatrix} \right]^{-1}$$

$$= \phi (1-\rho^2) \left[k \begin{pmatrix} 2(1-\rho) & 1-\rho \\ 1-\rho & 1 \end{pmatrix} \right]^{-1}$$

$$= \frac{\phi}{k} \begin{bmatrix} 1 & \rho-1 \\ \rho-1 & 2(1-\rho) \end{bmatrix} .$$

Hence,

$$Var (\hat{\gamma}_1) = \frac{2\phi(1-\rho)}{k} ,$$

and therefore the Var $(\hat{\gamma}_1)$ correctly accounts for the within-subjects correlation.

To solve (6.24) the correlation parameter α must be known. However, regardless of the true correlation structure (e.g., serial, exchangeable, no correlation), Liang and Zeger (1986) use a "working" correlation matrix, which is data driven, as an estimate of V_i. The resulting estimating equations are given by

$$\sum_{i=1}^{k} D_i^T \hat{V}_i^{-1} (y_i - \mu_i) = 0 \tag{6.32}$$

where $\mathbb{V}_i$ is an estimate of V_i in (6.24) and is given by

$$\hat{V}_i = (y_i - \mu_i)(y_i - \mu_i)^T . \tag{6.33}$$

The solution of (6.26) gives asymptotically multivariate normal estimates with covariance matrix given by

$$Cov(\hat{\beta}) = \phi M^{-1} \Sigma M^{-1} \tag{6.34}$$

where

$$M = \sum_{i=1}^{k} D_i^T \hat{V}_i^{-1} D_i$$

$$\Sigma = \sum_{i=1}^{k} D_i^T \hat{V}_i^{-1} V_i \hat{V}_i^{-1} D_i .$$

If $\hat{V}_1 = V_i$, then (6.28) reduces to (6.25). The useful feature of the GEE is that β and Cov $(\hat{\beta})$ are consistently estimated:

(1)   for non-normal data,
(2)   for misspecified correlation structure, and
(3)   for unbalanced data.

However, one should be able to correctly specify the link function; otherwise, the obtained estimates are no longer consistent.

The issue of robustness, of the GEE approach, against misspecification of the correlation structure was the subject of investigation by many authors. In particular, if the correlation structure is misspecified as "independence" which assumes that the within cluster responses are independent, the GEE has been shown to be nearly efficient relative to the maximum likelihood in a variety of settings. When the correlation between responses in not too high Zeger (1988) suggested that this estimator should be nearly efficient. McDonald (1993), focussing on the case of clusters of size $n_i = 2$ (i.e. the bivariate case) concluded that the estimator obtained under specifying independence may be recommended whenever the correlation between the within cluster pair is nuisance. This may have practical implications since the model can be implemented using standard software packages. In a more recent article, Fitzmaurice (1995) investigated this issue analytically. He has confirmed the suggestions made by Zeger (1988) and McDonald (1993). Furthermore, he showed that when the responses are strongly correlated and the covariate design includes a within-cluster covariate, assuming independence can lead to a considerable loss of efficiency if the GEE is used in estimating the regression parameters associated with that covariate. His results demonstrate that the degree of efficiency depends on both the strength of the correlation between the responses and the covariate design. He recommended that an effort should be made to model the association between responses, even when this association is regarded as a nuisance feature of the data and its correct nature is unknown.

Example 6.5

The following data are the results of a longitudinal investigation that aimed at assessing the long-term effect of two types of treatments (1≡excision arthroplasty; 2≡triple pelvic osteotomy) for Hip-Dysplasia in dogs. The owner's assessment was the response variable and was recorded on a binary scale at weeks 1,3,6,10, and 20. Using the generalized estimating equations approach (GEE) we test for treatment effect, controlling for laterality, and age as possible confounders.

Dog No.	Laterality	Age	Type of Surgery	Owner's assessment +				
				Week 1	Week 3	Week 6	Week 10	Week 20
1	U	6	1	1	1	1	1	1
2	U	4	1	1	1	1	1	1
3	U	7	2	1	1	1	1	1
4	B	7	2	1	1	1	1	1
5	U	4	2	1	1	1	1	1
6	U	8	2	1	1	0	1	1
7	U	7	2	1	0	1	1	1
8	U	6	2	1	1	1	0	1
9	U	8	1	1	1	1	1	1
10	U	5	2	1	1	1	1	1
11	U	6	1	1	1	1	1	1
12	U	6	2	0	0	0	1	1
13	U	7	1	1	1	1	1	1
14	U	7	1	1	1	1	1	1
15	B	7	2	1	1	1	1	1
16	B	7	2	1	1	1	0	1
17	B	5	1	1	0	1	0	1
18	B	6	1	0	0	1	1	0
19	B	8	2	1	1	0	1	0
20	B	6	1	1	1	1	0	1
21	U	8	1	1	1	1	1	1
22	B	2	2	1	1	0	0	0
23	B	1	1	0	1	1	1	1
24	U	1	1	1	1	1	1	1
25	U	1	1	1	1	1	1	1
26	B	2	2	1	0	1	0	1
27	B	2	2	0	0	0	0	0
28	B	2	2	0	0	0	0	0
29	U	1	1	1	0	1	1	1
30	B	2	2	1	1	0	1	1
31	B	2	2	0	0	0	0	0
32	B	2	2	0	0	0	1	1
33	U	1	1	1	1	1	1	1
34	U	2	2	1	1	0	1	1
35	U	1	1	1	1	1	1	1
36	U	2	2	1	0	1	1	1
37	U	1	1	0	1	1	1	1
38	U	2	2	1	1	1	1	1
39	U	6	1	1	1	1	1	1
40	U	8	1	0	0	1	0	0
41	U	8	2	1	1	0	0	0
42	U	8	1	0	1	0	0	0
43	U	2	2	1	0	0	1	0
44	U	1	1	0	0	1	1	1
45	U	1	1	1	1	1	1	1
46	U	2	2	0	0	0	1	0

*   U≡unilateral, B≡bilateral
**   1≡excision arthroplasty
        2≡triple pelvic osteotomy (TPO)
+   1≡good; 0≡poor

The data in example 6.5 have been fitted under 3 different specifications of the correlations among the scores in weeks 1, 3, 6, 10 and 20. The SAS programs required to fit the GEE model are given below:

Model 1 : This model specifies an autocorrelation with lag 1. The rationale being that adjacent responses would be correlated while responses that are separated by more than one time unit are uncorrelated.

```
proc genmod;
        class dog lateral typsurg;
                model score = lateral typsurg age/
                    dist = binomial;
                repeated subject = dog / type = AR(1);
    run;
```

Model 2 : This model specifies an exchangeable or compound symmetry correlation. The SAS program is similar to that of model 2, except the TYPE option in the "repeated" statement is type = CS.

Model 3 : This model is fitted for comparative purposes and the program is similar to the above, with TYPE = IND included in the "repeated" statement.

The results of fitting the three models are summarized in Table 6.13.

Table 6.13
Fitting of the Data in Example 6.4

Parameter Estimate	Model 1 AR(1)	Model 2 CS	Model 3 Independence
Intercept	0.821 (0.555)	0.755 (0.529)	0.755 (0.529)
Lateral b	-1.013 (0.450)	-1.009 (0.436)	-1.009 (0.436)
Lateral u	0.000	0.000	0.000
Typsurg 1	0.701 (0.490)	0.802 (0.475)	0.802 (0.475)
Typsurg 2	0.000	0.000	0.000
age	0.058 (0.102)	0.059 (0.099)	0.059 (0.099)
ϕ (scale)	1.010	1.011	1.011
scaled deviance	1.1008	1.1008	1.1008

The scale parameter is considered known. It is estimated as the square root of the normalized Pearson's chi-square. The scaled deviance is the ratio of the deviance to the number of degrees of freedom carried by the model (df= 230 - 4 = 226) since we have 5x46=230 observations, and estimated 4 parameters.

Comments

1. All the covariates in the model are measured at the cluster (subject) level, and no time varying covariates are included in the study. With this type of covariate design one should expect little or no difference between Model 3 and the other two models which specify some degree of correlation.

2. Relative to the unilateral, bilaterality (lateral) seems to have significant negative effect on the condition of the subject. However, neither age nor the type of surgery (Typsurg) have significant effect.

3. The bracketed numbers, are the empirical standard errors obtained from the sandwich estimator (see Chapter 4).

Chapter 7

SURVIVAL DATA ANALYSIS

I. INTRODUCTION

Studies with survival data as the response involve observing units until failure is experienced. In the case of medical studies, the units are humans or animals and failure may be broadly defined as the occurrence of a pre-specified event. Events of this nature include times of death, disease occurrence or recurrence and remission.

Although survival data analysis is similar to the analysis of other types of data discussed previously (continuous, binary, time series) in that information is collected on the response variable as well as any covariates of interest, it differs in one important aspect : the anticipated event may not occur for each subject under study. Not all subjects will experience the outcome during the course of observation, resulting in the absence of a failure time for that particular individual. This situation is referred to as *censoring* in the analysis of survival data, and a study subject for which no failure time is available is referred to as *censored*.

Unlike the other types of analysis discussed previously, censored data analysis requires special methods to compensate for the information lost by not knowing the time of failure of all subjects. In addition, survival data analysis must account for highly skewed data. Often one or two individuals will experience the event of interest much sooner or later than the majority of individuals under study, giving the overall distribution of failure times a skewed appearance and preventing the use of the normal distribution in the analysis. Thus the analysis of survival data requires techniques which are able to incorporate the possibility of skewed and censored observations.

The above paragraphs referred to censoring in a broad sense, defining censored survival data as data for which the true failure time is not known. This general definition may be broken down for three specific situations, resulting in three types of censoring :

Type I Censoring : Subjects are observed for a fixed period of time, with exact failure times recorded for those who fail during the observation period. Subjects not failing during the observation period are considered censored. Their failure times become the time at which they were last observed or the time at which the study finished.

Type II Censoring: Subjects are observed until a fixed number of failures occur. As with Type I Censoring, those failing during the observation period are considered uncesnsored and have known failure times. Those not failing are considered censored and have failure times which become the time at which they were last observed or the time at which the largest uncensored failure occurred.

Random Censoring : Often encountered in clinical trials, random censoring occurs due to the accrual of patients gradually over time, resulting in unequal times under study. The study takes place over a fixed period of time, resulting in exact failure times for those failing during the period of observation and censored failure times for those lost to follow up or not failing before study termination. All failure times reflect the period under study for that individual.

The information presented for the three censoring situations is summarized in Table 7.1:

Table 7.1
Summary Information for Three Types of Censoring

	Type I	Type II	Random
Study Characteristics	-study continues for a fixed period of time	-study continues until a fixed number/ proportion of failures	-study continues for a fixed period of time -unequal periods of observation possible
Uncensored Failure Time	-equal to the exact failure time which is known	-equal to the exact failure time which is known	-equal to the exact failure time which is known
Censored Failure Time	-equal to the length of the study	-equal to the largest uncensored failure time	-calculated using time of study completion and time of subject enrollment
Lost to Follow up Failure Time	-calculated using time at which subject is lost and time at which study starts	-calculated using time at which subject is lost and time at which study starts	-calculated using time at which subject is lost and time of subject enrollment

Reproduced from *Biometrics,* 52, 328–334, 1996. With permission from David Santini, International Biometrics Society.

If the survival time is denoted by the random variable T, then the following definitions are useful in the context of survival analysis:

1. Cumulative Distribution Function (CDF) - denoted by $F(t)$, this quantity defines the probability of failure before time t :

$$F(t) = \Pr(\text{individual fails before time } t)$$
$$= \Pr(T < t).$$

2. Probability Density Function (PDF) - denoted by $f(t)$, this quantity is the derivative of the Cumulative Distribution Function and defines the probability that an individual fails in a small interval per unit time :

325

$$f(t) = \lim_{\Delta t \to 0} \frac{\Pr(\text{an individual dies in } (t, t + \Delta t))}{\Delta t}$$

$$= \frac{d}{dt} F(t).$$

As with all density functions, $f(t)$ is assumed to have the following properties:

a) The area under the density curve equals one

b) The density is a non-negative function such that

$$\begin{aligned} f(t) \quad &\geq 0 \quad \text{for all } t \geq 0 \\ &= 0 \quad \text{for } t < 0. \end{aligned}$$

3. Survival Function - denoted by $S(t)$, this function gives the probability of survival longer than time t :

$$\begin{aligned} S(t) &= \Pr(\text{an individual survives longer than } t) \\ &= \Pr(T > t) \end{aligned}$$

so that the relationship with the probability density function is

$$\begin{aligned} S(t) &= 1 - \Pr(T < t) \\ &= 1 - F(t). \end{aligned}$$

The survival function is assumed to have the following properties:

a) The probability of survival at time zero is one, $S(t) = 1 \quad$ for $t = 0$

b) The probability of infinite survival is zero, $S(t) = 0 \quad$ for $t = \infty$

c) The survival function is non-increasing

4. Hazard Function - denoted by $h(t)$, this function gives the probability an individual fails in a small interval of time conditional on their survival at the beginning of the interval :

$$h(t) = \lim_{\Delta t \to 0} \frac{\Pr(\text{an individual dies in } (t, t + \Delta t) | T > t)}{\Delta t}.$$

326

In terms of the previously defined quantities, the hazard function may be written as

$$h(t) = \frac{f(t)}{1 - F(t)}.$$

In practice, the hazard function is also referred to as the instantaneous failure rate or the force of mortality. It represents the failure risk per unit of time during a lifetime.

The cumulative hazard function is written as $H(t)$ and is the integral of the hazard function:

$$H(t) = \int_0^t h(x)dx.$$

Although the quantities $F(t)$, $f(t)$, $S(t)$, and $h(t)$ may be defined for any continuous random variable, $S(t)$ and $h(t)$ are usually seen in the context of survival data since they are particularly suited to its analysis.

Notice as well that given any one of the four quantities, the other three are easily obtained. Thus specifying the survival function, for instance, also determines what the cumulative distribution function, probability density function and hazard function are.

In addition, the following relationships hold:

$$h(t) = f(t) / S(t)$$
$$f(t) = -S'(t)$$
$$h(t) = -\frac{d}{dt} \ln S(t)$$
$$S(t) = \exp\left[-\int_0^t h(x)dx\right]$$
$$f(t) = h(t) \exp\left[-H(t)\right].$$

II. EXAMPLES

A. VENTILATING TUBE DATA

One-third of pediatric visits arise due to inflammation of the middle ear, also known as otitis media, resulting in a substantial health care burden. In addition, concerns have surfaced relating to long-term language, behavior and speech development. Unsuccessful treatment with various drug therapies often leads to surgical intervention, in which tubes are placed in the ear. It has been shown that ventilating tubes are successful in preventing otitis media as long as the tubes are in place and unblocked.

Le and Lindgren (1996) studied the time of tube failure (displacement or blockage) for 78 children. Each child was randomly assigned to be treated with placebo or prednisone and sulfamethoprim. The children were observed from February 1987 to January 1990, resulting in the possibility of censored failure times for children not experiencing tube failure before the completion of the study. The data are listed below (Table 7.2), where the following coding is used for brevity:

1 - child [4 digit ID] (data set variable)
2 - group [2 = medical, 1 = control] (data set variable treat)
3 - ear [1 = right, 2 = left] (data set variable)
4 - time, in months (data set variable)
5 - status [1 = failed, 0 = censored] (data set variable)

Table 7.2
Ventilating Tube Data (Le and Lindren, 1996)

1	2	3	4	5	1	2	3	4	5	1	2	3	4	5
1001	1	1	3.10	1	1065	1	1	0.80	1	1031	2	1	24.30	1
		2	4.80	0			2	0.80	1			2	18.80	1
1003	1	1	12.70	1	1066	1	1	13.30	1	1033	2	1	15.20	1
		2	6.00	1			2	13.30	1			2	12.50	1
1008	1	1	3.10	1	1071	1	1	0.80	1	1036	2	1	33.00	1
		2	6.40	1			2	8.10	1			2	12.10	1
1009	1	1	8.50	1	1073	1	1	3.00	1	1037	2	1	13.10	1
		2	12.70	1			2	15.80	1			2	0.70	1
1010	1	1	9.10	1	2003	1	1	9.40	1	1039	2	1	6.10	1
		2	9.10	1			2	9.40	1			2	17.10	0
1014	1	1	0.50	1	2005	1	1	3.10	1	1041	2	1	9.50	1
		2	5.10	1			2	3.10	1			2	3.40	1
1015	1	1	18.00	1	2009	1	1	7.60	1	1042	2	1	15.10	1
		2	15.00	1			2	10.10	1			2	17.80	1
1019	1	1	6.00	1	2011	1	1	5.50	1	1047	2	1	5.80	1
		2	6.00	1			2	5.50	1			2	5.80	1
1020	1	1	6.40	1	2015	1	1	0.70	1	1048	2	1	0.60	1
		2	6.40	1			2	0.70	1			2	3.00	1
1022	1	1	4.40	0	2016	1	1	7.00	1	1049	2	1	2.80	1
		2	1.30	1			2	7.00	1			2	1.60	1
1024	1	1	12.80	1	2018	1	1	11.70	1	1050	2	1	6.20	1
		2	12.80	1			2	3.10	1			2	9.00	1
1025	1	1	8.80	1	2020	1	1	14.30	1	1058	2	1	8.70	1
		2	8.80	1			2	3.20	1			2	3.40	1
1027	1	1	2.80	0	1002	2	1	15.40	1	1059	2	1	20.90	0
		2	2.80	0			2	9.20	1			2	3.40	1
1032	1	1	9.30	1	1004	2	1	9.30	1	1061	2	1	9.20	1
		2	27.10	1			2	9.30	1			2	6.00	1
1034	1	1	6.10	1	1006	2	1	15.00	1	1063	2	1	6.40	1

ID	Trt	Ear	Time	C	ID	Trt	Ear	Time	C	ID	Trt	Ear	Time	C
		2	6.10	1			2	0.90	1			2	14.30	0
1035	1	1	17.90	1	1007	2	1	15.00	1	1067	2	1	8.80	1
		2	20.90	1			2	11.90	1			2	8.80	1
1038	1	1	9.30	1	1011	2	1	17.80	1	1068	2	1	18.50	1
		2	3.10	1			2	12.20	1			2	13.30	1
1040	1	1	2.90	1	1012	2	1	5.90	1	1069	2	1	12.20	1
		2	1.00	1			2	8.70	1			2	12.20	1
1043	1	1	9.10	1	1013	2	1	8.90	1	1072	2	1	12.50	0
		2	9.10	1			2	12.60	1			2	8.80	1
1044	1	1	5.80	1	1016	2	1	0.60	1	2001	2	1	8.50	1
		2	9.30	1			2	5.70	1			2	21.70	1
1045	1	1	2.90	1	1017	2	1	6.00	1	2004	2	1	1.80	1
		2	1.10	1			2	9.40	1			2	20.70	1
1046	1	1	6.20	1	1018	2	1	14.60	1	2007	2	1	6.20	1
		2	9.20	1			2	9.90	1			2	9.00	1
1055	1	1	1.10	0	1021	2	1	12.10	1	2010	2	1	9.70	1
		2	1.10	0			2	2.90	1			2	11.10	0
1060	1	1	6.00	1	1023	2	1	3.00	1	2012	2	1	6.00	1
		2	10.70	1			2	3.00	1			2	6.00	1
1062	1	1	6.20	1	1026	2	1	24.90	1	2013	2	1	11.90	1
		2	6.20	1			2	8.70	1			2	8.80	1
1064	1	1	9.30	1	1028	2	1	5.20	1	2017	2	1	8.70	1
		2	19.30	0			2	9.00	1			2	8.70	1

In addition, it is anticipated that the failure times of the two ears from one child will be similar, and in fact, correlated. This particular aspect of the Ventilating Tube Data will be further discussed when the topic of Correlated Survival Data arises.

B. CYSTIC OVARY DATA

This study examined the effectiveness of hormonal therapy for treatment of cows with cystic ovarian disease. Two groups of cows were randomized to hormonal treatment and one group received placebo. The time of cyst disappearance was then recorded, with the possibility of censored data due to not all cysts disappearing. The data are as follows (Table 7.3):

Table 7.3
Cystic Ovary Data (a)

Treatment 1	Treatment 2	Placebo
4,6,8,8,9,10,12	7,12,15,16,18,22*	19,24,18*,20*,22*,27*,30*

(* means censored observation)

For statistical analyses and computer analyses with SAS, the data are conveniently re-expressed as shown (Table 7.4):

Table 7.4
Cystic Ovary Data (b)

Cow	Treatment	Time	Censor	Cow	Treatment	Time	Censor
1	1	4	1	11	2	16	1
2	1	6	1	12	2	18	1
3	1	8	1	13	2	22	0
4	1	8	1	14	3	19	1
5	1	9	1	15	3	24	1
6	1	10	1	16	3	18	0
7	1	12	1	17	3	20	0
8	2	7	1	18	3	22	0
9	2	12	1	19	3	27	0
10	2	15	1	20	3	30	0

(censor = 1 if failure, censor = 0 if no failure)

C. BREAST CANCER DATA

An increase in breast cancer incidence in recent years has resulted in a substantial portion of health care dollars being directed towards research in this area. The aims of such research include early detection through mass screening and recurrence prevention through effective treatments. The focus of this investigation was to determine which prognostic measures were predictive of breast cancer recurrence in female patients.

At the start of the study, 73 patients were enrolled and measurements were taken for the following variables:

> 1-patient identification number (data set variable id)
> 2-age at the start of the study in years (data set variable age)
> 3-size of original tumor [small=0 or large=1] (data set variable tsize)
> 4-mean AgNOR count [0=low, 1=high] (data set variable mag)
> 5-proliferative AgNOR index [0=low, 1=high] (data set variable pag)
> 6-time until breast cancer recurrence in months (data set variable time)
> 7-breast cancer recurrence [0=no, 1=yes] (data set variable censor).

The data are as follows (Table 7.5):

Table 7.5
Breast Cancer Data

1	7	6	5	4	2	3	1	7	6	5	4	2	3
1	0	130	0	0	57	0	37	0	92	1	1	79	0
2	0	136	1	1	67	0	38	0	48	1	1	41	1
3	0	117	0	0	65	1	39	0	89	0	0	69	0
4	1	50	1	1	45	0	40	0	95	0	1	58	0
5	0	106	0	1	63	0	41	0	91	0	0	37	1
6	0	103	0	0	63	0	42	1	47	1	1	47	1
7	1	86	1	1	61	1	43	0	75	0	0	47	1
8	0	63	1	1	67	1	44	0	49	0	0	74	0
9	0	120	0	0	43	0	45	0	66	1	1	67	0
10	0	121	1	1	49	1	46	0	65	0	1	45	0
11	1	108	1	1	66	1	47	0	22	0	0	67	1
12	0	121	0	0	68	0	48	0	73	0	0	66	0
13	0	109	0	1	68	1	49	0	67	0	0	69	0
14	0	111	0	0	52	0	50	0	75	1	1	53	1
15	1	60	1	1	68	1	51	0	71	1	1	38	1
16	0	106	0	1	50	0	52	0	80	0	0	61	0
17	0	108	1	1	70	1	53	0	25	0	1	75	1
18	0	105	0	0	50	0	54	0	67	0	0	73	1
19	0	98	1	1	56	1	55	0	74	0	1	45	1
20	1	108	1	1	70	0	56	1	64	0	0	45	0
21	0	62	0	1	65	0	57	0	64	0	0	65	1
22	0	106	1	1	63	0	58	1	41	1	1	73	0
23	1	95	1	0	66	1	59	1	70	1	1	48	0
24	0	94	1	1	44	1	60	1	57	1	1	53	1
25	0	19	0	1	57	0	61	0	59	0	0	45	1
26	1	103	1	1	71	0	62	0	53	0	0	73	0
27	0	60	0	0	59	1	63	0	69	0	0	35	0
28	0	91	1	1	66	1	64	0	55	0	1	47	1
29	0	70	1	1	57	1	65	1	58	1	1	66	0
30	0	65	0	0	54	0	66	0	68	0	0	46	1
31	0	91	1	1	64	1	67	0	60	0	0	60	0
32	1	86	1	1	45	0	68	1	126	1	1	55	0
33	0	90	0	0	39	1	69	0	127	1	1	60	0
34	0	87	0	1	51	0	70	0	126	0	0	60	1
35	1	89	0	1	57	1	71	0	102	0	0	64	1
36	1	89	1	1	27	0	72	0	122	1	1	49	0
							73	0	100	1	1	50	1

The above three examples will be used throughout the chapter to demonstrate methods of survival data analysis using SAS.

III. THE ANALYSIS OF SURVIVAL DATA

Two distinct methodologies exist for the analysis of survival data : non-parametric approaches in which no distributional assumptions are made for the previously defined probability density function $f(t)$ and parametric approaches in which distributional restrictions are imposed. Each methodology will be discussed separately.

A. NON-PARAMETRIC METHODS

1. Methods for Non-censored Data

Estimates of the survival function, probability density function and hazard function exist for the specific case of non-censored data. They are given as follows:

i) Estimate of the survival function for non-censored data

$$\hat{S}(t) = \frac{\text{number of patients surviving longer than } t}{\text{total number of patients}}$$

ii) Estimate of the probability density function for non-censored data

$$\hat{f}(t) = \frac{\text{number of patients dying in the interval beginning at time } t}{(\text{total number of patients})(\text{interval width})}$$

iii) Estimate of the hazard function for non-censored data

$$\hat{h}(t) = \frac{\text{number of patients dying in the interval beginning at time } t}{(\text{number of patients surviving at } t)(\text{interval width})}$$

$$= \frac{\text{number of patients dying per unit time in the interval}}{\text{number of patients surviving at } t}$$

It is also possible to define the average hazard rate, which uses the average number of survivors at the interval midpoint to calculate the denominator of the estimate:

$$\hat{h}^{*}(t) = \frac{\text{number of patients dying per unit time in the interval}}{(\text{number of patients surviving at } t) - .5(\text{number of deaths in the interval})}$$

The estimate given by $\hat{h}^{*}(t)$ results in a smaller denominator and thus a larger hazard rate. $\hat{h}^{*}(t)$ is used primarily by actuaries.

Obviously different methods are required in the presence of censored data. These methods are now discussed.

2. Methods for Censored Data

The Kaplan-Meier (1958) estimate of the survival function in the presence of censored data, also known as the Product-Limit estimate, is given by

$$\hat{S}(k) = p_1 \times p_2 \times p_3 \times \cdots \times p_k$$

where $k \geq 2$ years and p_i denotes the proportion of patients surviving the ith year conditional on their survival until the $(i-1)^{th}$ year. In practice, $\hat{S}(k)$ is calculated using the following formula:

$$\hat{S}(t) = \Pi_{t_{(r)} \leq t} \frac{n-r}{n-r+1} \tag{7.1}$$

where the survival times have been placed in ascending order so that $t_{(1)} \leq t_{(2)} \leq \cdots \leq t_{(n)}$ for n the total number of individuals under study and r runs through the positive integers such that $t_{(r)} \leq t$ and $t_{(r)}$ is uncensored.

Table 7.6 is used in calculation of the product-limit survival estimate:

Table 7.6
General Calculations for Product-Limit Survival Estimate

Ordered Survival Times (censored (*) and uncensored)	Rank	Rank (r) (uncensored observations)	Number in Sample (n)	$\dfrac{n-r}{n-r+1}$	$\hat{S}(t)$
$t_{(1)}$	1	1	n		
$t_{(2)}$ *	2	/	$n-1$	/	/
$t_{(3)}$	3	3	$n-2$		
$\vdots$	$\vdots$	$\vdots$	$\vdots$		
$t_{(n)}$	n	n	1		

333

The last column of the table is filled in after calculation of $\frac{n-r}{n-r+1}$. Notice that the estimate of the survival function at time t is available only for those times at which the failure was not censored. In Table 7.2, the second failure time was deemed censored to exemplify this.

Example 7.1: The Kaplan-Meier (product-limit) estimate of the survival function for the cystic ovary data is calculated within each treatment group as follows (Table 7.7):

Table 7.7
KM Survival Estimate for Cystic Ovary Data

Treatment	Ordered Survival Times *censored	Rank	Rank of uncensored observations	Number in Sample	$\frac{n-r}{n-r+1}$	$\hat{S}(t)$
1	4	1	1	7	6/7	0.85
	6	2	2	6	5/6	(0.83)(0.85)=0.71
	8	3	3	5	4/5	(0.80)(0.71)=0.56
	8	4	4	4	3/4	(0.75)(0.56)=0.42
	9	5	5	3	2/3	(0.66)(0.42)=0.28
	10	6	6	2	1/2	(0.50)(0.28)=0.13
	12	7	7	1	0	0
2	7	1	1	6	5/6	0.83
	12	2	2	5	4/5	(0.80)(0.83)=0.66
	15	3	3	4	3/4	(0.75)(0.66)=0.49
	16	4	4	3	2/3	(0.66)(0.49)=0.33
	18	5	5	2	1/2	(0.50)(0.33)=0.16
	22*	6	/	1	/	/
control	18*	1	/	7	/	/
	19	2	2	6	5/6	0.83
	20*	3	/	5	/	/
	22*	4	/	4	/	/
	24	5	5	3	2/3	(0.66)(0.83)=0.55
	27*	6	/	2	/	/
	30*	7	/	1	/	/

Example 7.2: It is informative to determine the product limit estimates of the survival function for the breast cancer data using the SAS computer package.

The following SAS statements produce the chart of product limit survival estimates for the breast cancer data:

334

```
proc lifetest data = agnor;
      time time*censor(0);
      run;
```

With output as shown below:

The LIFETEST Procedure

Product-Limit Survival Estimates

TIME	Survival	Failure	Survival Standard Error	Number Failed	Number Left
0.000	1.0000	0	0	0	73
19.000*	.	.	.	0	72
22.000*	.	.	.	0	71
25.000*	.	.	.	0	70
41.000	0.9857	0.0143	0.0142	1	69
47.000	0.9714	0.0286	0.0199	2	68
48.000*	.	.	.	2	67
49.000*	.	.	.	2	66
50.000	0.9567	0.0433	0.0245	3	65
53.000*	.	.	.	3	64
55.000*	.	.	.	3	63
57.000	0.9415	0.0585	0.0284	4	62
58.000	0.9263	0.0737	0.0317	5	61
59.000*	.	.	.	5	60
60.000	0.9109	0.0891	0.0348	6	59
60.000*	.	.	.	6	58
60.000*	.	.	.	6	57
62.000*	.	.	.	6	56
63.000*	.	.	.	6	55
64.000	0.8943	0.1057	0.0379	7	54
64.000*	.	.	.	7	53
69.000*	.	.	.	7	46
70.000	0.8749	0.1251	0.0417	8	45
70.000*	.	.	.	8	44
80.000*	.	.	.	8	38
86.000	.	.	.	9	37
86.000	0.8288	0.1712	0.0507	10	36
87.000*	.	.	.	10	35
89.000	.	.	.	11	34
89.000	0.7815	0.2185	0.0578	12	33
89.000*	.	.	.	12	32
94.000*	.	.	.	12	26
95.000	0.7514	0.2486	0.0629	13	25
95.000*	.	.	.	13	24
102.000*	.	.	.	13	21
103.000	0.7156	0.2844	0.0693	14	20
103.000*	.	.	.	14	19

335

106.000*	.	.	.	14	15
108.000	.	.	.	15	14
108.000	0.6202	0.3798	0.0869	16	13
108.000*	.	.	.	16	12
120.000*	.	.	.	16	8
122.000*	.	.	.	16	5
126.000	0.4962	0.5038	0.1309	17	4
126.000*	.	.	.	17	3
136.000*	.	.	.	17	0

* Censored Observation

NOTE: The last observation was censored so the estimate of the mean is biased.

Notice that the survival function is only estimated at death times which are not censored. Some of the censored observations have been deleted from the chart for brevity.

In addition, summary statistics are provided concerning sample size and percent of censored observations:

Summary of the Number of Censored and Uncensored Values

Total	Failed	Censored	%Censored
73	17	56	76.7123

So far, the breast cancer data analysis has not distinguished between the AgNOR groups when looking at survival. Adding a 'strata' statement to the SAS program as shown below results in the calculations shown above being performed within each level of the variable specified (in this case, pag) in the strata statement:

```
proc lifetest data = agnor plots = (s) graphics;
      time time*censor(0);
      strata pag;
      symbol1  v = none  color = black  line = 1;
      symbol2  v = none color = black line = 2;
      run;
```

The product limit survival estimates are then calculated for failures within each pag group as shown (note that some of the censored observations not contributing estimates of survival have been removed from the charts for brevity):

Product-Limit Survival Estimates
PAG = 0

TIME	Survival	Failure	Survival Standard Error	Number Failed	Number Left
0.000	1.0000	0	0	0	40
19.000*	.	.	.	0	39
62.000*	.	.	.	0	30
64.000	0.9667	0.0333	0.0328	1	29
64.000*	.	.	.	1	28
87.000*	.	.	.	1	17
89.000	0.9098	0.0902	0.0632	2	16
89.000*	.	.	.	2	15
130.000*	.	.	.	2	0

* Censored Observation

NOTE: The last observation was censored so the estimate of the mean is biased.

Product-Limit Survival Estimates
PAG = 1

TIME	Survival	Failure	Survival Standard Error	Number Failed	Number Left
0.000	1.0000	0	0	0	33
41.000	0.9697	0.0303	0.0298	1	32
47.000	0.9394	0.0606	0.0415	2	31
48.000*	.	.	.	2	30
50.000	0.9081	0.0919	0.0506	3	29
57.000	0.8768	0.1232	0.0577	4	28
58.000	0.8455	0.1545	0.0636	5	27
60.000	0.8141	0.1859	0.0685	6	26
63.000*	.	.	.	6	25
66.000*	.	.	.	6	24
70.000	0.7802	0.2198	0.0736	7	23
70.000*	.	.	.	7	22
86.000	.	.	.	8	19
86.000	0.7022	0.2978	0.0844	9	18
89.000	0.6632	0.3368	0.0883	10	17
91.000*	.	.	.	10	16
94.000*	.	.	.	10	13
95.000	0.6122	0.3878	0.0951	11	12
98.000*	.	.	.	11	11
100.000*	.	.	.	11	10
103.000	0.5510	0.4490	0.1034	12	9
106.000*	.	.	.	12	8
108.000	.	.	.	13	7
108.000	0.4132	0.5868	0.1146	14	6
108.000*	.	.	.	14	5

337

```
121.000*          .              .                .        14        4
122.000*          .              .                .        14        3
126.000       0.2755         0.7245          0.1360        15        2
127.000*          .              .                .        15        1
136.000*          .              .                .        15        0
                          * Censored Observation
NOTE: The last observation was censored so the estimate of the mean is biased.


          Summary of the Number of Censored and Uncensored Values
             PAG           Total      Failed     Censored     %Censored
             0              40           2           38        95.0000
             1              33          15           18        54.5455
             Total          73          17           56        76.7123
```

The plots statement in the above SAS program produces a graph depicting survival in each pag group, as shown in Figure 7.1

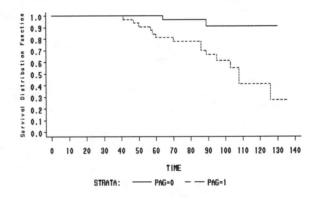

Figure 7.1   Survival function curve for breast cancer recurrence data using Kaplan-Meier Product Limit estimator

Due to the fact that the survival is always higher in the group pag = 0, it appears that this group experiences a much more favorable outcome.

The survival curves for the ventilating tube data showing the Kaplan-Meier cancer recurrence time estimates within the treatment and medical groups are shown in Figure 7.2.

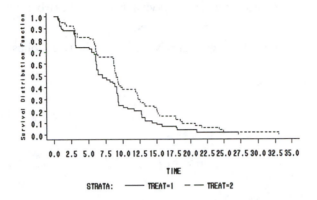

Figure 7.2   Survival function curve for ventilating tube data using Kaplan-Meier Product Limit estimator

Once again, the time until ventilating tube failure is longer for the medical group, implying greater effectiveness than the control in preventing tube displacement.

Part of the output from the above SAS commands includes calculation of the Log-Rank test. This aspect will be discussed subsequently.

At this point we have summarized measures describing survival for non-censored and censored data. Our graph indicated longer times until breast cancer recurrence in the low proliferation AgNOR group for the breast cancer data and longer times until tube displacement in the medical group for the ventilating data. However, in order to determine whether there is a significant difference between groups in a data set, a statistical test must be used for comparison of time until failure. Without any prior knowledge of the distribution which may be appropriate for the data, a non-parametric (distribution-free) test is preferable. A test designed for this purpose is the Log-Rank test (Peto and Peto (1972)).

B. NON-PARAMETRIC SURVIVAL COMPARISONS BETWEEN GROUPS

1. The Log-Rank Test for Comparisons Between Two Groups

Suppose that we have two groups, A and B, each given different treatments and it is of interest to compare the survival in these two groups. For a hypothesis test of

H_0: no difference in survival between the two groups

versus

H_A: difference in survival between the two groups

the test statistic is given by

$$\chi^2 = \frac{U_L^2}{V_L}$$
(7.2)

where

$$U_L = \sum_{j=1}^{r} (d_{1j} - e_{1j})$$

$$V_L = \sum_{j=1}^{r} \frac{n_{1j} n_{2j} d_j (n_j - d_j)}{n_j^2 (n_j - 1)} = \sum_{j=1}^{r} v_j$$

for n_{1j}, n_{2j}, d_{1j}, d_{2j}, e_{1j}, e_{2j} defined as follows:

$t_{(j)}$ = j^{th} death time (regardless of group)

n_{1j} = number at risk in group A just before time $t_{(j)}$

n_{2j} = number at risk in group B just before time $t_{(j)}$

d_{1j} = number of deaths in group A at time $t_{(j)}$

d_{2j} = number of deaths in group B at time $t_{(j)}$

e_{1j} = expected number of individuals dying in group A at time $t_{(j)}$

e_{2j} = expected number of individuals dying in group B at time $t_{(j)}$

where

$$e_{kj} = \frac{n_{kj} d_j}{n_j}$$

for $k = 1,2$ so that

n_j = total number at risk in both groups just before $t_{(j)}$

and

d_j = total number of deaths in both groups at $t_{(j)}$.

The test statistic is calculated by constructing a table of the following nature (Table 7.8):

Table 7.8
Calculations for Log-Rank Test Statistic

Death Time	d_{1j}	n_{1j}	d_{2j}	n_{2j}	d_j	n_j	$e_{1j} = n_{1j}d_j / n_j$	(1) $d_{1j} - e_{1j}$	(2) v_{1j}
$t_{(1)}$									
$t_{(2)}$									
$\vdots$									
$t_{(n)}$									

The squares of the sum of column (1) and the sum of column (2) are then used in calculation of the statistic. Under the null hypothesis of no differences between groups A and B, the test statistic is distributed as $\chi^2_{(1)}$.

Example 7.3: The SAS program described above (with the strata = pag statement) calculates the log-rank test to determine whether differences are present in survival for patients with a high proliferative AgNOR index (pag = 1) and those with a low index (pag = 0). The output is as shown:

```
            Test of Equality over Strata

                                       Pr >
        Test       Chi-Square   DF   Chi-Square

        Log-Rank     10.3104     1     0.0013
        Wilcoxon      8.6344     1     0.0033
        -2Log(LR)    12.7162     1     0.0004
```

The p-value for the Log-Rank test (p=0.0013) indicates that there are significant differences in the two AgNOR groups with respect to survival. This is in agreement with the survival graph previously examined, which implied substantially better survival for the low index group (pag = 0).

Notice that the above SAS output also provides output for the Wilcoxon and -2 Log(LR) tests of equality over the strata AgNOR. The Wilcoxon test statistic has a similar form to the Log-Rank; however the Wilcoxon test weights each term in the summation over the various death times by the number of individuals alive at that death time, thus giving less weight to terms where few women had not experienced breast cancer recurrence. For general use, the Log-Rank test is most appropriate when the assumption of proportional hazards between treatment groups holds. The topic of proportional hazards will be discussed subsequently.

2. The Log-Rank Test for Comparisons Between Several Groups

Often it is desirable to make comparisons of the survival between three or more groups; in this case, an extension of the Log-Rank test is used.

If there are q groups we wish to make comparisons between, then the following are calculated for each group $i = 1,...,q$:

$$U_{Li} = \sum_{j=1}^{r} \left(d_{ij} - \frac{n_{ij} d_j}{n_j} \right)$$

$$U_{Wi} = \sum_{j=1}^{r} n_j \left(d_{ij} - \frac{n_{ij} d_j}{n_j} \right).$$

The vectors

$$U_L = \begin{bmatrix} U_{L1} \\ U_{L2} \\ \vdots \\ U_{Lq} \end{bmatrix}$$

and

$$U_W = \begin{bmatrix} U_{W1} \\ U_{W2} \\ \vdots \\ U_{Wq} \end{bmatrix}$$

are then formed.

In addition, the variances and covariances are needed and are given by the formula

$$V_{Lii'} = \sum_{j=1}^{r} \frac{n_{ij} d_j (n_j - d_j)}{n_j (n_j - 1)} \left(\delta_{ii'} - \frac{n_{ij}}{n_j} \right)$$

where $\delta_{ii'}$ is such that

$$\delta_{ii'} = 1 \quad \text{if } i = i'$$
$$0 \quad \text{otherwise.}$$

The variance-covariance matrix is then given by

$$V = \begin{bmatrix} V_{L11} & V_{L12} & \cdots & V_{L1i} \\ V_{L21} & V_{L22} & \cdots & V_{L2i} \\ \vdots & \vdots & \ddots & \vdots \\ V_{Li1} & V_{Li2} & \cdots & V_{Lii} \end{bmatrix}$$

where $V_{Lij} = Cov(U_{Li}, U_{Lj})$ and $V_{Lii} = Var(U_{Li})$.

To test a null hypothesis of

H_0: no difference in survival between all groups

versus

H_A: difference in survival between at least two groups

the test statistic given by

$$\chi^2 = U_L' V_L^{-1} U_L \qquad (7.3)$$

has a $\chi^2_{(q-1)}$ distribution under the null hypothesis, where q is the number of strata.

For calculation of the stratified log-rank test in SAS, the same strata statement is used as for calculation of the Log-Rank test between two groups. A stratified test will be performed automatically for variables having more than two levels.

Example 7.4: The following SAS program (with the strata = group statement) calculates the Log-Rank test to determine whether differences are present in time until cyst disappearance for cows treated with hormone treatment 1, hormone treatment 2 and the control. The program and Log-Rank portion of the output are as shown:

```
proc lifetest;
        time weeks*censor(0);
        strata group;
run;
```

SAS output:

```
                 Test of Equality over Strata
                                          Pr >
        Test        Chi-Square    DF    Chi-Square

        Log-Rank      21.2401      2      0.0001
        Wilcoxon      17.8661      2      0.0001
        -2Log(LR)     10.6663      2      0.0048
```

343

In this case, the p-value for the Log-Rank test (p=0.0001) indicates that there are significant differences in the three groups with respect to time until cyst disappearance.

Addition of the statement `plots = (s)` at the end of the `proc lifetest` statement shown above would produce a survival curve with each of the three groups graphed separately.

C. PARAMETRIC METHODS

All parametric methods involve specification of a distributional form for the probability density function, $f(t)$. This in turn specifies the survival function $S(t)$ and the hazard function $h(t)$ using the relationships defined previously. The two parametric models to be discussed are the Exponential and the Weibull. Their survival and hazard functions are given and their properties reviewed.

1. The Exponential Model for Survival Analysis

An exponential density function is given by

$$
\begin{aligned}
f(t) &= \lambda \exp(-\lambda t) & t \geq 0, \lambda > 0 \\
&= 0 & t < 0
\end{aligned}
\tag{7.4}
$$

so that

$$
S(t) = \exp(-\lambda t) \qquad t \geq 0
$$

and

$$
h(t) = \lambda \qquad t \geq 0.
$$

Notice that the hazard function is independent of time, implying the instantaneous conditional failure rate does not change within a lifetime. This is also referred to as the memoryless property of the Exponential distribution, since the age of an individual does not affect the probability of future survival. When $\lambda = 1$, the distribution is referred to as the unit exponential.

In practice, most failure times do not have a constant hazard of failure and thus the application of the Exponential model for survival analysis is limited. The Exponential model is in fact a special case of a more general model which is widely applicable: the Weibull model.

2. The Weibull Model for Survival Analysis

The Weibull density function is given by

$$
f(t) = \lambda \gamma t^{\gamma-1} \exp(-\lambda t^{\gamma}) \qquad t \geq 0, \ \gamma, \lambda \geq 0
\tag{7.5}
$$

344

giving

$$S(t) = \exp(-\lambda t^\gamma)$$

and

$$h(t) = \lambda\gamma t^{\gamma-1}.$$

In the above, λ and γ are the scale and shape parameters, respectively. The specific case of $\gamma = 1$ defines the Exponential model with constant hazard previously discussed, $\gamma > 1$ implies a hazard which increases with time and $\gamma < 1$ yields a hazard decreasing with time. It is evident that the Weibull distribution is widely applicable since it allows modeling of populations with various types of failure risk.

Example 7.5: It is desired to fit a) an exponential model and b) a Weibull model to the breast cancer data discussed. This is done using the 'proc lifereg' statement in SAS with specification of exponential and Weibull models:

a) proc lifereg data = agnor;
      model time*censor(0) = pag age tsize / dist = exponential;

b) proc lifereg data = agnor;
      model time*censor(0) = pag age tsize / dist = weibull;

Note that the model statement specifies the dependent variable as time*censor(0), where time is the variable in the analysis recording the failure or censoring time, censor is an indicator variable denoting whether or not a failure time is censored, and the number in brackets indicates the coded value of the censor variable indicating an observation was censored.

The SAS output is as shown:

a) Exponential Model

```
Log Likelihood for EXPONENT -36.73724081

Lifereg  Procedure

Variable  DF   Estimate    Std Err   ChiSquare  Pr>Chi  Label/Value

INTERCPT  1    6.63279679  1.352221  24.06015   0.0001  Intercept
PAG       1   -2.1810958   0.755017   8.345181  0.0039
AGE       1    0.01069086  0.020383   0.275112  0.5999
TSIZE     1    0.39384413  0.494546   0.634214  0.4258
SCALE     0        1          0                         Extreme value scale
                                                        parameter
Lagrange Multiplier ChiSquare for Scale       . Pr>Chi is  .      .
```

345

b) Weibull Model

```
Log Likelihood for WEIBULL -26.88658984


Lifereg   Procedure

Variable   DF   Estimate   Std Err   ChiSquare   Pr>Chi   Label/Value

INTERCPT   1    5.10839008   0.473726   116.2824   0.0001   Intercept
PAG        1   -0.6610495    0.281155   5.528123   0.0187
AGE        1    0.00576767   0.007038   0.671549   0.4125
TSIZE      1    0.03470751   0.162151   0.045815   0.8305
SCALE      1    0.32047388   0.065261                        Extreme value scale
                                                             parameter
```

As was noted using the log-rank test and the survival curves, a significant difference exists in the failure times between those with a high AgNOR proliferative index and those with a low index, both when the failure times are modeled to be exponential and when they are modeled as Weibull. This is reflected in the p-values of 0.0039 and 0.0187 for the exponential and Weibull models, respectively. The other variables included in the model, age and tumor size, appear to have little effect on the time of breast cancer recurrence. The log-likelihood values of -36.74 and -26.89 indicate that the Weibull distribution is slightly better at modeling the breast cancer recurrence times. This may be due to the fact that the previously discussed restrictions imposed by an exponential model (i.e. a constant hazard rate) may not be valid when examining disease recurrence. In such a situation, a distribution with a hazard function that changes over time is preferable.

Graphical Assessment of Model Adequacy

The fit of the exponential and Weibull models to the breast cancer data may also be assessed graphically using the following SAS program:

```
proc lifetest data = agnor  outsurv = a;
        time time*censor(0);
run;

data graph;
        set a;
        s = survival;
        logs = log(s);
        loglogs = log(-log(s));
        logtime = log(time);
run;
```

```
proc gplot;
        symbol value = none   i = join;
        plot logs*time loglogs*logtime;
run;
```

The first program is to calculate the product limit estimates of the survival function and to output them to the data set a. The second program calculates the log of survival, the log of (-log(survival)) and the log of the failure time variable. The third SAS program produces two graphs, with the survival times joined to form a linear plot for the variables log of survival versus time and log(-log(survival)) versus log time. The graphs (Figures 7.3 and 7.4) look as follows:

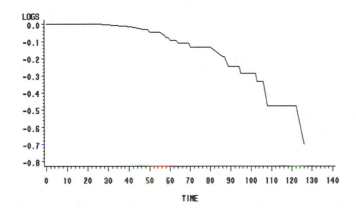

Figure 7.3   Plot showing the goodness of fit of the exponential model for breast cancer data.

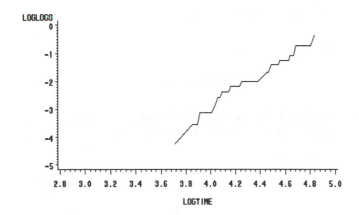

Figure 7.4   Plot showing the goodness of fit of the Weibull model for breast cancer data.

347

A straight line indicates no departures from model adequacy. The above graphs appear to indicate a better fit to the Weibull model for the breast cancer data, as did the log-likelihood value.

The reason we expect to see a straight line relationship between log(survival) and time if the exponential model is adequate and log(-log(survival)) and log(time) if the Weibull model is adequate is as follows:

Recall the survival functions for the exponential and Weibull models were given by

$$S(t) = \exp(-\lambda t)$$

and

$$S(t) = \exp(-\lambda t^{\gamma}),$$

respectively.

We rearrange the equations to obtain linear functions of t. Taking the log of each side in the exponential model gives

$$\log(S(t)) = -\lambda t$$

which is now a linear function of t, so that a graph of the log of the survival estimates versus time should be linear if the exponential model is adequate.

Similarly, rearranging the Weibull survival function gives

$$\log(S(t)) = (-\lambda t^{\gamma})$$

so that upon taking the negative log, we have

$$\log[-\log(S(t))] = \log \lambda + \gamma \log t$$

which is now a linear function of t, so that a graph of the log of the -log(survival) estimates versus log of time should be linear if the Weibull model is adequate.

The exponential and Weibull model adequacy graphs (Figures 7.5 and 7.6) are given below for the ear data:

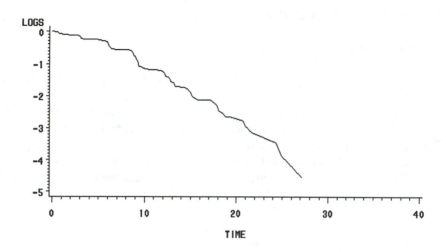

Figure 7.5 Plot showing the goodness of fit of the exponential model for ventilating tube data using lambda = 0.1527.

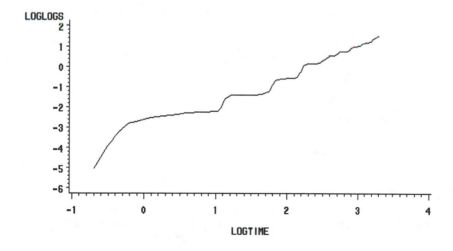

Figure 7.6 Plot showing the goodness of fit of the Weibull model for ventilating tube data using log(lambda)=-3.144 and gamma=1.36.

The above graphs imply the exponential model is an adequate choice for the ventilating tube data.

D. SEMI-PARAMETRIC METHODS

1. The Cox Proportional Hazards Model

One positive feature of the parametric methods previously discussed is that specification of a form for the probability density function allows the likelihood to be constructed. Maximum likelihood estimates and standard errors may then be obtained for all parameters in the model. However, the drawback in parametric modeling lies in the fact that it may not be desirable to specify a probability distribution function for a particular set of data, making non-parametric calculations more attractive. The ideal situation would involve no distributional restrictions on the density, yet maximum likelihood estimates of regression parameters (and thus treatment effects) would be readily available. An analysis with such properties may be performed using the Cox Proportional Hazards Model (Cox, 1972).

As its name implies, the Proportional Hazards Model is constructed by assuming that the hazard function of the ith individual is the product of a baseline hazard common to all individuals, denoted $h_0(t)$ and a function of the covariate vector, x_i for that individual, $\varphi(x_i)$:

$$h_i(t) = h_0(t)\varphi(x_i) \quad . \tag{7.6}$$

Since rearrangement yields the hazard ratio or relative hazard to be the non-negative function

$$\varphi(x_i) = \frac{h_i(t)}{h_0(t)},$$

the covariate function is specified to be

$$\varphi(x_i) = \exp(x_i\beta),$$

which is also non-negative.

Writing the covariate vector of the ith individual $x_i = \left(x_{i1}, x_{i2}, ..., x_{ip}\right)$, further rearrangement shows

$$\log\left(\frac{h_i(t)}{h_0(t)}\right) = \beta_1 x_{i1} + \beta_2 x_{i2} + ... + \beta_p x_{ip},$$

so that the proportional hazards model is a linear model for the log hazard ratio.

The partial likelihood for the proportional hazards model is given by

$$L(\beta) = \prod_{j=1}^{r} \frac{\exp(\beta'x_{(j)})}{\sum\limits_{l \in R(t_{(j)})} \exp(\beta'x_{(l)})},$$

where $R(t_{(j)})$ is the risk set (those individuals alive) at the jth ordered death time, $t_{(j)}$. Thus the likelihood takes the product over the jth ordered death times of terms of the form

$$L_i(\beta) = \frac{\exp(\beta'x_{(i)})}{\sum\limits_{l \in R(t_{(i)})} \exp(\beta'x_{(l)})} \qquad (7.7)$$

$$= \Pr \begin{pmatrix} \text{failure occurs to observed} \\ \text{individual given the risk set} \\ R(t_{(i)}) \text{ and a failure at } t_{(i)} \end{pmatrix}$$

so that $L_i(\beta)$ is the ratio of the hazard for the individual who died at the ith ordered death time divided by the sum of hazards for the individuals who were at risk when the ith ordered death occurred. Notice that individuals with censored failure times do not contribute a term to the likelihood, however, they are included in the risk sets. Hence $L(\beta)$ may be expressed as

$$L(\beta) = \prod_{i=1}^{n} \left(\frac{\exp(\beta'x_{(i)})}{\sum\limits_{l \in R(t_{(i)})} \exp(\beta'x_{(l)})} \right)^{\delta_i} \qquad (7.8)$$

where

$$\delta_i = 1 \text{ if the } i^{th} \text{ individual fails during the study}$$
$$= 0 \text{ if the failure time of the } i^{th} \text{ individual is censored.}$$

Taking the product over the j uncensored failure times in (7.7) is equivalent to taking the product over the n censored and uncensored failure times in (7.8) due to the indicator variable δ_i.

Notice that the likelihood (7.8) is referred to as a partial likelihood due to the fact that it is based on the product of probabilities of failure rather than official density functions. Maximum likelihood estimation of the regression parameters β occurs by treating the partial likelihood given in (7.8) as a true likelihood, so that differentiation of the log and subsequent maximization is possible. Variance estimates are found using the matrix of partial second derivatives. Newton-Rhapson techniques are required for the maximization.

2. Treatment of Ties in the Proportional Hazards Model

The proportional hazards model previously discussed implicitly assumes (due to its continuous nature) that the exact time of failure or censoring is known for each individual under study. For this situation, there are no ties in failure times. Such accuracy is not usually encountered in practice, where frequently survival times are only available for the nearest day, week or month. This may result in tied failure times in the data. Several methods exist for dealing with tied failure times in the Cox Proportional Hazards model. Two will be discussed here.

Note that in the treatment of ties, it is assumed that

i) there are r distinct deaths at $t_{(j)}, j = 1,...,r$ and

ii) there may be d_j deaths at $t_{(j)}$ and

iii) $s_j = x_1 + x_2 + ... + x_{d_j}$ $j = 1,...,r$, where $x_1, x_2, ..., x_{d_j}$ are the covariate vectors of the individuals dying at $t_{(j)}$.

In addition, it is assumed that if an uncensored failure and a censored failure (i.e. a death and a drop-out) occur at the same time, the uncensored failure occurs first, so that the discussion below focuses on ties in uncensored failure times.

i) Breslow's Method for Tied Failure Times

Breslow (1972) suggested the following approximation to the partial likelihood function to account for tied failure times in the data:

$$\prod_{j=1}^{r} \frac{\exp(\beta's_j)}{\left[\sum_{l \in R(t_{(j)})} \exp(\beta'x_l)\right]^{d_j}},$$

where all possible sequences of deaths are summed for the respective component of the likelihood function. Breslow's approximation has the advantage of being quite accurate when the number of ties at a given death time is small and is simple to compute.

ii) Cox's Method for Tied Failure Times

To account for the presence of tied data, Cox (1972) suggested using a likelihood of the form

$$\prod_{j=1}^{r} \frac{\exp(\beta's_j)}{\displaystyle\sum_{l \in R(t_{(j)};d_j)} \exp(\beta's_l)} \;,$$

where $R(t_{(j)};d_j)$ denotes a set of d_j individuals drawn from $R(t_{(j)})$. The denominator sum is over all possible sets of d_j individuals sampled from the risk set without replacement. Notice that if $d_j = 1$ (meaning no ties in the data), Cox's approximation reduces to the original partial likelihood proposed.

Example 7.6: We fit the Cox proportional hazards model to the breast cancer data with the following SAS program:

```
proc phreg data = agnor;
        model time*censor(0) = pag age tsize;
run;
```

The program produces the output shown below:

```
                      The PHREG Procedure

    Data Set: WORK.AGNOR
    Dependent Variable: TIME
    Censoring Variable: CENSOR
    Censoring Value(s): 0
    Ties Handling: BRESLOW


                  Summary of the Number of
                  Event and Censored Values

                                              Percent
               Total       Event    Censored  Censored

                73          17          56      76.71


           Testing Global Null Hypothesis: BETA=0

                  Without       With
    Criterion     Covariates    Covariates    Model Chi-Square

    -2 LOG L      121.174       108.867       12.306 with 3 DF (p=0.0064)
    Score            .             .          11.087 with 3 DF (p=0.0113)
    Wald             .             .           8.225 with 3 DF (p=0.0416)
```

```
                Analysis of Maximum Likelihood Estimates

                    Parameter     Standard      Wald        Pr >        Risk
     Variable   DF   Estimate      Error    Chi-Square   Chi-Square     Ratio

     PAG        1    2.086512     0.75574    7.62241      0.0058       8.057
     AGE        1   -0.018556     0.02149    0.74557      0.3879       0.982
     TSIZE      1   -0.175210     0.50803    0.11894      0.7302       0.839
```

Once again, the pag variable for proliferative AgNOR index implies there is a significant difference (p = 0.0058) between survival for those with high and low scores, and that the variables age and tumor size are not predictive of breast cancer recurrence (p = 0.982 and p = 0.839, respectively). In addition, the positive parameter estimate of pag implies the hazard of breast cancer recurrence becomes larger as the level of pag increases, meaning a greater hazard for the pag = 1 group versus pag = 0. The risk ratio of 8.057 for the pag variable indicates the hazard of breast cancer recurrence is 8 times greater in the high AgNOR proliferative index group.

SAS uses the Breslow method as the default for tie handling in the Cox proportional hazard model.

Example 7.7: We fit a stratified (by ear) Cox proportional hazards model to the ventilating tube data with the following SAS program:

```
proc phreg data = ear;
        model time*censor(0) = treat;
        strata ear;
run;
```

The program produces the output shown below:

```
The PHREG Procedure

        Data Set: WORK.EAR
        Dependent Variable: TIME
        Censoring Variable: CENSOR
        Censoring Value(s): 0
        Ties Handling: BRESLOW


                        Summary of the Number of
                        Event and Censored Values

                                                        Percent
            Stratum    Ear     Total     Event   Censored   Censored

               1        1       78        73        5        6.41
               2        2       78        71        7        8.97
   Total                       156       144       12        7.69
```

```
              Testing Global Null Hypothesis: BETA=0

                    Without        With
     Criterion      Covariates     Covariates     Model Chi-Square

     -2 LOG L        985.448         981.182         4.266 with 1 DF  (p=0.0389)
     Score              .               .            4.336 with 1 DF  (p=0.0373)
     Wald               .               .            4.293 with 1 DF  (p=0.0383)


               Analysis of Maximum Likelihood Estimates

                       Parameter    Standard      Wald         Pr >       Risk
    Variable    DF     Estimate     Error     Chi-Square    Chi-Square    Ratio

    TREAT       1      -0.357598    0.17259     4.2932        0.0383      0.699
```

Notice that the stratified proportional hazards model implies there is a significant difference in time to ventilating tube failure between the medical and control groups ($p = 0.0383$). The negative coefficient of the parameter estimate implies the hazard of tube failure becomes lower as the treatment category becomes higher, meaning the medical group (treat = 1) has a lower hazard than the control (treat = 0). The risk ratio implies the hazard for children receiving medical treatment is 70% of the hazard for children receiving the control. Note that the correlation between the two ears in the sample for each child has not yet been accounted for in the analyses.

3. Time Dependent Covariates in the Cox Proportional Hazards Model

Until this point in the discussion, we have assumed that the covariates, x_i included in the model did not change with time. This assumption is quite important because it ensured that an at risk individual's contribution to the denominator of the partial likelihood would be the same regardless of the time of the failure, meaning that the covariate value could be assessed solely at the beginning of the study. Covariates of this nature include sex, height and race. However, often it may be of interest to examine covariates that can change over time, such as blood pressure, weight, white blood cell count and cigarette smoking. In this case, several modifications of the hazard and partial likelihood functions are required. They are discussed subsequently.

Recall the original definition of the proportional hazard function (7.6) as giving the hazard of individual   as the product of a baseline hazard common to all individuals and a function of the covariates for that individual. For the case of time dependent covariates, the proportional hazard model is

$$h_i(t) = h_0(t) \exp(\beta x_i(t))$$

where $h_0(t)$ now becomes the baseline hazard for an individual with all covariate values equal to zero at time zero and throughout the study. In addition, the relative hazard is

$$\exp(\beta x_i(t)) = \frac{h_i(t)}{h_0(t)},$$

which is a function of time (t), so that the hazard ratio over time is not a constant function. This means the hazards can no longer be proportional.

The partial likelihood as a function of time dependent covariates becomes

$$L(\beta) = \prod_{i=1}^{n} \left(\frac{\exp(\beta'x_{(i)}(t_i))}{\sum\limits_{l \in R(t_{(i)})} \exp(\beta'x_{(l)}(t_i))} \right)^{\delta_i}, \qquad (7.9)$$

so that covariate values must be known for each individual in the risk set at the time of the respective death i.e. the blood pressure of individual k must be known at the time of death of individual j in order to include them in the risk set at that time.

Although use of time dependent covariates in this fashion complicates the study design by requiring measurements on all individuals at each death time, one can think of many situations where inferences made using time dependent covariates would not only be useful but absolutely necessary. For example, blood pressure may be considered to be a required covariate when examining time of stroke or heart attack, or glucose level when examining time of diabetic seizure. A more detailed discussion of analysis of survival data with time dependent covariates is given by Collett (1994) and Lawless (1982).

IV. CORRELATED SURVIVAL DATA

As with all types of response variables, techniques must be available for analyses performed on correlated time to event data. The complexity of studies involving multiple treatment centres, family members and measurements repeatedly made on the same individual requires methods to account for correlation in the data. Such is the case for any type of response, be it continuous, binary or a time of event. Use of the multivariate normal distribution allows correlation to be accounted for in continuous data, where techniques are well established. For the situation of binary responses, work over the past few decades has resulted in tests adjusted for correlation in the data. However, for the time to event case, methods of accounting for correlation in the data have only recently been developed, reflecting the fact that time to event models are themselves quite new.

As mentioned earlier, correlation is anticipated between outcomes in certain situations. Correlation due to the three most common types of studies will be discussed. For example, in multi-centre clinical trials, the outcomes for groups of patients at several centres are examined. In some instances, patients in a centre might exhibit similar responses due to uniformity of surroundings and procedures within a centre. This would result in correlated outcomes at the level of the treatment centre. For the situation of studies of family members or litters, correlation in outcome is likely for genetic reasons. In this case, the outcomes would be correlated at the family or litter level. Finally, when one person or animal is measured repeatedly over time, correlation will most definitely exist in those responses. Within the context of correlated data, the observations which are correlated for a group of individuals (within a treatment centre or a family) or for one individual (because of repeated sampling) are referred to as a *cluster*, so that from this point on, the responses within a cluster will be assumed to be correlated.

These three types of studies are becoming increasingly common in biomedical study designs. Thus it is essential to have methods of accounting for the resulting correlated data. Two methods with existing programs in computer software packages are currently available: the Marginal Approach and Frailty Models. They will be discussed subsequently and their properties contrasted. Situations in which each method is desirable are discussed.

A. MARGINAL METHODS

i) The GEE Approach

Similar to the Cox Proportional Hazards Model, a marginal proportional hazard function is adopted for the jth individual in the ith cluster:

$$\lambda_{ij}(t) = \lambda_0(t)\exp(\beta' X_{ij})$$

The likelihood function of the jth individual in the ith cluster is then

$$
\begin{aligned}
L_{ij}(\alpha,\beta) &= f_{ij}(t)^{\delta_{ij}} (S_{ij}(t))^{1-\delta_{ij}} \\
&= (\lambda_0(t)\exp(\beta' X_{ij}))^{\delta_{ij}} \exp(-\Lambda_0(t)\exp(\beta' X_{ij})) \\
&= (\mu_{ij}^{\delta_{ij}} \exp(-\mu_{ij}))(\lambda_0(t)/\Lambda_0(t))^{\delta_{ij}}
\end{aligned}
\qquad (7.10)
$$

where $\Lambda_0(t) = \int_0^t \lambda_0(u)du$ is the cumulative baseline hazard, $\mu_{ij} = \Lambda_0(t)\exp(\beta' X_{ij})$ and α represents baseline hazard parameters. The fact that the censoring indicator variable, δ_{ij} takes only the values 0 or 1 implies that $\mu_{ij}^{\delta_{ij}} \exp(-\mu_{ij})$ is a Poisson likelihood in the random variable δ_{ij} with mean μ_{ij}. Regression parameter estimates are obtained using Poisson regression of δ_{ij} on X_{ij} with offset $\log(\Lambda_0(t))$. The resulting MLE $\hat{\beta}$ is used to calculate the MLE of the baseline hazard parameter,

α, using the score equation. A subsequent Poisson regression yields an updated $\hat{\beta}$, which may be used in calculation of an updated baseline hazard parameter estimate. Iteration continues between the two processes until convergence.

The method described above wherein iteration occurs between Poisson regression and score equation calculations results in consistent regression parameter estimates, even when the data are correlated. However, the variance estimates obtained using such methods are not robust when the data occur in a clustered form and correlation is present within the cluster. It is for this reason that the well known 'sandwich estimates' are used to compute robust variance estimates. The following discussion focuses on the GEE (generalized estimating equation) approach to obtaining robust variance estimates for clustered survival data.

Define the covariance structure within the ith cluster to be $V_i = Cov(Y_i)$. Then the generalized estimating equations are of the form

$$\sum_{i=1}^{I} D_i^T V_i^{-1} S_i = 0$$

where

$\quad\quad$ D_i is the matrix of derivatives of $E(Y_{ij} | X_i)$ with respect to β

$\quad\quad$ S_i is the vector of residuals $\{Y_{ij} - E(Y_{ij} | X_i)\}$

$\quad\quad$ $V_i = A_i^{1/2} R(\gamma) A_i^{1/2} / \phi$

$\quad\quad$ A_i diagonal with elements $Var(Y_{ij})$

$\quad\quad$ $R(\gamma)$ the working correlation matrix depending on γ and ϕ

where ϕ is a scale factor.

The procedure is then as follows:

For initial estimates, $\hat{\gamma}$ and $\hat{\phi}$, define the block diagonal matrix $\tilde{V}$ with diagonal elements $\tilde{V}_i = V_i[\beta, \hat{\gamma}(\beta, \hat{\phi}(\beta))]$. Let $Z = D\beta - S$, where $D = (D_1, ..., D_I)'$ and $S = (S_1, ..., S_I)'$. The GEE method involves iteratively reweighted linear regression of Z on D with weight $\tilde{V}^{-1}$. Given β, one then estimates γ and ϕ using the current Pearson residuals,

$$\hat{r}_{ij} = [y_{ij} - \hat{E}(Y_{ij})] / \sqrt{\hat{var}(Y_{ij})},$$

where

$$\hat{E}(Y_{ij}) = \hat{\mu}_{ij} = g^{-1}(\hat{\beta}' X_{ij}),$$

for g the link function. An exchangeable correlation structure characterized by

$$corr(Y_{ij}, Y_{ik}) = \gamma, \quad j \neq k$$

gives a moment estimate of

$$\hat{\gamma} = \hat{\phi} \sum_{i=1}^{I} \sum_{j>k}^{I} \hat{r}_{ij} \hat{r}_{ik} / [1/2 \sum_{i=1}^{I} n_i(n_i - 1) - p]$$

for

$$\hat{\phi}^{-1} = \sum_{i=1}^{I} \sum_{j=1}^{I} \hat{r}_{ij}^2 / (N - p)$$

where $N = \sum_{i=1}^{I} n_i$.

Example 7.8: Recall the Ventilating Tube Data, in which each child under study contributed two failure times, one for each ear. Proper statistical analysis of the data requires the correlated nature of the observations within a child be accounted for. Although parameter estimates are not directly affected by the within cluster (child) correlation, variance estimates are. Thus the SAS macro GEE is utilized to obtain parameter estimates (regression, for given estimates of the parameter of the baseline hazard function) and robust variance estimates.

```
data poisson;
        set ear;
        lam = .15267;
        ttime = (time)**lam;

proc genmod data = poisson;
        class treat patient;
        model censor = treat / noint
        dist = poisson
        link = log
        offset = ttime
        pscale
        type3;
repeated subject = patient / type = exch;
run;
```

```
/* GJE macro */

data temp1;
        set ear;
        %let xx=treat;
        %include 'f:surv.cox';
        %phlev (data=temp1,
                time=time,
                event=censor,
                xvars=&xx,
                        id=patient,
                collapse=y,
                outlev=temp2 outvar=temp3);

proc print data=temp3;
run;
```

Discussion of SAS program:

Recall lambda = 0.15267 was found to be the estimate of λ for the time variable when the exponential model was fit to the ear data. The option noint in the model statement specifies that a regression model with no intercept should be fitted. In addition, the Poisson specification requires that the Poisson distribution be used, and the log link statement forces the relationship between the mean and the covariates to be log-linear. The repeated statement at the end specifies the correlation structure within the cluster to be exchangeable, meaning equal correlation is expected between any two individuals within the cluster.

ii) The GJE Approach

The Generalized Jackknife Estimator (GJE) (Therneau, 1993) is similar to the Generalized Estimating Equation approach in that both result in robust parameter estimate variances. The score residual matrix B and Fisher's Information Matrix are required to obtain these types of robust variance estimates. The motivation for the GJE approach is now discussed.

The partial likelihood function discussed earlier

$$L(\beta) = \prod_{i=1}^{n} \left(\frac{\exp(\beta'x_{(i)})}{\sum_{l \in R(t_{(i)})} \exp(\beta'x_{(l)})} \right)^{\delta_i} \qquad (7.8)$$

is differentiated with respect to β_r, the rth parameter. This yields

$$\frac{\partial \log L(\beta)}{\partial \beta_r} = \sum_{i=1}^{N} \delta_i (x_{ri} - a_{ri})$$

$$= \sum_{i=1}^{N} \delta_i S_{ri}, \ r = 1,\dots,p$$

where

$$a_{ri} = \frac{\sum_{t \in R_i} x_{rt} \exp(x_t'\hat{\beta})}{\sum_{t \in R_i} \exp(x_t'\hat{\beta})}.$$

The implication is that $\dfrac{\partial \log L(\beta)}{\partial \beta_r}$ is the difference between the rth covariate, x_{ri}, and a weighted average of values of the explanatory variable over individuals at risk at the failure time of the ith individual. The information matrix is obtained through differentiation a second time and is seen to be

$$-\frac{\partial^2 \log L(\beta)}{\partial \beta_r \partial \beta_s} = \sum_{i=1}^{N} \delta_i \left(\frac{\sum_{t \in R_i} x_{rt} x_{st} \exp(x_t'\hat{\beta})}{\sum_{t \in R_i} \exp(x_t'\hat{\beta})} - a_{ri} a_{si} \right) \qquad r, s = 1,\dots,p \, .$$

If B is the vector with r^{th} component $B_r = \partial \log L(\beta) / \partial \beta_r$ and A is the matrix with the entry in the rth row, s^{th} column being $-\partial^2 \log L(\beta) / \partial \beta_r \partial \beta_s$, then the traditional sandwich estimate is given by

$$V(\hat{\beta}) = H'H$$

where

$$H = BA^{-1} \, .$$

For the GJE, the matrix B (containing one row per patient per cluster) is collapsed into $\tilde{B}$ with one row per cluster, where the row for the cluster has been summed over patients in that cluster. The result is then the GJE robust variance estimate, given by

$$\tilde{V}(\hat{\beta}) = \tilde{H}'\tilde{H}$$

where

$$\tilde{H} = \tilde{B}A^{-1} \, .$$

Note that $\tilde{V}$ underestimates V when the number of clusters is small.

The collapse statement tells SAS to perform the analysis using only unique values of the id variable (defined as patient). Thus only one observation per child is included. Outlev is the name of the SAS data set containing the leverage residuals and outvar the name of the SAS output data set containing the robust variance estimate.

The GEE portion of the SAS output for the correlated ear data is as shown:

```
                      GEE Model Information

         Description                         Value
         Correlation Structure            Exchangeable
         Subject Effect                Patient (78 levels)
         Number of Clusters                   78
         Correlation Matrix Dimension          2
         Maximum Cluster Size                  2
         Minimum Cluster Size                  2

                Analysis of GEE Parameter Estimates
                Empirical Standard Error Estimates

                        Emp. Std.   95%
Parameter    Estimate   Error       Confidence Limits    Z        Pr>|Z|
                                    Lower     Upper


INTERCEPT    0.000        .           .          .         .         .
TREAT 1     -1.4189     0.0444     -1.5058    -1.3319   -33.99    0.0000
TREAT 2     -1.4612     0.0354     -1.5305    -1.3918   -41.31    0.0000
Scale        0.3253       .           .          .         .         .

Note: the scale parameter was held fixed.
```

The GJE portion of the SAS output for the correlated ear data is as shown:

```
Comparison of Cox model Beta, SE and chi-square to robust estimates

Wald chi-square is based on the robust estimates
Robust SE is based on the collapsed (summed within patient) L matrix


Variable  Parameter   SE    Robust  Chi-Square  Robust      Wald        df    p
          Estimate           SE                 Chi-Square  Chi-Square


treat     -0.35292  0.168  0.186    4.379       3.600        .          .    0.0578
wald         .        .      .        .           .         3.600       1    0.0578
```

The results indicate that the medical group experiences a significantly (p=0.0578) lower tube failure (displacement) rate than does the control group.

Marginal models (such as the GEE) treat the correlation within a cluster as a nuisance, meaning that it is indirectly accounted for in the analyses. Parameter estimates obtained using the marginal approach have the desirable interpretation of applying to the whole population. This means that an estimate of treatment effect, for example, would pertain to all clusters, or to all children in the Ventilating Tube Data.

Although parameter estimates under marginal models are easily calculated using available computer software, they have one large drawback. Treatment of within cluster correlation as a nuisance is an acceptable approach for situations in which the within cluster dynamics are not of interest, such as within a hospital or farm. In these cases, we simply wish to account for the correlation within patients in hospitals (treatment centres) or animals in farms. The correlation within these units is not of interest on its own. This is in contrast to the situations involving family members or repeated measuring on an individual. In these cases, the within cluster correlation represents disease relationships within a family and propensity of an individual to experience multiple disease events, respectively. This type of information is usually of primary or secondary importance for studies of this nature, making treatment of within cluster correlation as a nuisance unjustified and inaccurate. For example, knowledge of the tendency for a child to experience ventilating tube failure in both ears would definitely be of interest due to the fact that the majority of children are bilateral (have tubes in both ears). Another approach is required for this type of situation in which within cluster dynamics are of interest: Frailty Models.

B. FRAILTY MODELS

Quite the opposite to marginal models, frailty models directly account for and estimate within cluster correlation. A parameter estimate of within cluster propensity for events is obtained directly.

The random effects approach to frailty models involves the assumption that there are unknown factors within a cluster causing similarity (homogeneity) in failure times within the cluster and thus differences (heterogeneity) in failure times between different clusters. The reason such factors are referred to as unknown is that if they were known to the investigator, they could be included in the analysis, resulting in independence within a cluster. Frailty modeling (known as such because it examines the tendency for individuals within a cluster to fail at similar times, or experience similar frailties) involves specification of independence within a cluster, conditional on a random effect. This random effect for the ith cluster, v_i, is incorporated (conditionally) into the proportional hazard function previously examined:

$$h(t|v_i) = v_i h_0(t) \exp(\beta x_{ij})$$

which may be re-expressed as

$$h(t|v_i) = h_0(t)\exp(\beta x_{ij} + \eta_i),$$

showing v_i actually behaves as an unknown covariate for the ith cluster in the model.

Using previous relationships between the survival and hazard function, we have the conditional survival function as

$$S(t|v_i) = \exp[v_i\Lambda_0(t)\exp(\beta X_{ij})]$$

and the conditional likelihood as

$$L(\gamma,\beta|v_i) = \prod_{i=1}^{I}\prod_{j=1}^{n_i}\left(h(t_{ij}|v_i)^{\delta_{ij}}S(t_{ij}|v_i)\right),$$

where there are I clusters, the ith one being of size n_i and γ and β represent baseline hazard and regression parameters, respectively. Substitution gives

$$L(\gamma,\beta|v_i) = \prod_{i=1}^{I}\prod_{j=1}^{n_i}\left([h_0(t)v_i\exp(\beta X_{ij})]^{\delta_{ij}}\exp[-v_i\Lambda_0(t)\exp(\beta X_{ij})]\right) \qquad (7.11)$$

$$= \prod_{i=1}^{I}\prod_{j=1}^{n_i}\left(\frac{\phi}{\Phi}\right)^{\delta_{ij}}\prod_{i=1}^{I}\exp(-\Phi)(\Phi)^{\delta_{i.}}.$$

where

$$\phi = v_i\exp(\beta'X_i)\exp(\alpha'W_{ij})\gamma t_{ij}^{\gamma-1}$$

$$\Phi = v_i\exp(\beta'X_i)\sum_{j=1}^{n_i}\exp(\alpha'W_{ij})t_{ij}^{\gamma} = v_i\exp(\beta'X_i)e_{i.}.$$

The marginal (i.e. independent of v_i) likelihood, $L(\gamma,\beta)$, is obtained through integration of the random effect distribution. A common assumption is for the random effect to follow a Gamma distribution with mean 1 and variance τ, i.e.,

$$f(v_i) = \frac{v_i^{1/\tau-1}\exp(-v_i/\tau)}{\Gamma(1/\tau)\tau^{1/\tau}}.$$

The marginal likelihood is then obtained to be

$$L(\gamma,\beta,\alpha,\tau) = \prod_{i=1}^{I} \prod_{j=1}^{n_i} \int_0^{\infty} L(\gamma,\beta,\alpha|v_i)dG(v_i)$$

$$= \prod_{i=1}^{I} \prod_{j=1}^{n_i} \left(\frac{\exp(\alpha w_{ij})\gamma t_{ij}^{\gamma-1}}{e_{i.}} \right)^{\delta_{ij}} \prod_{i=1}^{I} \left(\exp(\beta x_i)e_{i.}\tau \right)^{\delta_{i.}} \frac{\Gamma(\delta_{i.}+1/\tau)}{\Gamma(1/\tau)} \left(\frac{1}{1+\exp(\beta x_i)e_{i.}\tau} \right)^{\delta_{i.}+1/\tau} \qquad (7.12)$$

Inference on the regression parameters, baseline hazard parameter and dispersion parameter is then possible using maximum likelihood procedures. Newton Rhapson methods are required for estimation.

Example 7.9: A macro was designed to obtain the maximum likelihood estimates shown in the marginal likelihood above (7.12) for the ventilating tube data. It was desired to examine the significance of treatment in delaying time to tube failure after accounting for correlation within ears. Notice that maximum likelihood estimates are obtained for the cluster level treatment effect, β, the Weibull baseline hazard parameter, γ, and the dispersion parameter τ, but not for within cluster covariates due to the fact that each child received the same medicine in each ear. The results are as shown:

	β	γ	τ
Parameter Estimate	-4.19	2.00	2.75

To examine the significance of treatment effect, the estimate of the standard error of β is required and was found to be 0.479, so that $\hat{\beta}/S_{\hat{\beta}} = -8.75$, implying the treatment substantially decreases time to tube failure after adjusting for the correlation between ears (p = 0.000).

Frailty models have a great deal of potential in accounting for correlation arising in clustered survival data (Hougaard, 1995). Although the Gamma frailty distribution has been examined here, other possibilities include the Inverse-Gaussian and Log-Normal. The Inverse-Gaussian appears to be particularly well-suited to the situation in which survival times are positively skewed as well as correlated. However, these types of models have the downfall of being difficult to fit due to complex distributional structure and divergence is not uncommon when attempting to maximize the likelihood. On the positive side, a great deal of research is under way in the area of frailty models (Liang et al., 1995, Sastry, 1997) and their introduction into commercial software (such as SAS) is not far in the future.

Appendix I

Average Milk Production per Month (kg) for 10 Ontario Farms

Farm	Milk Yield		Farm	Milk Yield		Farm	Milk Yield
1	32.33		4	27.25		8	21.73
1	29.47		4	29.69		8	21.90
1	30.19		4	28.29		8	25.07
1	28.37		4	28.08		8	23.49
1	29.10		5	29.69		8	26.65
1	28.19		5	31.92		8	27.08
1	30.28		5	29.51		8	25.23
1	29.28		5	30.64		8	27.20
1	30.37		5	30.75		9	23.88
1	31.37		5	30.96		9	22.31
1	34.38		5	29.95		9	23.19
1	31.66		5	29.10		9	22.53
2	30.09		5	27.51		9	22.98
2	31.55		5	27.12		9	27.12
2	31.06		5	26.74		9	27.09
2	32.01		5	26.93		9	25.93
2	28.28		6	30.05		9	25.90
2	22.27		6	24.80		9	25.99
2	25.24		6	26.84		9	26.07
2	26.77		6	26.22		9	25.62
2	29.42		6	23.79		10	27.78
2	31.04		6	26.89		10	26.57
2	29.89		6	26.84		10	23.64
2	30.87		6	30.29		10	21.03
3	26.33		6	29.17		10	18.77
3	26.32		6	27.64		10	16.33
3	26.74		6	27.51		10	15.42
3	22.49		6	30.36		10	18.33
3	23.16		7	26.35		10	20.02
3	19.95		7	26.41		10	21.92
3	19.70		7	27.51		10	21.70
3	19.09		7	26.45		10	24.12
3	24.57		7	26.07			
3	28.74		7	26.78			
3	28.36		7	29.18			
3	23.84		7	30.45			
4	35.73		7	30.68			
4	31.78		7	30.78			
4	25.60		7	30.58			
4	23.29		7	29.89			
4	28.14		8	24.72			
4	25.92		8	23.56			
4	26.07		8	24.43			
4	25.69		8	22.62			

REFERENCES

Abraham, B. and Ledolter, J. (1983). *Statistical Methods for Forecasting*, John Wiley, New York.

Abramowitz, M. and Stegum, I. (1972). Handbook of Mathematical Functions with Formulas, Graphs and Mathematical Tables, U.S. Government Printing Office, Washington, D.C.

Agresti, A. (1990). *Categorical Data Analysis*, John Wiley, New York.

Ahn, C., and Odom-Maryon,T. (1995). Estimation of a common odds ratio under binary cluster sampling. *Statistics in Medicine*, 14, 1567-1577.

Albers, W. (1978). Testing the mean of a normal population under dependence, *The Annals of Statistics*, 6, 6, 1337-1344.

Alsawalmeh, Y., and Feldt, L. (1994). Testing the equality of two related intraclass reliability coefficient. *Applied Psychological Measurements,* 18, 2, 183-190.

Andersen, A., Jensen, E. and Schou, G. (1981). Two-way analysis of variance with correlated errors, *International Statistical Review*, 49, 153-167.

Anderson, D. and Aitken, M. (1985). Variance component models with binary response: interviewer variability, *J. Stat. Soc. B*, 47, 203-210.

Andrews, D., Gnanadesikan, R. and Warner, J. (1971). Transformations of multivariate data, *Biometrics*, 27, 825-840.

Bahadur, R. (1961). A representation of the joint distribution of responses to n dichotomous items, in Solomon, H. (Ed.), *Studies in Item Analysis and Prediction*, pp. 158-176. Stanford University Press, Palo Alto, California.

Bartko, J. J. (1966). The intraclass correlation coefficient as a measure of reliability, *Psychological Reports*, 19, 3-11.

Bartlett, M. (1937). Properties of sufficiency and statistical tests, *Proceedings of the Royal Society, A*, 160, 268-282.

Bartlett, M. (1946). On the theoretical justification of sampling properties of an autocorrelated time series, *J. Stat. Soc. B*, 8, 27-41.

Bland, M., and Altman, D. (1986). Statistical methods for assessing agreement between two methods of clinical measurement. *Lancet*,1, 307-310.

368

Bloch, D. and Kraemer, H. (1989). 2x2 kappa coefficient: Measure of agreement or association, *Biometrics*, 45, 269-287.

Boos, D.D. (1992). On the generalization of the score test. *The American Statistician*, 46, 4, 327-333.

Box, G. (1954a). Some theorems on quadratic forms applied in the study of analysis of variance problems II. Effects of inequality of variances and of correlation between errors in the two-way classification, *Annals of Mathematics and Statistics*, 25, 484-498.

Box, G. (1954b). Some theorems on quadratic forms applied in the study of analysis of variance problems, *Annals of Mathematical Statistics*, 25, 290-302.

Box, G. and Cox, D. (1964). An analysis of transformation (with discussion), *J. Stat. Soc. B*, 26, 211-252.

Box, G. and Jenkins (1970). *Time Series Analysis, Forecasting, and Control*, Holden Day, San Francisco.

Box, G.E.P. and Pierce, D.A. (1970). Distribution of residual autocorrelations in autoregressive-integrated moving average time series models. *Journal of the American Statistical Association*, 70, 1509-1526.

Breslow, N. (1972). Contribution to the discussion of a paper by D. R. Cox. *Journal of the Royal Statistical Society, B*, 34, 216-217.

Breslow, N., Day, N., Halvorsen, K., Prentice, R. and Sabai, C. (1978). Estimation of multiple relative risk functions in matched case-control studies, *American Journal of Epidemiology*, 108, 299-307.

Brown, B. and Forsythe, A. (1974a). Robust tests for the equality of variances, *Journal of the American Statistical Association*, 69, 364-367.

Brown, B. and Forsythe, A. (1974b). The small sample behavior of some statistics which test the equality of several means, *Technometrics*, 16, 129-132.

Chen, C. and Swallow, W. (1990). Using group testing to estimate a proportion and to test the binomial model, *Biometrics*, 46, 1035-1046.

Cicchetti, D. and Fleiss, J. (1977). Comparison of the null distributions of weighted kappa and the C ordinal statistic, *Appl. Psychol. Meas.*, 1, 195-201.

Cicchetti, D. and Allison, T. (1971). A new procedure for assessing reliability of scoring EEG sleep recordings, *Am. J. EEG Techol.*, 11, 101-109.

Cohen, J. (1960). A coefficient of agreement for nominal scale, *Educational and Psychological Measurement*, 20, 37-46.

Cohen, J. (1968). Weighted kappa: Nominal scale agreement with provision for scaled disagreement or partial credit, *Psychological Bulletin*, 70, 213-220.

Collett, D. (1995). *Modelling Survival Data in Medical Research.* Chapman and Hall, London.

Collier, R., Baker, F., Mandeville, G. and Hayes, T. (1967). Estimates of test size for several test procedures based on conventional variance ratios in the repeated measures design, *Psychometrika*, 32, 339-353.

Connolly, M. and Liang, K-Y. (1988). Conditional logistic regression models for correlated binary data, *Biometrika*, 75, 501-506.

Cox, D. R. (1970). *Analysis of Binary Data*, Chapman and Hall, London.

Cox, D. R. (1972). Regression Models and Life-Tables (with discussion), *Journal of the Royal Statistical Society, Series B*, 34, 187-220.

Cox, D. R. and Snell, E. (1989). *Analysis of Binary Data*, Chapman and Hall, London.

Crouch, E. and Spiegelman, D. (1990). The evaluation of integral of the form $\int_{-\infty}^{\infty} f(t) \exp(-t^2)$:

Application to Logistic-Normal models, *Journal of the American Statistical Association*, 85, 464-469.

Crowder, M. (1978). Beta-binomial ANOVA for proportions, *Applied Statistics*, 27, 34-37.

Crowder, M. and Hand, D. (1990). *Analysis of Repeated Measures*, Chapman and Hall, London.

Cryer, J. (1986). *Time Series Analysis*, Duxbury Press, Boston.

D'Agostino, R., Belanger, A. and D'Agostino, J.R. (1990). A suggestion for powerful and informative tests of normality, *The American Statistician*, 44, 316-321.

D'Agostino, R. and Pearson, E. (1973). Testing for the departures from normality. I. Fuller experimental results for the distribution of b_2 and $\sqrt{b_1}$, *Biometrika*, 60, 613-622.

D'Agostino, R. and Stephens, M. (1986). *Goodness-of-fit Techniques*, Marcel Dekker, New York.

Davies, M. and Fleiss, J. (1982). Measuring agreement for multinomial data, *Biometrics*, 38, 1047-1051.

Diggle, P. (1989). Testing for random dropouts in repeated measurement data, *Biometrics*, 45, 1255-1258.

Diggle, P. (1990). *Time Series: A Biostatistical Introduction*, Oxford Science Publication, Oxford.

Diggle, P., Liang, K-Y and Zeger, S. (1994). *The Analysis of Longitudinal Data*, Oxford Science Publications, Oxford.

Donald, A. and Donner, A. (1987). Adjustments to the Mantel-Haenszel chi-squared statistic and odds ratio estimator when the data are clustered, *Statistics in Medicine*, 6, 491-499.

Donner, A. (1989). Statistical methods in ophthalmology: An adjusted chi-squared approach, *Biometrics*, 45, 605-611.

Donner, A. and Bull, S. (1983). Inferences concerning a common intraclass correlation coefficient, *Biometrics*, 39, 771-776.

Donner, A. and Eliasziw, M. (1992). A goodness-of-fit approach to inference procedures for the kappa statistic: Confidence interval construction, significance testing and sample size estimation, *Statistics in Medicine*, 11, 1511-1519.

Donner, A. Eliasziw, M., and Klar, N. (1994). A comparison of methods for testing homogeneity of proportions in teratologic studies. *Statistics in Medicine,* 13, 1253-1264.

Donner, A. and Wells, G. (1985). A comparison of confidence interval methods for the intraclass correlation coefficient, *Biometrics*, 41, 401-412.

Dorfman, R. (1943). The detection of defective members of large populations, *Annals of Mathematical Statistics*, 14, 436-440.

Draper, N. and Smith, H. (1981). *Applied Regression Analysis* (2nd edition), John Wiley, New York.

Dunn, G. (1992). Design and analysis of reliability studies, *Statistical Methods in Medical Research*, 1, 123-157.

Dunn, G. and Everitt, B. (1982). *An Introduction to Mathematical Taxonomy*, Cambridge University Press, Cambridge.

Durbin, J. and Watson, G. (1951). Testing for serial correlation in least squares regression II. *Biometrika*, 38, 159-178.

Elston, R.C. (1977). Response to query: estimating "heritability" of a dichotomous trait. *Biometrics*, 33, 232-233.

Firth, D. (1987). On the efficiency of quasi-likelihood estimation, *Biometrika*, 74, 233-245.

Firth, D. (1990). Generalized linear models, in *Statistical Theory and Modelling,* D. Hinkley, N. Reid, E. Snell, Eds., p. 55. Chapman and Hall, London.

Fisher, R. (1932). *Statistical Methods for Research Workers* (4th edition), Oliver and Boyd, Edinburgh.

Fleiss, J. (1979). Confidence intervals for the odds ratio in case-control studies: the state of the art, *Journal of Chronic Diseases*, 32, 69-77.

Fleiss, J. (1981). *Statistical Methods for Rates and Proportions,* John Wiley, New York.

Fleiss, J. (1986). *The Design and Analysis of Clinical Experiments,* John Wiley, New York.

Fleiss, J. and Cohen, J. (1973). The equivalence of weighted kappa and the intraclass correlation coefficient as measures of reliability, *Educational and Psychological Measurement*, 33, 613-619.

Fleiss, J., Cohen, J. and Everitt, B. (1969). Large sample standard errors of kappa and weighted kappa, *Psychological Bulletin*, 72, 323-327.

Friedman, M. (1937). The use of ranks to avoid the assumption of normality implicit in the analysis of variance. *Journal of the American Statistical Association*, 32, 675-701.

Gart, J. and Buck, A. (1966). Comparison of a screening test and a reference test in epidemiologic studies. II. A probabilistic model for the comparison of diagnostic tests, *American Journal of Epidemiology*, 83, 593-602.

Gart, J. and Thomas, D. (1982). The performance of three approximate confidence limit methods for the odds ratio, *American Journal of Epidemiology*, 115, 453-470.

Gastwirth, J. (1987). The statistical precision of medical screening tests, *Statistical Science*, 2, 213-238.

Geisser, S. (1963). Multivariate analysis of variance of a special covariance case, *Journal of the American Statistical Association*, 58, 660-669.

Geisser, S. and Greenhouse, S. (1958). An extension of Box's results on the use of the F distribution in multivariate analysis, *The Annals of Mathematical Statistics*, 29, 885-891.

Godambe, V. and Kale, B. (1991). Estimating functions: An overview, in *Estimating Functions*, V.P. Godambe, Ed., Oxford Science Publications, Oxford.

Greenhouse, S. and Geisser, S. (1959). On methods in the analysis of profile data, *Psychometrika*, 24, 95-112.

Grubbs, F. E. (1948). On estimating precision of measuring instruments and product variability, *Journal of the American Statistical Association*, 43, 243-264.

Haldane, J. (1956). The estimation and significance of the logarithm of a ratio of frequencies, *Annals of Human Genetics*, 20, 309-311.

Harris, R. (1975). *A Primer of Multivariate Statistics*, Academic Press, New York.

Hauck, W. (1979). The large sample variance of the Mantel-Haenszel estimator of a common odds ratio, *Biometrics*, 35, 817-819.

Henderson, C.R. (1975). Best linear unbiased estimation and prediction under a selection model. *Biometrics*, 31, 423-447.

Hosmer, D. and Lemeshow, S. (1989). *Applied Logistic Regression*, John Wiley, New York.

Hougaard, P. (1995). Frailty models for survival data, *Lifetime Data Analysis*, 1, 255-273.

Hui, S. and Walter, S. (1980). Estimating the error rates of diagnostic tests, *Biometrics*, 36, 167-171.

Hwang, F. (1976). Group testing with a dilution effect, *Biometrika*, 63, 671-673.

Imhof, J. (1962). Testing the hypothesis of no fixed main-effects in Scheffe's mixed model, *The Annal of Mathematical Statistics*, 33, 1085-1095.

Jennrich, R.H. and Schluchter, M.D. (1986). Unbalanced repeated measures models with structured covariance matrices. *Biometrics*, 42, 805-820.

Jewell, N. (1984). Small-sample bias of point estimators of the odds ratio from matched sets, *Biometrics*, 40, 421-435.

Jewell, N. (1986). On the bias of commonly used measures of association for 2x2 tables, *Biometrics*, 42, 351-358.

Jones, R.H. (1993). *Longitudinal Data with Serial Correlation: A State-Space Approach*. Chapman and Hall, London.

Kalman, R. (1960). A new approach to linear filtering and prediction problems, *Trans. ASME J. Basic Eng.*, 82, 35-45.

Kaplan, E. L., and Meier, P. (1958). Nonparametric estimation from incomplete observations. *Journal of the American Statistical Association*, 53, 457-481.

Karim, M. and Zeger, S. (1988). *GEE: A SAS Macro for Longitudinal Data Analysis*, Technical Report #674 from the Department of Biostatistics, Johns Hopkins University, Baltimore.

Katz, J., Carey, V., Zeger, S., and Sommer, L. (1993). Estimation of design effects and Diarrhea clustering within household and villages. *American Journal of Epidemiology*, 138, 11, 994-1006.

Kendall, M. and Ord, K. (1990). *Time Series* (3rd edition), Edward Arnold, London.

Kish, L. (1965). *Survey Sampling*, John Wiley, New York.

Kleinbaum, D., Kapper, L. and Muller, K. (1988). *Applied Regression Analysis and Other Multivariable Methods*, PWS-Kent, Boston.

Kogan, L. (1948). Analysis of variance: Repeated measurements, *Psychological Bulletin*, 45, 131-143.

Kolmogorov, A. (1939). Sur L'interpolation et L'extrapolation des suites stationnaires, *C.R. Acad. Sci.*, Paris, 208, 2043-2045.

Kolmogorov, A. (1941). Interpolation and extrapolation von stationären Zufälligen Folgen, *Bull. Acad. Sci. (Nauk), USSR, Ser. Math.* 5, 3-14.

Kreppendorff, K. (1970). Bivariate agreement coefficient for relability of data, *Sociological Methodology*, F. Borgatta and G.W. Bohrnstedt, Eds., pp. 139-150. Jossey-Bass, San Francisco.

Laird, N. (1988). Missing data in longitudinal studies, *Statistics in Medicine*, 7, 305-315.

Landis, J. and Koch, G. (1977). The measurement of observer agreement for categorical data, *Biometrics*, 33, 159-174.

Lawless, J.F. (1982). *Statistical Models and Methods for Lifetime Data*, John Wiley, New York.

Le, C.T. and Lindgren, B.R. (1996). Duration of Ventilating Tubes: A Test for Comparing Two Clustered Samples of Censored Data, *Biometrics*, 52, 328-334.

Leugrans, S. (1980). Evaluating laboratory measurement techniques: in *Biostatistics Casebook*, R. Miller, B. Efron, B. Brown, and L. Moses, Eds., pp. 190-219. John Wiley, New York.

Levene, H. (1960). Robust tests for equality of variances, in *Contributions to Probability and Statistics*, I. Olkin, Ed., pp. 278-292. Stanford University Press, Palo Alto, California.

Levinson, N. (1947). The Wiener RMS error criterion in filter design and prediction, *J. Math. Physics*, 25, 261-278.

Liang, K-Y and Zeger, S. (1986). Longitudinal data analysis using Generalized Linear Models, *Biometrika*, 73, 13-22.

Liang, K-Y, Self, S.G., Bandeen-Roche, K., Zeger, S.L. (1995). Some recent developments for regression analysis of multivariate failure time data, *Lifetime Data Analysis*, 1, 403-415.

Lin, L. I. (1989). A concordance correlation coefficient to evaluate reproducibility, *Biometrics*, 45, 255-268.

Lindsay, J. (1993). *Models for Repeated Measurements*, Oxford Science Publications, Oxford.

Lipsitz, S., Laird, N. and Harrington, D. (1990). Multivariate regression models and analyses for categorical data, *Technical Report 700*, Department of Biostatistics, Johns Hopkins University, Baltimore.

Littell, R.C., Milliken, G.A., Stroup, W.W., and Wolfinger, R.D. (1996). SAS system for mixed models. SAS Institute, Cary, N.C.

Little, R. and Rubin, D. (1987). *Statistical Analysis with Missing Data*, John Wiley, New York.

Ljung, G.M. and Box, G.E.P. (1978). On the measure of lack of fit in time series models. *Biometrika*, 65, 297-304.

Looney, S. and Stanley, W. (1989). Exploratory repeated measures analysis for two or more groups: Review and update, *The American Statistician*, 43, 4, 220-225.

MacDonald, B. (1993). Estimating logistic regression parameters for bivariate binary data, *J. Stat. Soc. B*, 55 No. 2, 391-397.

Mantel, N. (1973). Synthetic retrospective studies and related topics, *Biometrics*, 29, 479-486.

Mantel, N. and Haenszel, W. (1959). Statistical aspects of the analysis of data from retrospective studies of disease, *J. Natl. Cancer Inst.*, 22, 719-748.

McCullagh, P. and Nelder, J. (1989). *Generalized Linear Models*, 2nd edition, Chapman and Hall, London.

McDermott, J., Allen, O., and Martin, W. (1992). Culling practices of Ontario cow-calf producers, *Canadian Journal of Veterinary Research*, 56, 56-61.

McGilchrist, C.A. (1994). Estimation in generalized mixed models. *Journal of the Royal Statistical Society,* B, 56, No.1, 61-69.

Meyer, N. (1964). Evaluation of screening tests in medical diagnosis, *Biometrics*, 20, 730-755.

Miller, R. (1986). *Beyond ANOVA: Basics of Applied Statistics*, John Wiley, New York.

Milliken, G. and Johnson, D. (1984). *Analysis of Messy Data*, Vol. I. Van Nostrand Reinhold, New York.

Morrison, D. (1976). *Multivariate Statistical Methods* (2nd edition), McGraw-Hill, New York.

Mosteller, F. and Tukey, J. (1977). *Data Analysis and Regression: A Second Course in Statistics*, Addison-Wesley, Reading, MA.

Oakes, M. (1986). *Statistical Inference*, Epidemiology Resources Inc., Chestnut Hills, MD.

Paul, S.R. (1982). Analysis of proportions of affected foetuses in teratological experiments. *Biometrics*, 38, 361-370.

Peto, R. and Peto, J. (1972). Asymptotically efficient rank invariant procedures. *Journal of the Royal Statistical Society, A,* 135, 185-207.

Potthoff, R.F. and Roy, S.N. (1964). A generalized multivariate analysis of variance model useful especiallly for growth curve problems. *Biometrika*, 51, 313-326.

Prentice, R. (1976). Use of the logistic model in retrospective studies, *Biometrics*, 32, 599-606.

Prentice, R. (1988). Correlated binary regression with covariates specific to each binary observation, *Biometrics*, 44, 1033-1048.

Quan, H. and Shih, W.J. (1996). Assesssing reproducibility by the within- subject coefficient of variation with random effects models. *Biometrics*, 52, 1196-1203.

Quenouille, M. (1949). Approximate tests of correlation in time series, *J. Stat. Soc. B*, 11, 68-84.

Rao, J. N. K. and Scott, A. J. (1992). A simple method for the analysis of clusted binary data, *Biometrics*, 48, 577-585.

Robins, J., Breslow, N. and Greenland, S. (1986). Estimators of the Mantel-Haenszel variance consistent in both sparse data and large-strata limiting models, *Biometrics*, 42, 311-323.

Rogan, W. and Gladen, B. (1978). Estimating prevalence from the results of a screening test, *American Journal of Epidemiology*, 107, 71-76.

Rosenblatt, R., Reinken, J. and Shoemack, P. (1985). Is obstetrics safe in small hospitals? Evidence from New Zealand's regionalized perinatal system, *Lancet*, 24, 429-431.

Rubin, D. (1976). Inference and missing data, *Biometrika*, 63, 581-592.

Rubin, D. (1994). Modelling the drop-out mechanism in repeated measures studies. 9th International Workshop on Statistical Modelling, Exeter, U.K.

Sastry, N. (1997). A Nested Frailty Model for Survival Data, with an Application to the Study of Child Survival in Northeast Brazil, *Journal of the American Statistical Association*, 92, 438, 426-435.

Schall, R. (1991). Estimation in generalized linear models with random effects. *Biometrika*, 78, 4, 719-727.

Schlesselman, J. (1982). *Case-Control Studies Design, Conduct, Analysis*, Oxford University Press, Oxford.

Schouten, H. J. (1985). *Statistical Measurement of Interobserver Agreement*. Unpublished doctoral dissertation. Erasmus University, Rotterdam.

Schouten, H. J. (1986). Nominal scale agreement among observers, *Psychometrika*, 51, 453-466.

Scott, W. A. (1955). Reliability of content analysis: The case of nominal scale coding, *Public Opinion Quarterly*, 19, 321-325.

Shoukri, M. M., Martin, S. W. and Mian, I. (1995). Maximum likelihood estimation of the kappa coefficient from models of matched binary responses, *Statistics in Medicine*, 14, 83-99.

Shukla, G. K. (1973). Some exact tests on hypothesis about Grubbs' estimators, *Biometrics*, 29, 373-377.

Shumway, R. (1982). Discriminant analysis for time series, in *Handbook of Statistics*, Vol. II, *Classification, Pattern Recognition and Reduction of Dimensionality*, P. R. Kirshnaiah, Ed., 1-43. North Holland, Amsterdam.

Smith, C. A. B. (1956). On the estimation of intraclass correlation, *Annals of Human Genetics*, 21, 363-373.

Snedecor, G. and Cochran, W. G. (1980). *Statistical Methods* (8th edition), Iowa State University Press, Ames, Iowa.

Sobel, M. and Elashoff, R. (1975). Group testing with a new goal, estimation, *Biometrika*, 62, 181-193.

Stanish, W. and Taylor, N. (1983). Estimation of the intraclass correlation coefficient of rthe anlaysis of covariance model, *The American Statistician*, 37, 3, 221-224.

Sutradhar, B., MacNeill, I. and Sahrmann, H. (1987). Time series valued experimental designs: One-way ANOVA with autocorrelated errors, in *Time Series and Econometric Modelling*, I. B. MacNeill and G. J. Umphrey, Eds., D Reidel, Dordrecht, Holland.

Tarone, R. (1979). Testing the goodness of fit of the binomial distribution. *Biometrika*, 66, 3, 585-590.

Tenenbein, A. (1970). A double sampling scheme for estimating from binomial data with misclassifications, *Journal of the American Statistical Association*, 65, 1350-1361.

Therneau, T. (1993). Using a multiple-events Cox model. *Proceedings from the Biometrics Section of the American Statistical Association*, 1-14.

Thibodeau, L. (1981). Evaluating diagnostic tests, *Biometrics*, 37, 801-804.

Thomas, D. and Gart, J. J. (1977). A table of exact confidence limits for differences and ratios of two proportions and their odds ratios, *Journal of the American Statistical Association*, 72, 386-394.

Tukey, J. (1977). *Exploratory Data Analysis*, Addison-Wesley, Reading, MA.

Vacek, P. (1985). The effect of conditional dependence on the evaluation of diagnostic tests, *Biometrics*, 41, 959-968.

Walter, S. (1985). Small-sample estimation of log odds ratios from logistic regression and fourfold tables, *Statistics in Medicine*, 4, 437-444.

Walter, S. and Cook, R. (1991). A comparison of several point estimators of the odds ratio in a single 2x2 contingency table, *Biometrics*, 47, 795-811.

Walter, S., Hildreth, S. and Beaty, B. (1980). Estimation of infection rates in populations of organisms using pools of variable size, *American Journal of Epidemiology*, 112, 124-128.

Wedderburn, R. (1974). Quasi-likelihood functions, Generalized Linear Models and the Gauss-Newton Method, *Biometrika*, 61, 439-447.

Welch, B. (1951). On the comparison of several mean values: An alternative approach, *Biometrika*, 38, 330-336.

White, H. (1982). Maximum likelihood estimation of misspecified models, *Econometrica*, 50, 1-25.

Whittle, P. (1983). *Prediction and Regulations* (2nd edition), University of Minnesota Press, Minneapolis.

Wilk, M. and Gnanadesikan, R. (1968). Probability plotting methods for the analysis of data, *Biometrika*, 55, 1-17.

Willeberg, P. (1980). The analysis and interpretation of epidemiological data, *Veterinary Epidemiology and Economics*, Proceedings of the 2nd International Symposium, Canberra, Australia, W.A. Geering and L.A. Chapman, Eds.

Williams, D. (1982). Analysis of proportions of affected foetuses in teratological experiments, *Biometrics*, 38, 361-370.

Winer, B. J. (1972). *Statistical Principles in Experimental Design* (2nd edition), McGraw-Hill, New York.

Woolf, B. (1955). On estimating the relationship between blood group and disease, *Annals of Human Genetics*, 19, 251-253.

Yaglom, A. (1962). *An Introduction to the Theory of Stationary Random Functions*, Prentice-Hall, Englewood Cliffs, NJ.

Yule, G. (1927). On a method of investigating periodicities in disturbed series, with special reference to Wolfer's sunspot numbers, *Phil. Trans.*, A226, 267-298.

Zeger, S. and Liang, K-Y. (1986). Longitudinal data analysis for discrete and continuous outcomes, *Biometrics*, 42, 121-130.

Zhao, L. and Prentice, R. (1990). Correlated binary regression using a quadratic exponential model, *Biometrika*, 77, 642-648.

INDEX

A

additive seasonal variation, 206
Additive Seasonal Variation Model (ASVM), 189
Akaike-information criteria, 268, 310
Altman (see Bland), 41
Analysis of Variance, 1, 20, 23, 69, 103, 268, 271
approximate tests, 254
AR(1), 212, 230, 252, 256, 259
AR(1) model, 236, 245
AR(2), 215, 233, 261
AR(p), 233
ARIMA, 229, 244
ARMA, 237, 252, 253
ARMA (1,1), 219
ARMA (1,1) model, 237
ARMA (p,d,q), 229
autocorrelation, 204, 205, 208, 226, 230
autocovariance, 208, 226, 230
autoregression, 201, 226, 230, 251
autoregressive integrated moving average, 229
autoregressive process, 212

B

Bahadur's model, 176
Bartlett's test, 9
bias, 60, 112, 139
bivariate distribution, 207
Bland and Altman, 41
Bonferroni, 288
box plot, 5
Breslow, 352
Brown and Forsythe, 11

C

case-control, 45, 200
centered moving average, 194
cluster, 65, 163
Cochran, 12
Cochran Armitage, 171
Cochran and Orcutt procedure, 205, 221
Cochran's Q, 102
coefficient of variation, 41
Cohen's weighted kappa, 95, 96, 98
cohort, 45, 193
compound symmetry, 282
concordance correlation, 35, 37, 39
conditional likelihood, 197
confounding, 149
contrast, 288
correlation, 206
correlation function, 233
correlogram, 228, 232
covariance, 206
Cox proportional hazard, 350
cyclical variation, 198

D

deseasonalized observations, 195
deviance, 156, 157
deterministic trend model, 224, 240
differencing, 226, 227, 243
dilution effect, 134
Donald and Donner adjustment, 69, 77
dummy variables, 147, 199
Durbin-Watson statistic, 220

E

Estimation, 235
exchangeable correlation, 169

381

experimental time series, 251
exponential, 199
Exponential model, 344, 345
extra-binomial variation, 164

F

first order autoregression process, 204
Fisher's exact test, 51
forecast error, 249
forecasting, 245, 258
forecasts, 199
frailty, 363, 365

G

gamma, 364
Geisser-Greenhouse, 285
Generalized Estimating Equations (GEE), 184, 186, 192, 314
GLMM, 180
GLIMMIX, 181
gold standard, 113
goodness of fit, 155
group testing, 131
Grubbs, 31, 34

H

Hauck variance estimator, 59, 79, 81
hazard function, 326
Henderson, 180
heterogeneity, 9, 15
homogeneity, 10
homogeneity, test of, 55
homogeneous nonstationarity, 225

I

infinite-order AR process, 218
interaction, 149
interclinician agreement, 18, 85
inter-quartile range (IQR), 5

intraclass correlation, 17, 19, 69, 281
intracluster correlation, 15, 74, 163

K

Kaplan-Meier, 333
Kappa
    coefficient, 88
    weighted, 96

L

latent class models, 134
least squares, (LS) estimators, 242
Levine's test, 14
Likelihood Ratio Test (LRT), 155, 156
linear, 199
link function, 146
logistic
    regression, 141
    transformation, 143
logit, 146
log-rank test, 339

M

MA(1), 217, 218
MA(1) model, 237, 249
MA(2), 218
MA(q), 217, 233
Mantel-Haenszel, 58, 66
Marginal likelihood, 173
matching, 59, 196
McNemar's test, 64, 65, 93, 94
method of maximum likelihood, 239, 254
Method of moments, 235, 236, 251
missing:
    at random, 294
    observations, 293
mixed autoregressive moving average
    process, 219
Mixed effect, 22, 23, 162
Mixed model, 303

model specification, 232
moment estimate, 209
moment estimators, 257
moving average, 191, 237, 263, 268
moving average process, 217
multilevel, 179, 313
Multiplicative seasonal variation
model, 189, 190, 206, 207, 222

N

non-stationary time series, 224
nonstationary, 229

O

Odds ratio (OR), 46, 60, 145
test of common, 55
Mantel-Haenzel, 58
Omnibus test, 2, 6
one-way ANOVA, 268
Overdispersion, 164, 167

P

PAC, 232
partial autocorrelation, 205
partial autocorrelation function, 249
partial autocorrelation function (PAC), 233
Pearson's chi-square, 43, 155, 156
Pearson's product
moment correlation, 33, 36
phlev, 362
Poisson, 357
Power Transformation, 231
Precision, 19, 30
Predictive value:
negative, 115
positive, 115
Prevalence, 113
proc genmod, 193, 223, 359
proc lifereg, 345

proc lifetest, 34
proc phreg, 354

Q

Q-Q plot, 4
quadratic, 199
Quasi-likelihood, 185

R

random effect, 19, 21, 22, 26, 172
random walk model, 226
Rao and Scott's adjustment, 71, 74, 79, 84, 85
Regression:
linear, 141
logistic, 141
relative risk, 46
reliability, 18, 28, 29, 103
repeatability, 19
repeated measures, 277

S

sample correlogram, 209
Satterwaite, 290
Schall's algorithm, 181
Screening tests: medical, 112
seasonal coefficient, 210
seasonal pattern, 199
seasonal variation, 189, 222
seasonality, 189
sensitivity, 112, 124, 134
serial correlation, 15, 16, 254
skewness, 2, 4
specificity, 112, 124, 134
stationarity, 206
stationarity conditions, 215
stationary series, 225
stochastic process, 205, 224
stochastic trend model, 225, 241
stratification, 53